Biochar - Productive Technologies, Properties and Applications
http://dx.doi.org/10.5772/intechopen.100763
Edited by Mattia Bartoli, Mauro Giorcelli and Alberto Tagliaferro

Contributors
Yıldırım İsmail Tosun, Ana R.F. Rodrigues, Margarida R.G. Maia, Ana R.J. Cabrita, Hugo M. Oliveira, Inês M. Valente, José L. Pereira, Henrique Trindade, António J.M. Fonseca, Uplabdhi Tyagi, Neeru Anand, Stephen Okiemute Akpasi, Ifeanyi Michael Smarte Anekwe, Jeremiah Adedeji, Sammy Lewis Kiambi, Edward Kwaku Armah, Maggie Manimagalay Chetty, Denzil Erwin Estrice, Boldwin Mutsvene, Nikita Singh, Zikhona Tshemese, Dinesh Chandola, Smita Rana, Yalong Zhang, Yu Zhang, Dongdong Feng, Jiabo Wu, Jianmin Gao, Yudong Huang, Qian Du, Minister Obonukut, Sunday Alabi, Alexander Jock, Abdulkareem Ghassan Alsultan, Nurul-Asikin Mijan, Laith Kareem Obeas, Aminul Islam, Maadh Fawzi Nassar, Nasar Mansir, Robiah Yunus, Yun Hin Taufiq-Yap, Siti Zulaika Razali, Abdul Kadir Salam, Kingsley Ukoba, Tien-Chien Jen, Theophilus Olufemi Isimikalu, Obey Gotore, Tirivashe Phillip Masere, Vadzanayi Mushayi, Ramaraj Rameshprabu, Tomoaki Itayama, Yuwalee Unpaprom, Osamu Nakagoe, Run-Hua Zhang, Ling-Fang Shi, Zhi-Guo Li, Guo-Lin Zhou, Yan-Lan Xie, Xing-Xue Huang, An-Hua Ye, Chu-Fa Lin, Karl Williams, Ala Khodier, Peter Bentley, Ariharaputhiran Anitha, Nagarajan Ramila Devi

First published in London, United Kingdom, 2023 by IntechOpen
IntechOpen is the global imprint of INTECHOPEN LIMITED, registered in England and Wales, registration number: 11086078, 5 Princes Gate Court, London, SW7 2QJ, United Kingdom

British Library Cataloguing-in-Publication Data
A catalogue record for this book is available from the British Library

Additional hard and PDF copies can be obtained from orders@intechopen.com

Biochar - Productive Technologies, Properties and Applications
Edited by Mattia Bartoli, Mauro Giorcelli and Alberto Tagliaferro
p. cm.
Print ISBN 978-1-80356-251-3
Online ISBN 978-1-80356-252-0
eBook (PDF) ISBN 978-1-80356-253-7

For EU product safety concerns:
IN TECH d.o.o., Prolaz Marije Krucifikse Kozulić 3, 51000 Rijeka, Croatia,
info@intechopen.com or visit our website at intechopen.com.

Meet the editors

Dr. Mattia Bartoli's research output includes 82 published peer-reviewed papers in top journals. During his time at the Biorefinery Research Group hosted by the University of Alberta, which develops new materials and new technologies, he worked closely with ForgeHydrocarbon a spin-off company focusing on innovative lipid-to-hydrocarbon technology for renewable hydrocarbon production. In 2018, he joined the Carbon Group, hosted by the Polytechnic of Turin, Italy, where he studied the production and use of carbon from the thermochemical conversion of waste streams for materials science applications. He is particularly interested in advanced materials such as carbon nanodots and bismuth-based materials for biological and electrochemical applications. Since 2021, Dr. Bartoli has been working on CO_2 electrochemical and thermochemical conversion at the Center for Sustainable Future Technologies - CSFT@POLITO.

Mauro Giorcelli, Ph.D., is a professor of applied physics and a researcher in biochar materials. He is a member of the Carbon Group in the Department of Applied Science and Technologies (DISAT) at the Polytechnic of Turin, Italy. His main research interests are the morphological, electrical and mechanical characterizations and applications of carbon-based materials and composites. He has over 100 publications in international journals and has been an invited speaker at several conferences. Mauro Giorcelli has recently been a guest researcher at the University of Ontario Institute of Technology in Oshawa, Ontario. He also collaborates with several international universities and research centers.

Alberto Tagliaferro obtained his degree in nuclear engineering and his Ph.D. in physics at the Polytechnic of Turin. He is an Associate Professor in the Materials Science and Chemical Engineering Department at the Polytechnic of Turin. He is head of the Carbon Group, a research group founded in 2002. His research interests have moved in recent years to biochar production and applications. He is Chair Adjunct Professor at the University of Ontario Institute of Technology and editor of *Critical Reviews in Environmental Science and Technology* (CREST) by Taylor and Francis, *Reviews on Advanced Materials Science* by De Gruyter, and *Micromachines* by MDPI. He is Chair of the Education Committee of IUVSTA (International Union for Vacuum Science, Technique and Applications).

Contents

Preface

Biochar is the solid residue recovered from the thermal cracking of biomasses in an oxygen-poor atmosphere. Recently, biochar has been increasingly explored as a sustainable, inexpensive and viable alternative to traditional carbonaceous materials for the development of many cutting-edge applications. Biochar exhibits high thermal stability, high surface area and electrical conductivity, and its principal properties can be tuned appropriately by controlling the conditions of the pyrolysis process. Due to its intriguing characteristics, biochar is currently in competition with high-performing fillers in multifunctional approaches to environmental remediation, electrochemistry, energy application and materials science, and represents a promising aspect of the movement toward a biomass-based circular bioeconomy.

For this book, we selected crucial topics ranging from pollutants removal to the electrochemical and energetic applications of biochar, focusing both on contributions that could provide an introduction to the biochar sector and those that provide enlightening new perspectives on this subject and its future applications. The book promotes the spread of innovative rethinking of old technologies and problems that can merge the great potential of biochar-based technologies with respect for our world.

This book will be highly accessible to any reader with a strong scientific and technological background, from scientific advisors in private companies to academics. Students enrolled in graduate science programs may also find this text useful for deeper insights into the very complex field proposed by the authors. In view of its very strong scientific content, we believe that this book may come to be the reference text for any future study and application of biochar-based technologies. We hope that *Biochar - Productive Technologies, Properties and Applications* will contribute to focusing the attention of the scientific community on this emerging sector.

Mattia Bartoli
CSFT@POLITO,
Fondazione Istituto Italiano di Tecnologia,
Turin, Italy

Mauro Giorcelli
Department of Applied Science and Technology,
Polytechnic University of Turin,
Turin, Italy

Alberto Tagliaferro
Department of Applied Science and Technology,
Polytechnic University of Turin,
Turin, Italy

Section 1

Perspectives on the Biochar Future

Chapter 1

Review: Heads or Tails? Toward a Clear Role of Biochar as a Feed Additive on Ruminant's Methanogenesis

Ana R.F. Rodrigues, Margarida R.G. Maia, Ana R.J. Cabrita, Hugo M. Oliveira, Inês M. Valente, José L. Pereira, Henrique Trindade and António J.M. Fonseca

Abstract

The use of biochar has been suggested as a promising strategy in bio-waste management and greenhouse gases mitigation. Additionally, its use, as a feed additive, in ruminants has been reported to have contrasting effects on enteric methane production. Hence, this chapter intends to overview the most relevant literature that exploited the use of biochar as a mitigation strategy for methane. This includes the reported effects of biochar on methane production and rumen fermentation observed in *in vitro* and *in vivo* assays, as well as manure's methane emission. The information available about the biochar and the experimental conditions used in the different studies is still limited, which created additional challenges in identifying the biological mechanisms that potentially drive the contrasting results obtained. Nevertheless, it is clear from the current state-of-the-art that biochar may be a key player in the modulation of gut fermentation and in the reduction of greenhouse gases produced by ruminants that need to be consolidated by further research.

Keywords: biomass, biochar, enteric methane, *in vitro*, *in vivo*, ruminants

1. Introduction

The livestock sector was estimated to emit 14.5% of global anthropogenic greenhouse gases (GHG), mainly methane (CH_4), nitrous oxide (N_2O), and carbon dioxide (CO_2) [1], with enteric CH_4 corresponding to 40% of total livestock sector emissions, 77% of which emitted by cattle [1].

Ruminants are herbivorous animals that host a complex symbiotic microbial population composed of bacteria, protozoa, archaea, fungi, and bacteriophages in the two forestomach (reticulum and rumen) where feeds undergo fermentation, before entering the true stomach, the abomasum. Microbial population ferments structural and non-structural polysaccharides, and proteins originating volatile fatty acids (VFA) (mainly acetate, propionate, and butyrate), ammonia-N (NH_3-N), CO_2, and hydrogen (H_2) [2]. Volatile fatty acids are absorbed through the rumen wall and comprise the major energy source of the host animal. Hydrogen is mainly eliminated by the reduction of CO_2 by methanogenic archaea [3]. Enteric CH_4 represents a loss from 2 to 12% of total gross energy intake [4] and it is the second GHG contributor to climate change, with a global warming potential 28 times larger than CO_2, in a time horizon of 100 years. Mitigation of enteric CH_4 emissions is thus important not only to minimize the environmental impact of ruminant production but also to improve feed efficiency.

Several strategies have been evaluated to reduce enteric CH_4 production, including feeding management (e.g., ingredient selection, feed supplements, rate of passage, and better-quality ingredients), rumen modifiers (e.g., defaunation, bacteriocins, and immunization), and improvement of animal production through genetics (e.g., nutrient utilization, feed efficiency, and CH_4 production) [5], but effects are often transient [6] or conflicting [7]. Greenhouse gases (CH_4, N_2O) and ammonia (NH_3) are also produced during cattle manure decomposition in housing, storage, and treatment, and ultimately during land spreading [8]. Different strategies have been proposed to reduce gaseous emissions in each stage of manure management, from dietary manipulation to chemical application in slurry [9, 10]. One emerging strategy to cope with the mitigation of both enteric CH_4 and GHG from ruminants' manure is the use of biochar. Biochar is a stable porous carbon-rich material (between 65 and 90%), mainly produced by the pyrolysis method under oxygen-limited conditions, containing mineral elements whose physical and chemical characteristics are determined by feedstocks and technologies involved in the production process [11, 12]. Due to its characteristics, biochar has been studied for multiple uses, such as soil amending [13–15], mitigating GHG emissions from soil [16–19], recovering nutrients from wastewaters [20], and reducing GHG emissions from cattle manure during storage [21, 22]. Its porous structure promotes soil moisture retention, reduces bulk density, enhances the organic matter content, and can positively affect soil cation exchange capacity [23, 24]. Due to these properties, interest has emerged in biochar as a feed supplement to mitigate enteric and fecal CH_4, and manure gaseous emissions [25, 26], in a cascade approach, thus enhancing its effect along the cattle production system [27]. In this context, the European biochar foundation has developed guidelines for biochar production to be used as a feed additive [28] under the requirements of the European Food Safety Authority (EFSA) and respecting the commission regulation (EC) 178/2002 [29] and 834/2007 [30].

2. The role of biomass and production conditions on biochar characteristics

The biomass source and the type and conditions of production are key factors in biochar physicochemical properties resulting in different functional characteristics and applications [31], being pyrolysis the most common process for the production of biochar. The characteristics of biochar can be highly variable, especially in terms of elemental composition, surface chemical composition, structure, and stability. Each component's decomposition and depolymerization occurs through several reactions at different temperatures, contributing to the structural differences among biochars [32, 33].

In the works reviewed here, biochars were mainly produced by the pyrolysis of agriculture and forestry lignocellulosic biomasses, which are primarily composed of cellulose (40−45%), hemicellulose (25−35%) and lignin (20−30%), although their distribution varies among biomasses [34]. In terms of gaseous capture, the most relevant characteristics of biochar are the organic matter content (given by polarity and aromaticity), mineral content, cation exchange capacity, surface charge, and textural properties (surface area and pore size) [35].

The adsorption capacity of biochar related to the polarity and aromaticity is highly modulated by the pyrolysis conditions [35]. Due to the high carbon content and porous structure, adsorption is a valuable property of biochar, which has been used for environmental purposes, such as the reduction of GHG levels [35]. Therefore, the physical-chemical characteristics of the biochar have a strong influence on the capabilities of the materials for a particular application (**Figure 1**).

For example, the ash content that results from the decomposition of the inorganic matter of biomass [23] is expected to be low in wood-based biomass when compared to mineral-rich biomass, such as grass, manure, litter, and solid waste [36]. Wood, bamboo, corncob, corn stover, pellets (miscanthus, softwood, wheat straw, and oil-seed rape straw), rice straw, and potato peel biochar reported less than 25% of ash content, while rice husk presented higher than 40% [37–44]. The ash content has been demonstrated to be relevant for the surface polarity and distribution of pores, thus influencing the sorption capacity of the material. The mineral content in biochar (such as carbonates, oxides, phosphates, alkali, or alkaline earth metals) has been shown to

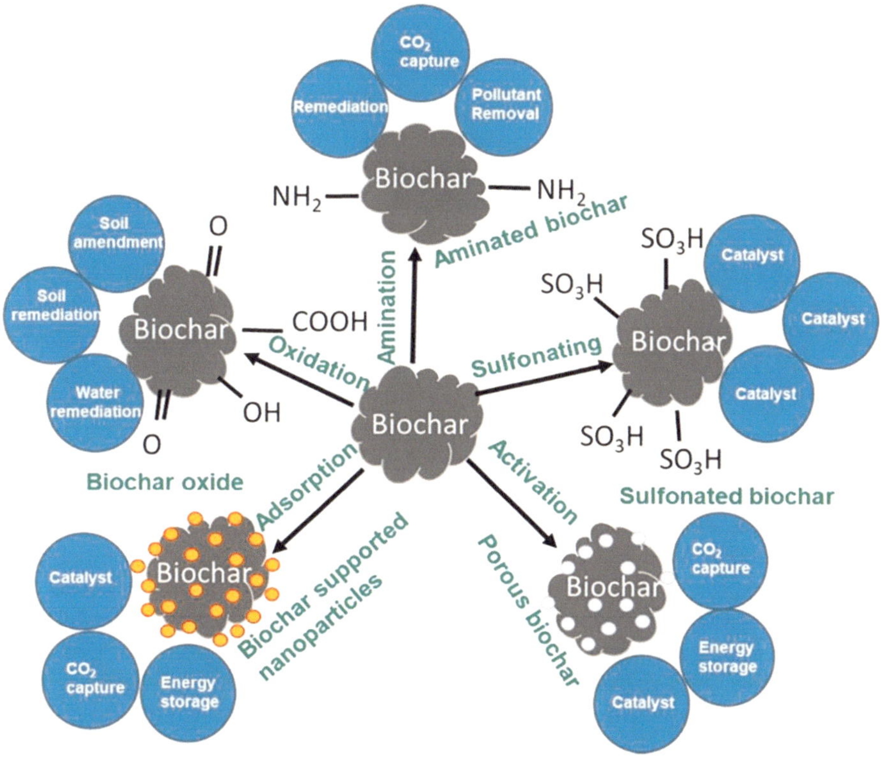

Figure 1.
Biochar post-production functionalization and potential applications. Reprinted with permission from Ghodake et al. [33].

increase the sorption capacity for acidic gases, such as sulfur dioxide, hydrogen sulfide, and CO_2 [12].

The surface area and pore size can also be modified by chemical and physical activation following the carbonization process [45]. The modification of biochar with a CO_2-NH_3 mixture resulted in a surface area increase besides improving the chemical properties of the surface by a nitrogen modification [46]. The microporous structure has a key role in CO_2 capture at low temperatures [47].

Using lignocellulosic biomasses (the main raw material present in the application herein described), a microporous structure is expected with higher cellulose and hemicellulose content, whereas mesoporous structures are expected with higher lignin content [27]. The increase in pyrolysis temperatures also increases the porosity, surface area, pH, ash, and carbon content of biochar due to the release of volatile components, while reducing biochar exchange capacity and yield [20, 48]. In the study of Calvelo Pereira *et al.* [38], an increase in surface area, carbon, nitrogen, and ash contents of biochar produced from the pyrolysis of pine chips and corn stover was observed. Also, biomass has been shown to highly influence the surface area, as demonstrated by other authors [40, 44].

3. Effects of biochar on *in vitro* rumen fermentation

There is a paucity of data on the effects of biochar on CH_4 production by short- and long-term *in vitro* studies. Therefore, these will be addressed separately.

3.1 *In vitro* short-term studies

Table 1 presents the results obtained in 14 studies evaluating the effects of biochar addition up to 16% on rumen fermentation and CH_4 production through *in vitro* short-term incubations (up to 48 h). No clear association is evident between effects on CH_4 production and biochar characteristics (e.g., biomass, temperature of pyrolysis) and level of inclusion. Increasing pyrolysis temperature increases surface area, which has the potential to improve biofilm formation and promote the adsorption capacity of microorganisms, nutrients, and gases, thus reducing CH_4 production [26, 49]. Indeed, some studies [44, 50–54] reported a decrease in CH_4 production with the addition of biochar produced at very high temperatures (700–1000°C), whereas in the studies using biochar produced at lower temperatures (350–700°C) no effect [38, 55] or an increased [42] CH_4 production was observed. However, Saenab *et al.* [56] reported a decrease in CH_4 production when biochar from cashew nutshell was produced at 300°C and Cabeza *et al.* [40] found higher CH_4 production with biochar produced at 700°C than 550°C. It must be realized that *in vitro* systems do not effectively reproduce the *in vivo* situation, particularly the adaptation of rumen microbiome to novel materials, and for this reason, effects *in vitro* might not be observed *in vivo* [5].

The information about biochar characteristics (besides pyrolysis temperature), is absent in the majority of the studies, making impossible any association between the results and the biochar characteristics and their respective effects on CH_4 production. Despite not having evaluated the effect on CH_4 mitigation, McFarlane *et al.* [39] found biochar particle size to affect rumen fermentation, being inhibited with large particles (>178 μm *vs.* <178 μm). Although without impact on gas production and

Biomass	Temperature	Time (h)	Incubation level	CH_4 production	Reference
Rice husk	900–1000	24	0, 1, 2, 3, 4, 5	↓with 1% biochar; no further benefits with 2–5% biochar	[51]
		24	0, 0.5, 1	↓with 0.5 and 1%; further reductions with addition of nitrate N and urea	[51]
		24	1.5	↓with adapted inoculum or biochar addition	[57]
	700–900	24, 48	1	↓ with higher reduction at 48 h	[50]
	1000	24	1	↓	[53]
	1000	6, 12, 18, 24	1	↓ at 18–24h	[54]
Pine wood chips	350	2, 6, 12, 24	16	Not affected	[38]
	550				
Corn stover	350				
	550				
Gasified	—	48	9	Not affected	[37]
Straw-based					
Wood-based					
Activated carbon					
Miscanthus straw pellets	550	24	1	↓ with biochar over the control; No differences between sources; ↑ with 700°C over 550°C	[40]
	700		10		
Oilseed rape straw pellets	550		1		
	700		10		
Rice husk	550		1		
	700		10		
Cashew nutshell	300	3, 6, 9, 12, 24, 30, 36, 48	0, 0.75, 1.5, 2.25, 3	↓ with biochar; ↓↓ with biochar and bio fat	[56]
Potato peel	500	24	0, 5, 10	↑ over the control	[42]
Agro-forestry	600	24			
Mixed species of green waste tree pruning	500	6, 12, 24	0, 0.5, 1, 2, 4	Not affected by inclusion level	[55]
Rice straw	300, 500, 700	4, 24, 48	3	↓ with rice straw and corncob in comparison to bamboo at 4 and 48 h; ↓ with increasing temperature	[44]
Corncob					
Bamboo					

Table 1.
Biochar biomass, temperature of pyrolysis (°C), and inclusion level (% dry matter basis) effects on methane (CH_4) production in short-term in vitro studies.

VFA proportions, these authors reported *in vitro* true digestibility of orchard grass hay to be increased by the inclusion of fine biochar particle size [39].

A comparison between studies is further complicated by the diversity of biomass sources used (e.g., rice husk, pine wood, corn stover, cashew nutshell, tree pruning, rice straw, corncob, bamboo) that might affect VFA profile, thus introducing a confounding effect on the mechanism of CH_4 reduction. Most studies that compared the impact of biomass sources on enteric CH_4 production [37, 38, 41, 42] observed no differences among biochar sources. Conversely, Van Dung *et al.* [44] found rice straw and bamboo biomass to reduce CH_4 production compared to corncob, at 4 and 48 h of incubation, but not at 24 h. Moreover, these authors observed an interaction effect between biomass source and pyrolysis temperature [44], supporting the need for a multi-aspect analysis of biochar's chemical and physical properties. The effects on VFA profile were further assessed [38, 40, 42, 55, 56]. In the study of Calvelo Pereira *et al.* [38], despite a decrease in propionate proportion found with some mixtures, which might indicate an increase in H_2 produced, the effects were insufficient to affect CH_4 production. In the study by Cabeza *et al.* [40], the addition of biochar slightly reduced CH_4 production, but it kept unchanged the amounts of total VFA or acetate produced and reduced those of propionate and butyrate. Saenab *et al.* [56] observed a reduction of CH_4 production by 11.5% with 3% [dry matter (DM) basis] cashew nutshell biochar supplementation, although total and individual VFA produced were unaffected. Rodrigues *et al.* [42] attributed the reduction of VFA production through biochar addition to a reduced energy supply for microbial growth. Supplementation of tree pruning biochar up to 4% (DM basis) did not affect CH_4 or VFA content and profile [55].

The study by Leng *et al.* [57] was the only one that evaluated the effect of rumen fluid adapted to biochar. The authors attributed the reduction in CH_4 production with rumen-adapted inoculum to a larger ruminal population that oxidizes CH_4. Indeed, adapted rumen inoculum is expected to present a higher density of methanotrophs [58], possible the effect of biochar on rumen CH_4 is solely due to the increase in potential habitat for this consortium. However, in the study by Leng *et al.* [57], CH_4 reduction was higher with biochar addition to unadapted rumen inoculum than without biochar addition to adapted rumen inoculum. Biochar addition promotes either the association of microorganisms that more efficiently ferment feed materials or facilitates CH_4 oxidation by bringing together methanogenic archaea and methanotrophic consortia [59].

However, from the available studies, the mechanism of CH_4 reduction through biochar is unclear. Although biochar favors methanotrophism in the soil [60], the anaerobic rumen precludes the growth of aerobic methanotrophs, thus the action of biochar is most possibly through the promotion of micro-environments by the large surface area of biochar [40].

3.2 *In vitro* long-term studies

The long-term effects of biochar supplementation on rumen fermentation and CH_4 production were further assessed *in vitro* using the rumen simulation technique system (**Table 2**). Despite differences among biochar biomass, pyrolysis temperature, and chemical and physical characteristics, only one study observed a CH_4 mitigation effect of biochar when compared to control [41]; supplementation levels (0.5, 1, and 2%, DM basis) having a quadratic effect, greatest with 0.5% inclusion. Jackpine biochar also improved most fermentation parameters (e.g., NH_3-N, total VFA,

Biomass	Temperature	Inclusion level	Substrate	Effects	Reference
Jackpine	600	0, 0.5, 1, 2	Barley silage: rolled barley grain: canola meal: concentrate (60:27:10:3)	Compared to control; ↓ CH_4 and ↑ VFA; = gas, pH, protozoa; linearly ↑ NH_3-N, DMD, CPD, NDFD, ADFD, total and LAB microbial N	[44]
Hardwood blackbutt, clay, and minerals	650	0, 3.6, 7.2	Oaten pasture: maize silage: concentrate (35:35:30)	= CH_4 and total gas, pH, NH_3-N, VFA, DMD, microbial richness, and diversity; 7.2% tended to ↓ CH_4 compared to 3.6%	[61]
Spruce steem	450	2	Barley silage: rolled barley grain: canola meal: premix (60:27:10:3)	Tended to ↓ CH_4 (% total gas); = total gas, pH, VFA, protozoa, microbial N, bacterial richness, diversity, and relative abundance	[62]
Jackpine/ yellow pine	400–600	2	Barley silage: rolled barley grain: canola meal: premix (60:27:10:3)	= CH_4, total gas, pH, VFA, protozoa, microbial N, bacterial richness, diversity, and relative abundance	[43]

NH_3-N- ammonia-N, DMD- dry matter digestibility, CPD- crude protein digestibility, NDFD- neutral detergent fiber digestibility, ADFD- acid detergent fiber, LAB- liquid associated bacteria, and VFA- volatile fatty acids.

Table 2.
Biochar biomass, temperature of pyrolysis (°C), and inclusion levels (% dry matter basis) effects on rumen fermentation and methane (CH_4) production in long-term in vitro studies.

acetate, propionate, butyrate, and branched-chain VFA yield), nutrient digestibility (DM, crude protein, neutral detergent fiber, acid detergent fiber), and microbial N of total and liquid associated bacteria while decreased that of loosely associated bacteria [41]. Conversely, mineral-activated blackbutt [61], jack/yellow pine [43] and spruce stem [62] biochar supplementation kept unaffected gas production, fermentation parameters (pH, NH_3-N, total and individual VFA yield), nutrient digestibility, microbial N produced, protozoa count, and bacterial diversity, richness, and relative abundance. Inconsistency of biochar effects has been attributed to variations in biochar chemical and physical properties, including particle size, adsorptive potential, electrical conductivity, and electron-mediation in redox reactions [37, 39]. Several modification methods have been used to improve biochar properties, such as acidification of surface area, to increase biochar adsorption [23]. Teoh *et al.* [61] further suggested that biochar pH could be of particular importance in enteric CH_4 reduction, based on the notable CH_4 reduction (25%, as mg/g DM incubated) of the acidic (pH 4.8) jack pine biochar used in Saleem *et al.* [41] study. Acidic biochar has been associated with improved carbon sequestrum and higher redox potential in soils, whereas neutral mineral-rich biochar lacked this ability [63]. However, acidic (pH 4.9) pine biochar failed to reduce enteric CH_4 production [43] similarly to observed with basic (pH 8.2) biochar supplementation [37, 38, 61].

Acidic biochar has also been suggested to improve the redox potential and thus increase biofilm development by the mediation of electrons among the microbial population [61, 64]. However, more developed biofilms were observed on readily digestible substrates than on biochar surfaces [62, 65]. Even though microbial diversity, richness, and relative abundance were not affected by long-term biochar supplementation, discriminant analysis unveiled biochar-type specific changes in rumen bacterial families [43, 61, 62]. Of particular interest, Teoh *et al.* [61] found a 19.8-fold reduction in the abundance of *Methanomethylophilaceae* with the supplementation of mineral-activated biochar. Members of *Methanomethylophilaceae* family are methanogenic archaea that use sources of hydrogen to reduce methylated compounds and produce CH_4 [66, 67], thus suggesting the potential mitigation effect of hardwood biochar [61].

4. Effects of biochar *in vivo*

The porous structure of biochar can adsorb gases and provide habitat for microbial biofilms [37, 68], which in addition to electron-mediation properties in biological redox reactions [69] suggest its potential to reduce enteric CH_4 production and promote rumen fermentation. As previously stated, *in vitro* studies present several advantages, but do not fully simulate the *in vivo* animal. Few studies have evaluated, *in vivo*, the effects of dietary biochar inclusion on ruminant performance and CH_4 production (**Table 3**). Globally, dietary supplementation with biochar from different sources increased or not affected ruminant performance and reduced or kept

Animals	Diet	Biochar level (source)	Observations	Reference
Cattle (80–100 kg)	Cassava root chips and fresh cassava foliage	0.6 (rice husks)	Live weight gain ↑ 25%; ↑ DM feed conversion; ↓ CH_4 production	[52]
Angus × Hereford heifers (565 ± 35 kg)	Barley silage-based diet	0, 0.5, 1, 2 (pine-enhanced biochar)	CH_4 emissions not affected; Specific rumen microbiota altered	[65]
Crossbred steers (529 ± 16 kg)	Growing diet: brome hay: wheat straw: corn silage: wet distillers' grains: supplement (21:20:30:22:7) Finishing diet: dry-rolled corn: corn silage: wet distillers: supplement (53:15:25:7)	0, 0.8, 3 (whole pine trees)	CH_4 tended to decrease in the growing animals; CH_4 is not affected in the finishing animals	[70]
Lambs (37.9 ± 0.8 kg)	Alfalfa and barley (60:40) *ad libitum*	0, 2 (Lodgepole pine and quaking aspen)	= feed intake and average daily gain; ↑ DM digestibility and digestible DM intake	[71]
Kermanian ram lambs (21.9 ± 2.24 kg)	Alfalfa: wheat straw: concentrate (30:10:60)	0, 1, 1.5 (Walnut shell and pistachio by-product at 1%, chicken manure at 1.5%)	= DM intake; ↑ average daily gain; ↑ feed conversion ratio	[72]
Bos taurus crossbred beef steers initial (286 ± 26 kg)	High-forage and high-grain diets	0, 0.5, 1, 2 (Yellow pine)	2% lean meat yield; = body weight and DM intake	[73]

Animals	Diet	Biochar level (source)	Observations	Reference
Milking dairy cows	Barley hay and compound (40:6) free-access to forage during the day	0.5 (powdered activated carbon)	↓ manure CH_4 by 30–40% and CO_2 emissions by 10%; ↑ milk production; ↓ manure methanogenic flora by 30%; ↑ nonmethanogenic species	[74]

DM- dry matter, and CO_2- carbon dioxide.

Table 3.
Effect of biochar biomass and inclusion level (% dry matter basis) on ruminant performance and methane (CH_4) production.

unaffected CH_4 production. Leng *et al.* [52] pointed out the need for CH_4 mitigation strategies to include alternative electron sinks rather than just focused on methanogens inhibition, due to the need for symbiotic associations in biofilm microbial colonies on feed particles for successful ruminal fermentation to occur. Rumen microbial biofilms are of particular importance for fiber fermentation, with microbial attachment to feed particles allowing pit formation as well as glycocalyx emission to fibrous amorphous material [75].

In Angus × Hereford heifers, Terry *et al.* [65] found that, although total tract digestibility, nitrogen balance, and CH_4 production were not affected by dietary biochar inclusion, the relative abundance of *Fibrobacter* and *Tenericutes* were reduced and that of *Spirochaetaes*, *Verrucomicrobia*, and *Elusimicrobia* increased. Modulation of the manure microbial population was also found to be affected by dietary biochar supplementation. Al-Azzawi *et al.* [74] reported decreased methanogenic population by 30% with a corresponding increase in the non-methanogenic archaeal species in manure, suggesting that formed CH_4 could be reduced by further utilization by methanotrophic species. Moreover, biochar was shown to affect nitrification by increasing ammonia-oxidizing organisms and reducing ammonooxygenase activity [76].

Although dietary biochar supplementation had variable effects on ruminant performance, these were overall promising and suggest potential benefits beyond methanogenesis. Indeed, 0.6% biochar increased the live weight gain of yellow cattle and DM feed conversion by 25% [52]. Terry *et al.* [73] found no effect on body weight gain or DM intake in beef steers up to 2% biochar, but lean meat yield increased with the highest biochar level tested (2%). In lambs, 2% biochar kept feed intake and average daily gain unaffected, and improved DM intake [71], while up to 1.5% biochar was found to maintain DM intake and increase average daily gain and feed conversion ratio [72]. In addition, milk production of cows fed 0.5% (DM basis) activated carbon was improved [74]. Furthermore, in an innovative solution for biochar utilization reported by Joseph *et al.* [77], biochar was mixed with molasses and fed directly to cows, the dung-biochar mixture being incorporated into the soil profile by dung beetles and the costs and benefits of integrating biochar with animal husbandry and improvement of pastures were assessed. These authors found that dung-biochar had an outer coating of mineral elements (P, K, Mg, Ca, Al, Si, and Fe) and nitrogen, adsorbed in the cow gut, that were available for soil, thus being an effective strategy to improve soil properties. In addition, increasing returns to farmers were calculated, suggesting the profitability of dietary biochar supplementation in ruminant production systems [77].

Notwithstanding, the inconsistent results in the literature on the effect of biochar on reducing CH_4 emissions, rumen *in vitro* fermentation, and *in vivo* rumen function limits the mechanistic understanding of the underlying mode of action. This is particularly difficult due to the use of different sources of biomass and production conditions, such as duration and temperature, of pyrolysis as well as post-treatment modifications, which alter the composition, porosity, and chemistry of biochar [65], but also to the poorly characterized biochar used in ruminant studies. These challenges make comparisons between studies difficult, and in addition to the lack of knowledge of the long-term effects of dietary biochar supplementation, could have limited its use in ruminant feeding practices on-farm.

5. Effects of biochar on manure CH_4 production

Ruminant production generates high amounts of manure that need to be stored until the land application. Manure is a rich source of nutrients, and its application is shown to improve soil quality, to reduce the use of mineral fertilizers and costs of production [21]. However, during manure storage and land application, malodorous compounds and GHG, such as CH_4, CO_2, and N_2O, as well as NH_3, are formed and emitted [78], with a detrimental impact on ecosystems [22]. Biochar application to manure can be an effective strategy to improve its environmental impact, as it can absorb and retain GHG, NH_3, and nutrients [79, 80]. Moreover, when applied to soils, biochar-enriched manure may provide nutrients, sequester carbon, and improve soil's structure [22, 79]. Although the already identified biochar potential in manure, differences have been reported among biochar biomass, production conditions, pH, hydrophobicity, and particle size [22, 68, 81]. Moreover, a life cycle assessment of the environmental implications of stored cattle slurry (a mixture of manure, split feed, and water) treatments revealed biochar to be one of the less effective approaches to suppress GHG emissions from liquid slurry, except for N_2O [21]. The inconsistent results from biochar application to manure pinpoint the need for more research in this field.

6. Conclusions

Biochar is undoubtedly a material with high potential to deal with ruminant methanogenesis due to its availability, stability, and large surface areas. Nevertheless, there is a significant knowledge gap about the mechanisms that govern the interactions between biochar and the plethora of microorganisms that are present in the ruminant's gut and manure. In this chapter, we addressed the most relevant literature on the topic, seeking additional clarification about the potential role of biochar in methanogenesis. The absence of detailed characterization of biochar used, and the diversity of the experimental conditions applied in the different studies, create additional challenges for a critical comparison of the past findings. Therefore, for future studies, some level of standardization and the detailed characterization of the biochar(s) used will have a significant impact on the clarification of its role in the mitigation of GHG emissions from ruminants.

Acknowledgements

The work received financial support from the Portuguese Foundation for Science and Technology (FCT) through projects UIDB/04033/2020, UIDB/50006/2020, UIDP/50006/2020, and OPTIMA: Optical monitoring of environmental emissions of ammonia by an integrated and autonomous membrane-based fluorescence platform (PTDC/CTA-AMB/31559/2017), and from project R&W Clean: new solutions for sensing environmental and biological parameters to help demedicalize the agricultural sector (POCI- 01-0247-FEDER-70109) supported by PORTUGAL 2020 program through the European Regional Development Fund. Ana R.F. Rodrigues thanks FCT and European Social Fund through Programa Operacional Capital Humano (POCH) and SANFEED Doctoral Programme, for funding her Ph.D. grant (PDE/BDE/114434/2016). Margarida R.G. Maia and Inês M. Valente thank FCT for funding through program DL 57/2016 – Norma transitória (Ref. SFRH/BPD/70176/2010 and SFRH/BPD/111181/2015, respectively).

Conflict of interest

The authors declare no conflict of interest.

Author details

Ana R.F. Rodrigues[1†], Margarida R.G. Maia[1], Ana R.J. Cabrita[1], Hugo M. Oliveira[2], Inês M. Valente[1,3], José L. Pereira[4,5], Henrique Trindade[5] and António J.M. Fonseca[1*]

1 REQUIMTE, LAQV, ICBAS, School of Medicine and Biomedical Sciences, University of Porto, Porto, Portugal

2 INL, International Iberian Nanotechnology Laboratory, Braga, Portugal

3 REQUIMTE, LAQV, Chemistry and Biochemistry Department, Faculty of Sciences University of Porto, Porto, Portugal

4 Centre for the Research and Technology of Agro-Environmental and Biological Sciences (CITAB), Inov4Agro, University of Trás-os-Montes and Alto Douro, Vila Real, Portugal

5 Agrarian School of Viseu, Polytechnic Institute of Viseu, Viseu, Portugal

*Address all correspondence to: ajfonseca@icbas.up.pt

† Present Address: FeedInov CoLab, Estação Zootécnica Nacional, R. Professor Doutor Vaz Portugal, 2005-424, Vale de Santarém.

References

[1] Gerber PJ, Steinfeld H, Henderson B, Mottet A, Opio C, Dijkman J, et al. Tackling Climate Change through Livestock – A Global Assessment of Emissions and Mitigation Opportunities. Rome: Food and Agriculture Organization of the United Nations (FAO); 2013

[2] Janssen PH. Influence of hydrogen on rumen methane formation and fermentation balances through microbial growth kinetics and fermentation thermodynamics. Animal Feed Science and Technology. 2010;**1601**(2):1-22. DOI: 10.1016/j.anifeedsci.2010.07.002

[3] Wang K, Xiong B, Zhao X. Could propionate formation be used to reduce enteric methane emission in ruminants? Science of the Total Environment. 2023; **855**:158867. DOI: 10.1016/j. scitotenv.2022.158867

[4] Johnson K, Johnson D. Methane emissions from cattle. Journal of Animal Science. 1995;**738**:2483-2492. DOI: 10.2527/1995.7382483x

[5] Hristov AN, Oh J, Firkins JL, Dijkstra J, Kebreab E, Waghorn G, et al. SPECIAL TOPICS—Mitigation of methane and nitrous oxide emissions from animal operations: I. A review of enteric methane mitigation options. Journal of Animal Science. 2013;**91**: 5045-5069. DOI: 10.2527/jas2013-6583

[6] Knapp JR, Laur GL, Vadas PA, Weiss WP, Tricarico JM. Invited review: Enteric methane in dairy cattle production: Quantifying the opportunities and impact of reducing emissions. Journal of Dairy Science. 2014;**976**:3231-3261. DOI: 10.3168/ jds.2013-7234

[7] Beauchemin KA, Ungerfeld EM, Eckard RJ, Wang M. Review: Fifty years of research on rumen methanogenesis: Lessons learned and future challenges for mitigation. Animal. 2020;**14**:S2-S16. DOI: 10.1017/s1751731119003100

[8] Zhuang M, Shan N, Wang Y, Caro D, Fleming RM, Wang L. Different characteristics of greenhouse gases and ammonia emissions from conventional stored dairy cattle and swine manure in China. Science of the Total Environment. 2020;**722**:137693. DOI: 10.1016/j.scito tenv.2020.137693

[9] Kavanagh I, Burchill W, Healy MG, Fenton O, Krol DJ, Lanigan GJ. Mitigation of ammonia and greenhouse gas emissions from stored cattle slurry using acidifiers and chemical amendments. Journal of Cleaner Production. 2019;**237**:117822. DOI: 10.1016/j.jclepro.2019.117822

[10] Zynda HM, Copelin JE, Weiss WP, Sun F, Lee C. Effects of reducing dietary cation-anion difference on lactation performance and nutrient digestibility of lactating cows and ammonia emissions from manure. Journal of Dairy Science. 2022;**1055**:4016-4031. DOI: 10.3168/ jds.2021-21195

[11] Yogalakshmi KN, Poornima DT, Sivashanmugam P, Kavitha S, Yukesh KR, Sunita V, et al. Lignocellulosic biomass-based pyrolysis: A comprehensive review. Chemosphere. 2022;**286**:131824. DOI: 10.1016/j.chemo sphere.2021.131824

[12] Xu X, Zhao Y, Sima J, Zhao L, Mašek O, Cao X. Indispensable role of biochar-inherent mineral constituents in its environmental applications: A review. Bioresource Technology. 2017;**241**: 887-899. DOI: 10.1016/j.biortech.2017. 06.023

[13] Xiu LQ, Zhang WM, Sun YY, Wu D, Meng J, Chen WF. Effects of biochar and straw returning on the key cultivation limitations of Albic soil and soybean growth over 2 years. Catena. 2019;**173**: 481-493. DOI: 10.1016/j.catena.2018.10.041

[14] Sun Y, Xiong X, He M, Xu Z, Hou D, Zhang W, et al. Roles of biochar-derived dissolved organic matter in soil amendment and environmental remediation: A critical review. Chemical Engineering Journal. 2021;**424**:130387. DOI: 10.1016/j.cej.2021.130387

[15] Huang J, Zhu C, Kong Y, Cao X, Zhu L, Zhang Y, et al. Biochar application alleviated rice salt stress via modifying soil properties and regulating soil bacterial abundance and community structure. Agronomy. 2022;**12**(2):409. DOI: 10.3390/agronomy12020409

[16] Pokharel P, Chang SX. Biochar decreases the efficacy of the nitrification inhibitor nitrapyrin in mitigating nitrous oxide emissions at different soil moisture levels. Journal of Environmental Management. 2021;**295**:113080. DOI: 10.1016/j.jenvman.2021.113080

[17] Wu Z, Song YF, Shen HJ, Jiang XY, Li B, Xiong ZQ. Biochar can mitigate methane emissions by improving methanotrophs for prolonged period in fertilized paddy soils. Environmental Pollution. 2019;**253**:1038-1046. DOI: 10.1016/j.envpol.2019.07.073

[18] Nan Q, Xin LQ, Qin Y, Waqas M, Wu WX. Exploring long-term effects of biochar on mitigating methane emissions from paddy soil: A review. Biochar. 2021; **32**:125-134. DOI: 10.1007/s42773-021-00096-0

[19] Liu J, Qiu H, Wang C, Shen J, Zhang W, Cai J, et al. Effects of biochar amendment on greenhouse gas emission in two paddy soils with different textures. Paddy and Water Environment. 2021;**191**:87-98

[20] Pan X, Gu Z, Chen W, Li Q. Preparation of biochar and biochar composites and their application in a Fenton-like process for wastewater decontamination: A review. Science of the Total Environment. 2021;**754**: 142104. DOI: 10.1016/j.scitotenv.2020.142104

[21] Miranda C, Soares AS, Coelho AC, Trindade H, Teixeira CA. Environmental implications of stored cattle slurry treatment with sulphuric acid and biochar: A life cycle assessment approach. Environmental Research. 2021;**194**:110640. DOI: 10.1016/j.envres.2020.110640

[22] Dougherty B, Gray M, Johnson MG, Kleber M. Can biochar covers reduce emissions from manure lagoons while capturing nutrients? Journal of Environmental Quality. 2017;**463**: 659-666. DOI: 10.2134/jeq2016.12.0478

[23] Zhang Y, Wang J, Feng Y. The effects of biochar addition on soil physicochemical properties: A review. Catena. 2021;**202**:105284. DOI: 10.1016/j.catena.2021.105284

[24] Blanco-Canqui H. Does biochar application alleviate soil compaction? Review and data synthesis. Geoderma. 2021;**404**:115317. DOI: 10.1016/j.geoderma.2021.115317

[25] Schmidt H-P, Hagemann N, Draper K, Kammann C. The use of biochar in animal feeding. PeerJ. 2019;7: e7373. DOI: 10.7717/peerj.7373

[26] Man KY, Chow KL, Man YB, Mo WY, Wong MH. Use of biochar as feed supplements for animal farming. Critical Reviews in Environmental

Science and Technology. 2021;**512**: 187-217. DOI: 10.1080/ 10643389.2020.1721980

[27] Osman AI, Fawzy S, Farghali M, El-Azazy M, Elgarahy AM, Fahim RA, et al. Biochar for agronomy, animal farming, anaerobic digestion, composting, water treatment, soil remediation, construction, energy storage, and carbon sequestration: A review. Environmental Chemistry Letters. 2022;**204**:2385-2485. DOI: 10.1007/s10311-022-01424-x

[28] European Biochar Certificate (EBC). Guidelines for a Sustainable Production of Biochar. Arbaz, Switzerland: Biochar Foundation (EBC); 2012-2022

[29] Council Regulation (EC) No 178/ 2002 of the European Parliament and of the Council of 28 January 2002 laying down the general principles and requirements of food law, establishing the European Food Safety Authority and laying down procedures in matters of food safety

[30] Council Regulation (EC) No 834/ 2007 of 28 June 2007 on organic production and labelling of organic products and repealing Regulation (EEC) No 2092/91

[31] Guo X-x, Liu H-t, Zhang J. The role of biochar in organic waste composting and soil improvement: A review. Waste Management. 2020;**102**:884-899. DOI: 10.1016/j.wasman.2019.12.003

[32] Seo MW, Lee SH, Nam H, Lee D, Tokmurzin D, Wang S, et al. Recent advances of thermochemical conversion processes for biorefinery. Bioresource Technology. 2022;**343**:126109. DOI: 10.1016/j.biortech.2021.126109

[33] Ghodake GS, Shinde SK, Kadam AA, Saratale RG, Saratale GD, Kumar M, et al. Review on biomass feedstocks, pyrolysis mechanism and physicochemical properties of biochar: State-of-the-art framework to speed up vision of circular bioeconomy. Journal of Cleaner Production. 2021;**297**:126645. DOI: 10.1016/j.jclepro.2021.126645

[34] Yaashikaa PR, Kumar PS, Varjani S, Saravanan A. A critical review on the biochar production techniques, characterization, stability and applications for circular bioeconomy. Biotechnology Reports. 2020;**28**:e00570. DOI: 10.1016/j.btre.2020.e00570

[35] Gwenzi W, Chaukura N, Wenga T, Mtisi M. Biochars as media for air pollution control systems: Contaminant removal, applications and future research directions. Science of the Total Environment. 2021;**753**:142249. DOI: 10.1016/j.scitotenv.2020.142249

[36] Premarathna KSD, Rajapaksha AU, Sarkar B, Kwon EE, Bhatnagar A, Ok YS, et al. Biochar-based engineered composites for sorptive decontamination of water: A review. Chemical Engineering Journal. 2019;**372**:536-550. DOI: 10.1016/j.cej.2019.04.097

[37] Hansen HH, Storm IMLD, Sell AM. Effect of biochar on *in vitro* rumen methane production. Acta Agriculturae Scandinavica, Section A — Animal Science. 2012;**624**:305-309. DOI: 10.1080/09064702.2013.789548

[38] Calvelo Pereira R, Muetzel S, Camps Arbestain M, Bishop P, Hina K, Hedley M. Assessment of the influence of biochar on rumen and silage fermentation: A laboratory-scale experiment. Animal Feed Science and Technology. 2014;**196**:22-31. DOI: 10.1016/j.anifeedsci.2014.06.019

[39] McFarlane ZD, Myer PR, Cope ER, Evans ND, Bone TC, Bliss BE, et al. Effect of biochar type and size on *in vitro*

rumen fermentation of orchard grass hay. Agricultural Sciences. 2017;**8**: 316-325. DOI: 10.4236/as.2017.84023

[40] Cabeza I, Waterhouse T, Sohi S, Rooke JA. Effect of biochar produced from different biomass sources and at different process temperatures on methane production and ammonia concentrations *in vitro*. Animal Feed Science and Technology. 2018;**237**:1-7. DOI: 10.1016/j.anifeedsci.2018.01.003

[41] Saleem AM, Ribeiro GO Jr, Yang WZ, Ran T, Beauchemin KA, McGeough EJ, et al. Effect of engineered biocarbon on rumen fermentation, microbial protein synthesis, and methane production in an artificial rumen (RUSITEC) fed a high forage diet. Journal of Animal Science. 2018; **968**:3121-3130. DOI: 10.1093/jas/sky204

[42] Rodrigues ARF, Maia MRG, Cabrita ARJ, Oliveira HM, Bernardo M, Lapa N, et al. Assessment of potato peel and agro-forestry biochars supplementation on *in vitro* ruminal fermentation. PeerJ. 2020;**8**:18. DOI: 10.7717/pearj.9488

[43] Tamayao P, Ribeiro GO, McAllister TA, Ominski KH, Saleem AM, Yang HE, et al. Effect of pine-based biochars with differing physiochemical properties on methane production, ruminal fermentation, and rumen microbiota in an artificial rumen (RUSITEC) fed barley silage. Canadian Journal of Animal Science. 2021;**1013**:577-589. DOI: 10.1139/cjas-2020-0129

[44] Van Dung D, Thao LD, Ngoan LD, Phung LD, Roubik H. Effects of biochar produced from tropical rice straw, corncob, and bamboo tree at different processing temperatures on *in vitro* rumen fermentation and methane production. Biomass Conversion and Biorefinery. 2022;**8**. DOI: 10.1007/s13399-022-02592-0

[45] Jung S, Park YK, Kwon EE. Strategic use of biochar for CO_2 capture and sequestration. Journal of CO_2 Utilization. 2019;**32**:128-139. DOI: 10.1016/j.jcou.2019.04.012

[46] Zhang X, Zhang SH, Yang HP, Feng Y, Chen YQ, Wang XH, et al. Nitrogen enriched biochar modified by high temperature CO_2-ammonia treatment: Characterization and adsorption of CO_2. Chemical Engineering Journal. 2014;**257**:20-27. DOI: 10.1016/j.cej.2014.07.024

[47] Zhang X, Wu J, Yang H, Shao J, Wang X, Chen Y, et al. Preparation of nitrogen-doped microporous modified biochar by high temperature CO_2–NH_3 treatment for CO_2 adsorption: effects of temperature. RSC Advances. 2016;**6100**: 98157-98166. DOI: 10.1039/C6RA23748G

[48] Tomczyk A, Sokolowska Z, Boguta P. Biochar physicochemical properties: Pyrolysis temperature and feedstock kind effects. Reviews in Environmental Science and Bio-Technology. 2020;**191**:191-215. DOI: 10.1007/s11157-020-09523-3

[49] Leng L, Xiong Q, Yang L, Li H, Zhou Y, Zhang W, et al. An overview on engineering the surface area and porosity of biochar. Science of the Total Environment. 2021;**763**:144204. DOI: 10.1016/j.scitotenv.2020.144204

[50] Leng RA, Inthapanya S, Preston TR. All biochars are not equal in lowering methane production in *in vitro* rumen incubations. Livestock Research for Rural Development. 2013;**25**:106

[51] Leng RA, Inthapanya S, Preston TR. Biochar lowers net methane production

from rumen fluid *in vitro*. Livestock Research for Rural Development. 2012; **24**:103

[52] Leng RA, Preston TR, Inthapanya S. Biochar reduces enteric methane and improves growth and feed conversion in local "Yellow" cattle fed cassava root chips and fresh cassava foliage. Livestock Research for Rural Development. 2012; **24**:199

[53] Vongsamphanh P, Napasirth V, Inthapanya S, Preston TR. Effect of biochar and leaves from sweet or bitter cassava on gas and methane production in an *in vitro* rumen incubation using cassava root pulp as source of energy. Livestock Research for Rural Development. 2015;**27**:73

[54] Vongkhamchanh B, Inthapanya S, Preston TR. Methane production in an *in vitro* rumen fermentation is reduced when the carbohydrate substrate is fresh rather than ensiled or dried cassava root, and when biochar is added to the substrate. Livestock Research for Rural Development. 2015;**27**:208

[55] O'Reilly GC, Huo YX, Meale SJ, Chaves AV. Dose response of biochar and wood vinegar on *in vitro* batch culture ruminal fermentation using contrasting feed substrates. Translational Animal Science. 2021;**5**(3):1-13. DOI: 10.1093/tas/txab107

[56] Saenab A, Wiryawan KG, Retnani Y, Wina E. Synergistic effect of biofat and biochar of cashew nutshell on mitigate methane in the rumen. Jurnal Ilmu Ternak Dan Veteriner. 2020;**253**: 139-146. DOI: 10.14334/jitv.v24i3.2475

[57] Leng RA, Inthapanya S, Preston TR. Methane production is reduced in an *in vitro* incubation when the rumen fluid is taken from cattle that previously received biochar in their diet. Livestock Research for Rural Development. 2012; **24**:211

[58] Kajikawa H, Valdes C, Hillman K, Wallace RJ, Newbold CJ. Methane oxidation and its coupled electron-sink reactions in ruminal fluid. Letters in Applied Microbiology. 2003;**366**: 354-357. DOI: 10.1046/ j.1472-765X.2003.01317.x

[59] Knittel K, Boetius A. Anaerobic oxidation of methane: Progress with an unknown process. Annual Review of Microbiology. 2009;**63**:311-334. DOI: 10.1146/annurev. micro.61.080706.093130

[60] Sonoki T, Furukawa T, Jindo K, Suto K, Aoyama M, Sanchez-Monedero MA. Influence of biochar addition on methane metabolism during thermophilic phase of composting. Journal of Basic Microbiology. 2013;**537**: 617-621. DOI: 10.1002/jobm.201200096

[61] Teoh R, Caro E, Holman DB, Joseph S, Meale SJ, Chaves AV. Effects of hardwood biochar on methane production, fermentation characteristics, and the rumen microbiota using rumen simulation. Frontiers in Microbiology. 2019;**10**:1534. DOI: 10.3389/fmicb.2019.01534

[62] Tamayao PJ, Ribeiro GO, McAllister TA, Yang HE, Saleem AM, Ominski KH, et al. Effects of post-pyrolysis treated biochars on methane production, ruminal fermentation, and rumen microbiota of a silage-based diet in an artificial rumen system (RUSITEC). Animal Feed Science and Technology. 2021;**273**:114802. DOI: 10.1016/j.anifeedsci.2020.114802

[63] Qi FJ, Dong ZM, Lamb D, Naidu R, Bolan NS, Ok YS, et al. Effects of acidic and neutral biochars on properties and

cadmium retention of soils. Chemosphere. 2017;**180**:564-573. DOI: 10.1016/j.chemosphere.2017.04.014

[64] Leng RA. Interactions between microbial consortia in biofilms: A paradigm shift in rumen microbial ecology and enteric methane mitigation. Animal Production Science. 2014;**545**: 519-543. DOI: 10.1071/AN13381

[65] Terry SA, Ribeiro GO, Gruninger RJ, Chaves AV, Beauchemin KA, Okine E, et al. A pine enhanced biochar does not decrease enteric CH_4 emissions, but alters the rumen microbiota. Frontiers in Veterinary Science. 2019;**6**:308. DOI: 10.3389/fvets.2019.00308

[66] Borrel G, Parisot N, Harris HMB, Peyretaillade E, Gaci N, Tottey W, et al. Comparative genomics highlights the unique biology of Methanomassiliicoccales, a Thermoplasmatales-related seventh order of methanogenic archaea that encodes pyrrolysine. BMC Genomics. 2014;**151**:679. DOI: 10.1186/1471-2164-15-679

[67] Lang K, Schuldes J, Klingl A, Poehlein A, Daniel R, Brune A. New mode of energy metabolism in the seventh order of methanogens as revealed by comparative genome analysis of "*Candidatus Methanoplasma termitum*". Applied and Environmental Microbiology. 2015;**814**:1338-1352. DOI: 10.1128/AEM.03389-14

[68] Lehmann J, Joseph S. Biochar for Environmental Management: Science, Technology and Implementation. 2nd ed. Routledge; 2015. DOI: 10.4324/9780203762264

[69] Yu L, Yuan Y, Tang J, Wang Y, Zhou S. Biochar as an electron shuttle for reductive dechlorination of pentachlorophenol by *Geobacter sulfurreducens*. Scientific Reports. 2015; **51**:16221. DOI: 10.1038/srep16221

[70] Winders TM, Jolly-Breithaupt ML, Wilson HC, MacDonald JC, Erickson GE, Watson AK. Evaluation of the effects of biochar on diet digestibility and methane production from growing and finishing steers. Translational Animal Science. 2019; **32**:775-783. DOI: 10.1093/tas/txz027

[71] McAvoy DJ, Burritt B, Villalba JJ. Use of biochar by sheep: impacts on diet selection, digestibility, and performance. Journal of Animal Science. 2020;**98**(12): 1-9. DOI: 10.1093/jas/skaa380

[72] Mirheidari A, Torbatinejad NM, Shakeri P, Mokhtarpour A. Effects of biochar produced from different biomass sources on digestibility, ruminal fermentation, microbial protein synthesis and growth performance of male lambs. Small Ruminant Research. 2020;**183**:106042. DOI: 10.1016/j.smallrumres.2019.106042

[73] Terry SA, Redman AAP, Ribeiro GO, Chaves AV, Beauchemin KA, Okine E, et al. Effect of a pine enhanced biochar on growth performance, carcass quality, and feeding behavior of feedlot steers. Translational Animal Science. 2020;**42**: 831-838. DOI: 10.1093/tas/txaa011

[74] Al-Azzawi M, Bowtell L, Hancock K, Preston S. Addition of activated carbon into a cattle diet to mitigate GHG emissions and improve production. Sustainability. 2021;**13**(15):8254. DOI: 10.3390/su13158254

[75] Rodrigues MAM, Cone JW, van Gelder AH, Sequeira JC, Fonseca AM, Ferreira LMM, et al. The effect of cellulose crystallinity on the *in vitro* digestibility and fermentation kinetics of meadow hay and barley, wheat and rice straws. Journal of the Science of Food

and Agriculture. 2003;**837**:652-657. DOI: 10.1002/jsfa.1338

[76] Deng L, Zhao M, Bi R, Bello A, Uzoamaka Egbeagu U, Zhang J, et al. Insight into the influence of biochar on nitrification based on multi-level and multi-aspect analyses of ammonia-oxidizing microorganisms during cattle manure composting. Bioresource Technology. 2021;**339**:125515. DOI: 10.1016/j.biortech.2021.125515

[77] Joseph S, Pow D, Dawson K, Mitchell DRG, Rawal A, Hook J, et al. Feeding biochar to cows: An innovative solution for improving soil fertility and farm productivity. Pedosphere. 2015;**255**: 666-679. DOI: 10.1016/S1002-0160(15) 30047-3

[78] Leytem AB, Dungan RS, Bjorneberg DL, Koehn AC. Emissions of ammonia, methane, carbon dioxide, and nitrous oxide from dairy cattle housing and manure management systems. Journal of Environmental Quality. 2011; **405**:1383-1394. DOI: 10.2134/ jeq2009.0515

[79] Ghezzehei TA, Sarkhot DV, Berhe AA. Biochar can be used to capture essential nutrients from dairy wastewater and improve soil physico-chemical properties. Solid Earth. 2014; **52**:953-962. DOI: 10.5194/se-5-953-2014

[80] Sarkhot DV, Ghezzehei TA, Berhe AA. Effectiveness of biochar for sorption of ammonium and phosphate from dairy effluent. Journal of Environmental Quality. 2013;**425**: 1545-1554. DOI: 10.2134/jeq2012.0482

[81] Zhang X, Wang H, He L, Lu K, Sarmah A, Li J, et al. Using biochar for remediation of soils contaminated with heavy metals and organic pollutants. Environmental Science and Pollution Research. 2013;**2012**:8472-8483. DOI: 10.1007/s11356-013-1659-0

Chapter 2

Biochar: Production, Application and the Future

Edward Kwaku Armah, Maggie Chetty, Jeremiah Adebisi Adedeji, Denzil Erwin Estrice, Boldwin Mutsvene, Nikita Singh and Zikhona Tshemese

Abstract

Biochar, or carbon obtained from biomass, is a particularly rich source of carbon created by thermal burning of biomass. There is a rise of interest in using biochar made from waste biomass in a variety of disciplines to address the most pressing environmental challenges. This chapter will provide an overview on the methods employed for the production of biochar. Biochar has been considered by a number of analysts as a means of improving their ability to remediate pollutants. Process factors with regards to biochar properties are mostly responsible for determining biomass production which is discussed in this present chapter. Several characterization techniques which have been employed in previous studies have received increasing recognition. These includes the use of the Fourier transform infrared spectroscopy and the Scanning electron microscope which duly presented in this chapter. This chapter also discusses the knowledge gaps and future perspectives in adopting biochar to remediate harmful contaminants, which can inform governmental bodies and law-makers to make informed decisions on adopting this residue.

Keywords: biochar, biomass, characterization, future perspective, pyrolysis, pretreatment

1. Introduction

The word char, is a common terminology used for the solid product of the combustion of carbonaceous material [1]. Generally, char product is rich in carbon content; an example is charcoal, which is almost the earliest invention of humans from fire or heat creation. Another vivid example of char is biochar. In this case, the study, is made from organic compounds such as forest, agricultural or animal products but in the absence/limited supply of oxygen compared to charcoal. Therefore, biochar is derived from biomass combustion in the presence of a limited oxygen supply and at relatively low temperatures below 700°C. The earliest known purpose for creating biochar was specifically for soil application such as carbon storage or sequestration in soil; improvement of soil performance such as increase in nutrient availability, reduction of compactness in soil, soil pH improvement; soil water filtration. Recent applications involve energy production, biochemical process stability and improvement, climate change mitigation, and construction additive [1–3]. The raw material determines carbonized organic matter properties and the operational parameters used during it production. Pyrolysis (slow or fast) and gasification are the main methods for the production of biochar. The physical nature of the biochar produced is directly affected by the chemical composition of the biomass feedstock. Most organic matter begin to thermally decompose at temperatures above 120°C. Hemicelluloses degrade between 200 and 260°C, cellulose between 240 and 350°C, and lignin between 280 and 500°C. As a result, the proportions of these components will affect the degree of reactivity and, as a result, the extent to which the physical structure is modified during processing [4]. Biochar is characterized with high porosity with pores ranging in size from micro to macropores. Large holes, which originate from the raw biomass's vascular bundles, are critical for increasing soil quality because they can serve as habitats for symbiotic microbes. Biochar major components are carbon, volatile matter, mineral matter (ash), and moisture. The percentage composition of each components varies based on the feedstock material and the operating parameters [1]. Biochar from plant-based materials have higher carbon composition which range from as low as 51% to as high. The understanding of the key mechanisms for changes in physicochemical properties of biochar during processing for various feedstock types and operating parameters is required to determine biochar's potential for application both now and in future. Therefore, this chapter explains biochar production techniques, factors affecting its properties and compositions and its application.

2. Biochar production techniques

An ever-growing appetency for using biochar for various applications has orchestrated an increase in converting it into biochar. Thermochemical conversion is a common technology for making biochar. Thermochemical conversion techniques are pyrolysis, hydrothermal carbonization (HTC), gasification, torrefaction, and hydrothermal liquefaction [5, 6].

2.1 Pyrolysis

Pyrolysis is a thermochemical technique that produces biochar, bio-oil, and syngas derived from biomass [7]. The process involves heating and thermally decomposing biomass under anaerobic conditions or limited oxygen supply (low stoichiometric oxygen atmosphere) with temperatures ranging between 400°C and 1200°C [2]. The absence of oxygen enables biomass heating beyond its thermal stability limit, causing the creation of more robust products, including solid residues. By creating an anaerobic atmosphere, it is also ensured that combustion will not occur when the biomass is heated. It is a highly complex process involving many distinct reactions in the reacting zone [8]. In another study, a low-temperature range for pyrolysis was recorded between 250°C and 900°C. Biomass from Agriculture comprises lignin, cellulose, hemicelluloses, and silica. Typically, cellulose pyrolyzes at 350°C, whereas the melting point of lignin is well above 350°C [6]. Although the product yield depends on various operating variables, char formation is generally favored by low temperatures and long residence times [9]. Therefore, it can be decoded that the effective temperature range for pyrolysis was between 300 and 700°C. The cracking of heavy chemicals happens in secondary pyrolysis and converts biomass into biochar or gases. **Figure 1** is a summary of the pyrolysis technique and the operating variables affecting pyrolysis.

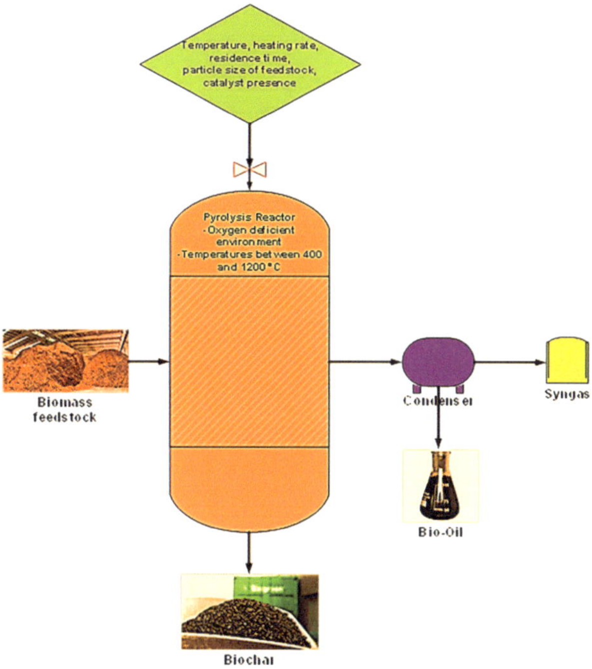

Figure 1.
Schematic representation of pyrolysis process [3].

In essence, this is an alternative way to valorize biomass into various products such as bio-oil, syngas and biochar. Depolymerization, fragmentation, and cross-linking are chemical mechanisms that occur during the process at specific temperature points, resulting in a different product state for lignocellulosic components, including cellulose and hemicellulose (solid, liquid and gas). Biochar and bio-oil are the solid and liquid products, whereas CO_2, CO, H_2, (collectively known as syngas) are evolved as the gaseous by-products (C_1-C_2 hydrocarbons) [3]. Biochar is made in a different type of reactors, such as paddle kiln, bubbling fluidized bed, wagon reactor, and agitated sand rotating kiln. The biomass nature and employed type determine the biochar yield during the pyrolysis route. The major operating parameter that impacts product efficiency is the temperature [10, 11]. When the pyrolysis temperature is increased, biochar's yield decreases and the generation of syngas increases. The gas yield is represented by the initial section of the product side (as shown in Eq. (1)), with various gases created during the process.

$$(C_6H_6O_6)_n \rightarrow (H_2 + CO + CH_4 + \ldots + C_5H_{12}) + (H_2O + CH_3OH + CH_3COOH + \ldots) + C \quad (1)$$

The mixture of multiple sorts of liquid outputs is shown in the second part of the products' side, and the solid yield is represented in the last component [12]. One of the most significant masteries of this technology is that it may be optimized to achieve the desired outcomes. Slow pyrolysis, for example, can be utilized to produce a considerable amount of biochar, whereas fast pyrolysis is better for dominantly producing bio-oil [13].

2.1.1 Types of pyrolysis

Pyrolysis is strongly dependent on the operating parameters, namely temperature, heating rate, and residence time [14]. These operating conditions further help to categorize pyrolysis into other six subclasses. These subclasses are slow pyrolysis, fast pyrolysis, flash pyrolysis, vacuum pyrolysis, intermediate pyrolysis, and hydro-pyrolysis [15]. Each classification of pyrolysis has its own documented benefits and drawbacks. The subclasses in question foster an environment for different reaction conditions and mechanisms to have various products. The pyrolysis technology mechanism is shown in **Figure 2**.

2.1.1.1 Slow pyrolysis

As indicated by the name, to complete the process, slow pyrolysis has a long residence time (more than 1 hour), and biochar is produced as a major product [16]. Slow pyrolysis is dubbed conventional pyrolysis, where biomass is heated at temperatures ranging between 300 and 600°C accompanied by a heating rate of 5–7°C/min [12, 17]. A lower heating rate and longer vapor residence time provide a suitable environment and adequate time for the secondary reactions to proceed. Furthermore, a prolonged residence period permits vapors created during the secondary reaction to be evacuated [15, 18]. This leads to the creation of solid carbonaceous biochar in the end. Slow pyrolysis favors char development, but liquid and gaseous products are also created in modest quantities. Biochar is formed as a primary product (35–45%) together with other products such as bio-oil (25–35%) and syngas (20–30%), as indicated in Eq. (1) [6, 19].

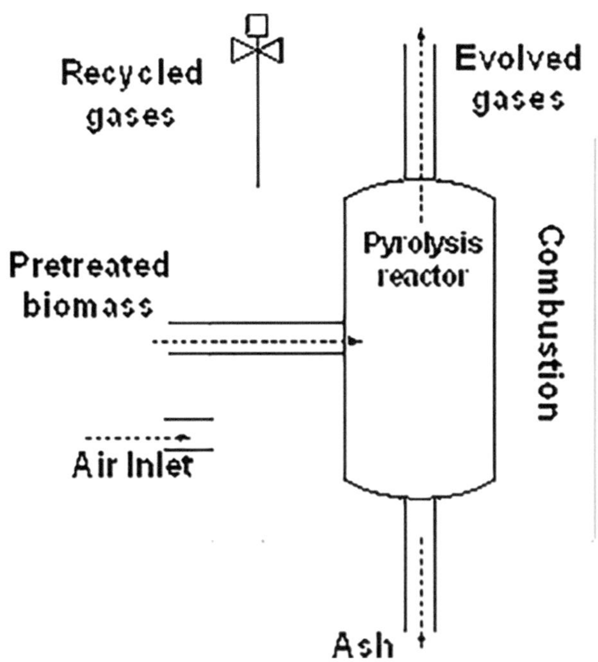

Figure 2.
Representation of a pyrolysis process [6].

2.1.1.2 Fast pyrolysis

Fast pyrolysis is a direct thermochemical process for converting solid biomass into high-energy liquid bio-oil. A high-efficiency thermochemical technique to produce biomass-derived biofuels, with reduced amounts of solids and gases produced [20, 21]. Fast pyrolysis is carried out without oxygen at temperatures above 500°C and a heating rate of over 300°C/min. Fast pyrolysis is a rapid biochar generation technique that takes only a few seconds. Fast pyrolysis produces 60% bio-oil, 20% biochar, and 20% syngas, as reported in other studies [21, 22]. Even higher temperatures in the range of 850–1250°C with a heating rate of 10–200°C for a short residence time ranging from 1 to 10 s have been reported in several experiments. 60%-75% of liquid products, 15%-25% of biochar and 10–20% of non-condensable gaseous products are produced by a typical pyrolysis process [23]. Fast pyrolysis takes biomass to temperatures in which thermal cracking can occur and minimizes the exposure time, which supports biochar production [24].

2.1.1.3 Flash pyrolysis

This is dubbed to be an enhanced and modified version of fast pyrolysis. Biomass decomposes quickly, usually in less than a minute, at 1000°C and even higher temperatures. Heating rates of above 1000°C/sec have been recorded on occasion. Flash pyrolysis is carried out at temperatures between 900 and 1200°C, which can be reached in less than one second (usually between 0.1 and 1 s) [25, 26]. A high bio-oil yield combines a high heating rate with a high temperature and a short vapor residence time. However, the yield of biochar is reduced because of the process [27, 28]. In flash pyrolysis, heat and mass transfer processes, reaction chemical kinetics and biomass phase transition behavior all play a role in product distribution. Although flash pyrolysis is performed in a fluidised bed reactor and a twin-screw mixing reactor, it has limited industrial applicability because of the reactor's architecture, which requires it to run at a high temperature with a very high heating rate [12].

2.1.1.4 Vacuum pyrolysis

This is the thermal decomposition of biomass under vacuum or relatively low pressure in an isolated oxygen environment [15, 29]. Pressure is usually regulated in the region between 0.5 and 2 bar, and temperature is maintained at 450–600°C [30]. Like slow pyrolysis, vacuum pyrolysis has comparably low heating rates. However, these two techniques, in comparison, yield significantly different products. This owes to the constant and effective discharge of the vapor produced during vacuum pyrolysis through condensation train. The rapid evacuation of organic vapors created during the primary pyrolysis also considerably minimizes the vapor residence time, which in turn minimizes the occurrence of secondary reactions and assures a high liquid product yield during the secondary pyrolysis [31]. As a result, only vacuum or low-pressure extraction is utilized to remove vapor evolved during pyrolysis, which substantially affects product quality and yield by preventing inorganic devolatilisation.

2.1.1.5 Intermediate pyrolysis

As the name suggests, this is a combination of slow and fast pyrolysis processes, and it is crucial when there is a need to balance solid and liquid products. This means that slow pyrolysis is more efficient at producing large amounts of char, but it also results in lower amounts of liquid products, while it is vice versa with fast pyrolysis. Generally, pressure is kept at 1 bar during the process. Intermediate pyrolysis has temperatures ranging between 500 and 650°C, with heating rates between 0.1 and 10°C/min and residence time between 5 and 17 mins [32]. 40–60% liquid, 20–30% non-condensable gases, and 15–25% biochar are typical constituents of finished products [33, 34]. Using intermediate pyrolysis conditions prevents the synthesis of high molecular reactive tars and results in dry biochar, which can be utilized for agricultural purposes or directly in boilers and engines in conjunction with high-quality bio-oil [2].

2.1.1.6 Hydropyrolysis

It relatively a novel technique that is used for the conversion of biomass into high quality products by injection of hydrogen or hydrogen based material into the reactor under high pressure, typically above the atmospheric pressure, stretching from 50 bar to 200 bar [15, 35]. The heating rate (10–300°C/s), residence time (over 15 sec) and temperature (350–600°C) are not highly deviated from fast pyrolysis [36]. In essence, hydropyrolysis can be considered a special type of fast pyrolysis subjected to high pressure in an atmosphere infused with hydrogen or hydrogen-based material. This method is not ideal for the production of biochar as the introduction of hydrogen under high temperature and pressure acts as a reducing agent, hence reducing oxygen content in the bio-oils produced and synchronously inhibiting the production of biochar [37, 38]. The employment of a catalyst to eradicate oxygen, water, and CO_x from the liquid product is typically linked with hydropyrolysis. Catalysts also reduce depolymerisation and coking reactions [39]. However, developing the catalyst for this intention remains a notable example of the difficult aspects of catalytic hydropyrolysis.

2.2 Carbohydrate decomposition

The majority of the material used in biochar production via pyrolysis contain carbohydrates in various forms (cellulose, hemicellulose and lignin), and these react differently based on the operating conditions they are subjected to, thus influencing the product yield of pyrolysis [15]. More specifically, lignin and cellulose are the major parts of biomass, making up its bulk [40]. On pyrolysis, cellulose mostly creates tar, a mixture of discrete ketones, aldehydes, organic liquids, and char, whereas lignin essentially produces char and a minimal amount of water. As the cellulose content grows but the char and tar content decreases, the yield of gaseous content increases. It's also been discovered that structural differences in biomass cause changes in the pyrolysis product's composition [41].

2.2.1 Cellulose decomposition

By lowering the extent of polymerization, the process of cellulose degradation is determined, which consists of two principal reactions:

i. Slow pyrolysis involves cellulose degradation over a prolonged period with a lower heating rate.

ii. Fast pyrolysis occurs at high heating rates through speedy volatilization and leads to levoglucosan formation.

In addition to producing the solid product biochar, levoglucosan is dehydrated to generate hydroxymethylfurfural, which can break down to produce liquid and gaseous products such as bio-oil and syngas, respectively. Furthermore, the hydroxymethylfurfural can undergo several processes, including aromatization, condensation, and polymerization, to generate solid biochar [42, 43]. At low temperatures, cellulose degrades to a reasonably stable anhydrocellulose that produces a lot of char, but it decomposes into volatiles [25, 44].

2.2.2 Hemicellulose decomposition

The hemicellulose degradation mechanism is like that of cellulose. Depolymerisation of hemicellulose leads to oligosaccharides production [45]. Decarboxylation, intramolecular rearrangement, depolymerisation, and aromatisation reactions can be used to synthesize biochar or the compound can degrade into syngas and bio-oil [46]. The volatile products and lignin are responsible for the char yield of the cellulose and hemicellulose components in biomass [40].

2.2.3 Lignin decomposition

Unlike the degradation of cellulose and hemicellulose, lignin decomposition is more complicated [47]. The creation of a more condensed solid structure and the shattering of relatively weak bonds result in the formation of char from lignin [48]. The β-O-4 lignin bond is broken and causes free radicals to be released. The protons emanating from other particles are captured by these free radicals, causing the production of degraded substances or compounds. Chain propagation is accomplished by free radicals moving to other molecules. Different amounts of lignin related to variable wood types bring about different breakdown rates. Coniferous lignin has been discovered to be more stable than deciduous lignin, and the former creates more char [49, 50].

2.3 Gasification

This is a thermochemical process that decomposes carbon-rich materials into gaseous products, including CO, CO_2, CH_4, H_2, and traces of hydrocarbons; these gases are referred to as syngas [51, 52]. Gasification happens at high temperatures between 700 and 900°C in an environment with restricted oxidizing agents such as oxygen, air, nitrogen, steam, carbon dioxide, or a mixture of these gases. It was discovered that when the temperature rose, carbon monoxide and hydrogen production increased, while other components such as methane, carbon dioxide, and hydrocarbons declined [53]. The main product of this process is syngas (mostly hydrogen), while char is referred to as a by-product (or waste) with a lower yield, along with ash, tar, and some oil [51]. Partial oxidation of biomass, unlike combustion, takes the energy available in the biomass and bundles it into chemical bonds in the form of gaseous products. The intrinsic chemical energy of carbon in biomass is transformed into combustible fuel gases, which are more efficient and convenient to utilize than raw biomass [54]. Commercial use of the gasification technique has also been documented. Because of its lower Levelised emissions and higher volume of syngas, gasification outperforms other traditional techniques including pyrolysis, combustion, and fermentation. The O/C ratio is critical to achieving high gasification efficiency. High gasification efficiency is achieved by using biomass with a low O/C ratio during gasification. Biomass can be reduced in its O/C ratio by the process of torrefaction. Before conventional gasification, torrefaction might be regarded as a pretreatment for better product quality. It is a low-temperature process between 200 and 300°C with a heating rate of roughly 50°C/min depending on the biomass composition and type [55, 56]. Pyrolysis and gasification are closely related processes. When gasification and pyrolysis are combined, there is no apparent separation between the two approaches [57, 58]. The little composition of oxygen used in gasification causes the biomass to undergo partial oxidation, changing the final product's characteristics. The product type is one of the most significant variations between pyrolysis and gasification. Gasification produces around 85% gaseous products, 10% solid char, and 5% liquid products [15, 58]. The schematic of the gasification process is shown in **Figure 3**.

The gasification mechanism can be sub-divided into many steps as follows [5]:

2.3.1 Drying

Biomass moisture is entirely removed from the material, and no energy is recovered in the process. Different types of biomass have varying moisture contents. When the biomass has a high moisture content, drying is used as a distinct step during gasification.

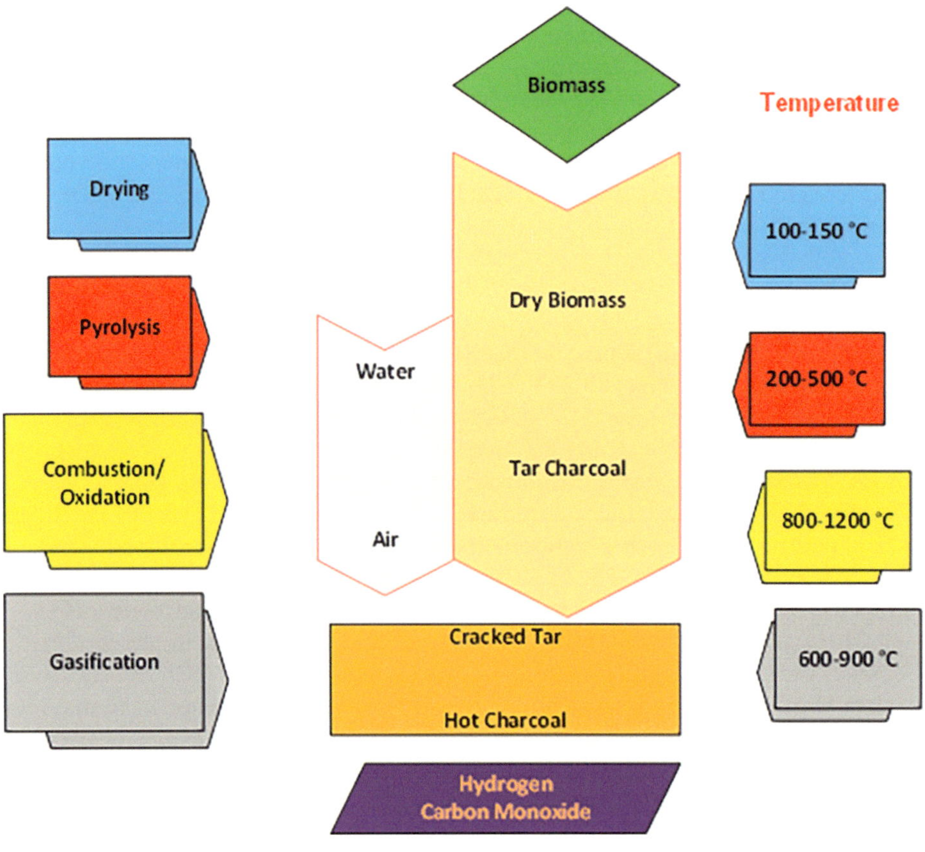

Figure 3.
Process diagram for gasification [54].

2.3.2 Pyrolysis

The biomass is heated from 200 to 700°C with restricted oxygen or air during the pyrolysis process. The volatile components of the biomass are evaporated under these circumstances. The volatile vapor contains CO, CO_2, CH_4, H_2, tar (heavier hydrocarbon) gases, and water vapor [59]. Tar and char are also formed [60].

2.3.3 Oxidation/combustion

The oxidation and combustion reactions of the gasification agents are the primary energy sources for the gasification process. These gasification agents react with the gasifier's combustible species to create CO_2, CO, and water.

2.3.4 Reduction

The CO_2 and H_2O are produced when the oxygen provided to the gasifier combines with the combustible elements. Upon contact with the char formed by pyrolysis, some of this CO_2 and H_2O are converted to CO and H_2 [60, 61]. Furthermore, the hydrogen in the biomass can be oxidized, resulting in the production of water. The reduction reactions that take place inside the gasifier are endothermic, and the energy necessary for them comes from the combustion of char and volatiles. Through a series of reactions, biomass reduction produces combustible gases such as hydrogen, carbon monoxide, and methane [62, 63].

2.3.5 Cracking

Furthermore, during the gasification process, the tar gases formed during the pyrolysis step are cracked, resulting in non-condensable gasses, light hydrocarbons, and unconverted tar [64]. The cracking stage follows more or less Eq. (2).

$$aC_nH_x \rightarrow bC_mH_y + CH_2 \tag{2}$$

Where C_nH_x is tar and C_mH_y is dehydrogenated hydrocarbons; a, b and c are mole ratios.

3. Factors affecting the properties of biochar

3.1 Feedstock

Biomass is a composite solid substance made up of organic, inorganic and biological material produced from living or non-living creatures/organisms. There are two main categories of biomass, namely Woody and Non-woody biomass. Woody biomass is mainly forestry and tree residue [1]. It is characterized by low moisture and ash content, high calorific and bulk density values, and low voidage; in contrast, Non-woody biomass is made up of agricultural crop residue, animal waste, and municipal and industrial solid waste [1]. Non-woody biomass is characterized by high moisture and ash content, decreased calorific value, low bulk density, and increased voidage compared to woody biomass [1]. The moisture content of the biomass has been shown to have a significant effect on the physicochemical characteristics of the derived biochar [2]. A study conducted by [3] comparing the pyrolytic charcoals produced from hard and softwood bark samples reported a direct correlation between initial sample moisture content and the surface chemistry derived charcoal; the study found that a decrease in the moisture content of maple bark resulted in charcoal surface becoming more graphite-like and polyaromatic attributed to prolonged pyrolysis time. The effect of feedstock lignin and cellulose content on biochar formation is a well-researched area [4]. Lignin is an amorphous, high molecular weight polymer that is hydrophobic in nature and has several aromatic functional groups in comparison; cellulose and hemicelluloses are made up of simple sugar monomers that disintegrate at temperatures below 450 degrees Celsius [5]. Studies conducted by Tripathi et al. 2016 and Yu et al. 2014 [2, 6] showed that the cellulose content of feedstock aided the formation of tar (which comprises aldehydes, organic liquids, ketones, and char); while a high lignin concentration is beneficial to the formation of char during pyrolysis. According to Demirba (2004) [7], high feedstocks lignin content will increase char formation. It has been shown that increased lignin content in plant biomass promotes carbonization and increases biochar carbon and ash content [8, 9].

3.2 Residence time

Residence (pyrolysis time) has been shown to affect the degree of carbonization and biochar yield of feedstock; this effect is particularly pronounced at low temperatures [18]. According to Zornoza et al. (2016), increased residence time during pyrolysis results in a higher degree of carbonization, reducing the liable organic matter mitigation the vulnerability of the biochar to microbial attack [19]. Residence time has also been shown to influence the specific surface area of biochar produced. A study conducted by Wang et al. (2019) found that the surface area of biochar's derived from the co-pyrolysis of sewage sludge and cotton stalks increased as residence time increased from 30 minutes to 90 minutes [20]. This was attributed to the formation and extension of pore structures of the biochar caused by the increased thermal decomposition of organic matter and volatiles released from etching pores during the increased residence time [21]. The same study noted a decrease in the surface area of the biochar's as the residence time was increased from 90 minutes to 150 minutes; this reduction was accounted for by the collapse of the pore structure of the biochar during the extended residence time [20]. Residence time has also been shown to affect the calorific value of the biochar produced; a study conducted by Ahmad et al. (2020) on coconut shell derived biochar showed an increase in calorific value from 25.99 MJ/kg to 29.54 MJ/kg as residence time increased for 45 minutes to 75 minutes [22].

3.3 Biomass pretreatment

The pre-treatment of biomass before the pyrolysis has been shown to influence biochar characteristics. Pre-treatment is primarily divided into four categories: physical, physiochemical/thermal, chemical, and biological. Physical pre-treatment describes methods (milling, grinding etc.) that use mechanical energy to alter biomass properties. The most common form of physical pre-treatment is particle size reduction via mechanical comminutions. The effect of particle size reduction and fractionation of ash content is well researched. A study conducted by Liu et al. showed that the ash content of switchgrass and pine bark varied considerably with particle size fractions [22]. The study also reported the potential 20% removal of inorganic constituents from switchgrass and a 30% removal of inorganic constituents from raw pine bark. A similar study conducted by Bridgeman et al. found that the ash content of switchgrass and reed canary greatly increased in fines with particle sizes smaller than 90 micrometers, increasing to 3.62 wt. % to 6.0 wt. % for reed canary grass and 3.12 wt. % to 6.88 wt. % (dry basis) for switchgrass [23]. Besides the ash content, feedstock particle size is also correlated to biochar particle size, with finer feedstocks producing finer biochar particle sizes [18]. Studies have found that biochar's derived from finer feedstocks exhibit lower nitrogen content as well as increased surface area, electrical conductivity, and pH [24, 25]. A study conducted by Sun et al. (2012) evaluating the properties of fine apple wood and corn stover-derived biochar (feedstock = 0.25 mm) reported a higher surface area when compared to applewood or corn stover-derived biochar stover-derived biochar of feedstock particle size = 1.5 mm [27]. Thermal pre-treatment describes methods that make use of thermal energy to produce changes in biomass properties; the most common forms of thermal pre-treatment are steam explosion, HTC and hot water extraction. Steam explosion involves the subjection of biomass to high temperatures and pressures between (160-260°C) and (0.69–4.83 MPa); the biomass subsequently undergoes sudden decompression scattering the fiber material and breaking the covalent bonds between the hemicellulose and lignin [28, 29]. Steam explosion increases the lignin content of the biomass by facilitating the depolymerisation of lignin into lower molecular weight molecules, which then condense with other degradation products [30]. A study conducted by Chen et al. 2017 [46] evaluating the effect of the steam explosion of crop straws before pyrolysis reported a change in the surface structure of the derived biochar; exhibiting a rougher surface when compared to the smoother, clearer and distinct pore structure of the untreated crop straw [31]. The same study also showed an approximate increase in the specific surface area of oil-rape straw-derived biochar 16 times greater than that on the untreated sample.

4. Biochar characterization

Properties of biochar produced depend on the composition, type of biomass and the conditions under which it is carbonized. Both physical and chemical characterizations are necessary when identifying the basic properties of biochar and predicting the various application uses. Biochar serves as a promising alternative to its surface area, charged surfaces and functional groups. **Figure 4** below displays the different physical and chemical methods used for biochar characterization, focusing on BET and FTIR, belonging to the chemical characterization and SEM as physical characterization.

The main aim of quantification to distinguish biochar from organic matter and other forms of black carbon produced. Majority of the potential technology is dependent on spectroscopic characteristics rather than physical separation or isolation.

Biochar being produced from a range of biomass that has different chemical and physical properties results in materials of different properties. Properties of each biomass are important during thermal conversion processes, proximate analysis (ash and moisture content); calorific value; fractions of fixed carbon; volatile components; fractions of lignin, cellulose and hemicellulose; inorganic substances; true density; particle size and moisture content.

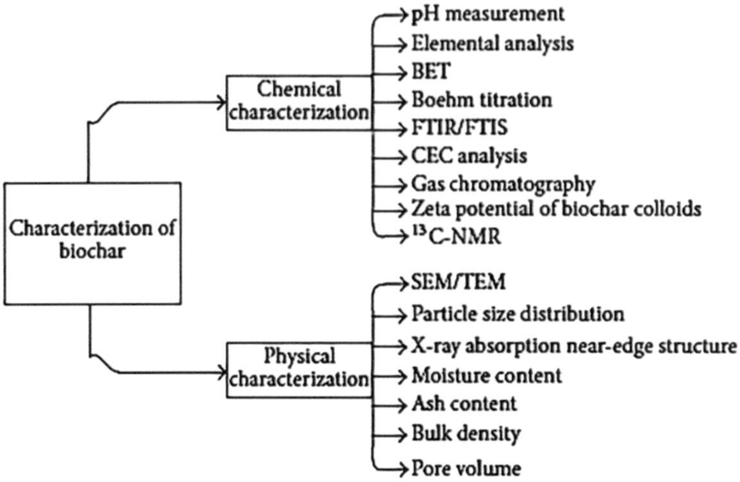

Figure 4.
Overview of a proposed characterization techniques for biochar [65].

4.1 Porosity and surface area

Chemical composition of biomass feedstock and biomass is subjected to a range of analyses to achieve the basic physicochemical characteristics of each raw material. **Figure 5** displays the physiochemical characteristics of biochar. Biochar production is often assessed through changes in the elemental concentrations of C, H, O, S and N and the associated ratios. The fixed carbon is the solid residue that remains after the particle size is carbonized and the volatile matter is expelled. The H/C and O/C ratios are used to determine the degree of aromaticity and maturation. Elemental ratios of O/C, O/H and C/H have been used to provide a reliable measure of the extent of pyrolysis and the level of oxidative adjustment of the biochar. Irrespective of the pyrolytic temperature, the BET areas increased with an increase in carbon burn off, indicating that the carbon burns off had a significant role in increasing pore volume and surface area while the average pore size increased with residence time and pyrolytic temperature. The BET surface area of biochar value of (1057 m2 .g – 1) has been reported, which appears slightly higher than that of activated carbon (970m^{2} .g^{-1}). Biochar micropore volume of (0.24 mL .g^{-1}) also appeared smaller than that of activated carbon, having a value of (0.32 mL .g^{-1}), however having an average pore diameter of (5.2 nm).

4.2 Scanning electron microscope (SEM)

Scanning electron microscopy is categorized as a physical characterization technique used to determine the samples macroporosity and the physical morphology of solid substance (**Figure 6**). A study by Amin 2016 [1] approximated that the biochar produced from cellulose plant materials had a pore diameter of 1 μm. This characteristic is highly dependable in the intrinsic architecture of the feedstock use.

SEM micrographs displayed that the biochar produced at different pyrolytic temperatures has a distinguishable and clear honeycomb structural appearance due to the original tubular structures present in plant cell materials (**Figure 6**). The well-developed pores have a direct impact on the high surface area. According to Cantrell et al. (2012), biochar produced at lower temperatures is appropriate for regulating

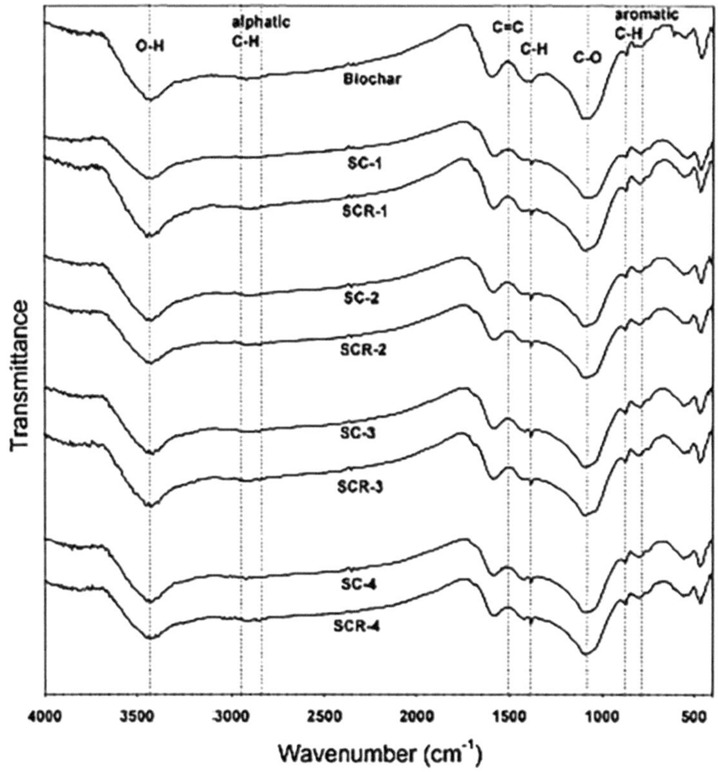

Figure 5.
Fourier-transform infrared spectra (FTIR) of the biochar samples [66].

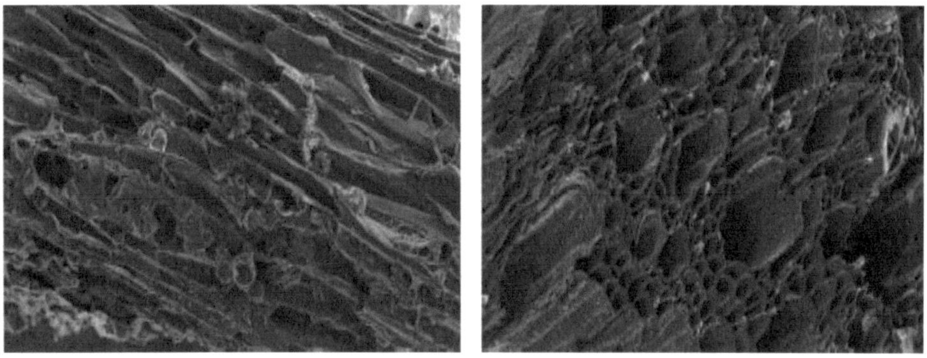

Figure 6.
SEM micrograph of biochar with magnification of 500x [67].

fertilizer nutrients and absorbing pollutants from the soil. Higher temperatures lead to material analogous to activated carbon and environmental remediation. SEM micrographs of biochar displayed a clean surface as the pyrolysis process had stabilized the volatile hydrocarbons, therefore smoothening the surface of the biochar. Pyrolysis at lower temperatures displays molded structures with small pores and uneven surface structure. In general, it is safe to say that since the biomass wastes contain lignin and high volatile matter content, the pore creation in biochar is directly affected.

4.3 Fourier transform infrared spectroscopy (FTIR)

FTIR spectroscopy serves as a great tool to observe the shift change of chemical compositions. The commonly used technique for biochar characterization using the FTIR is the pellet technique, which mixes 1 mg of dried biochar with 300 mg of pre-dried and pulverized spectroscopic grade KBr. Novak ae al. (2012) used the pellet technique to conclude 3400to 3410 cm^{-1}, H-bonded O–H stretching vibrations of hydroxyl groups from alcohols, phenols, and organic acids, 2850 to 2950 cm^{-1}, C–H stretching of alkyl structures; 1620–1650 cm^{-1}, aromatic and olefinic CDC vibrations, CDO in amide (I), ketone, and quinone groups; 1580 to 1590 cm^{-1}, COO- asymmetric stretching; 1460 cm^{-1}, C–H deformation of CH_3 group; 1280–1270 cm^{-1}, O–H stretching of phenolic compounds; and three bands around 460, 800, and 1000–1100 cm^{-1}, bending of Si–O stretching [68]. **Figure 5** illustrates the FTIR spectra of biochar collected during different stages of the production, i.e. (Biochar: Original, −1: pre-incubation, −2: jointing, −3: Heading; −4: Mature).

5. Applications of biochar and future perspective

Biochar is a product (together with bio-oil and gases) resulting from biomass pyrolysis. Biochar usage has increased because it reduces the negative impacts of biomass on the environment [69]. The physicochemical properties of biochar are what govern the applications of this material. Depending on the feedstock type, production technology and process conditions [70]; the quality, yield and toxicity of the resulting biochar differs (as shown in **Table 1**) [72, 73]. These applications (including potential applications) range from adsorption for water and air pollutants [74], activated carbon [75], anaerobic digestion promoter/catalyst [76], construction material [77], agriculture and horticulture use such as soil conditioning, compost additive [78], carbon sequestration, etc. [73]. **Figure 7** demonstrates these applications and how biochar contributes to the circular economy through its uses in agriculture and horticulture. Also, these numerous biochar benefits show a great potential to contribute to the economic sustainability of emerging cellulosic bioenergy production systems [79, 80]. It is worth noting that as the number of applications of biochar increases, so does the number of manufacturers, leading to a need for regulated standards and guidelines for the production of this material (see **Table 2**) [81, 82].

5.1 Biochar in agriculture and horticulture

Biochar application in agriculture and horticulture has been explored both on a laboratory scale and in the field. These applications include being used as a component of chemical fertilizer [83], soil microbial activity, soil amendment for crop productivity improvement through nutrient availability [84, 85] as well as water holding capacity [86]. Biochar has also been reported to alleviate heavy metals release in the soil while having a limiting effect that aids in increasing the pH of highly acidic soils [87, 88]. Though biochar is another soil conditioner type, it differs from compost by production pathways. Biochar is produced by thermal decomposition of food, horticultural and municipal solid waste in the absence of oxygen, while natural biodegradation of organic substrates produces compost by the microbial community under aerobic conditions. Another difference is that; compost degrades fast, making its benefits relatively short-lived compared to biochar which persists in the soil for more prolonged periods [78, 89].

Type of characterization	Determination method	Results and remarks
Elemental analysis	C, H, O, S and N associated ratios	The H/C, O/C and N ratios are used to determine the aromaticity and maturity of the biochar
BET	Surface area, pore structure, average pore diameter, pore volume and average pores of biochar	$1057 m^2 g^{-1}$; macroporosity and microporosity; 5.2 nm; $0.24 mLg^{-1}$; 3.3 nm. [71]
FTIR	Changes which occur in the biochar preparations as well as its functional groups present from the original biochar.	Changes include dehydration, pyrolysis, graphene nucleation, and finally carbonization; O–H (3600–3100 cm − 1), C=C and C=O stretching (1740–1600 cm − 1), C–O–C symmetric stretching (1097 cm − 1), –COOMe (1400–1500 cm − 1), and so on

Table 1.
List of notable chemical characterisations of biochar.

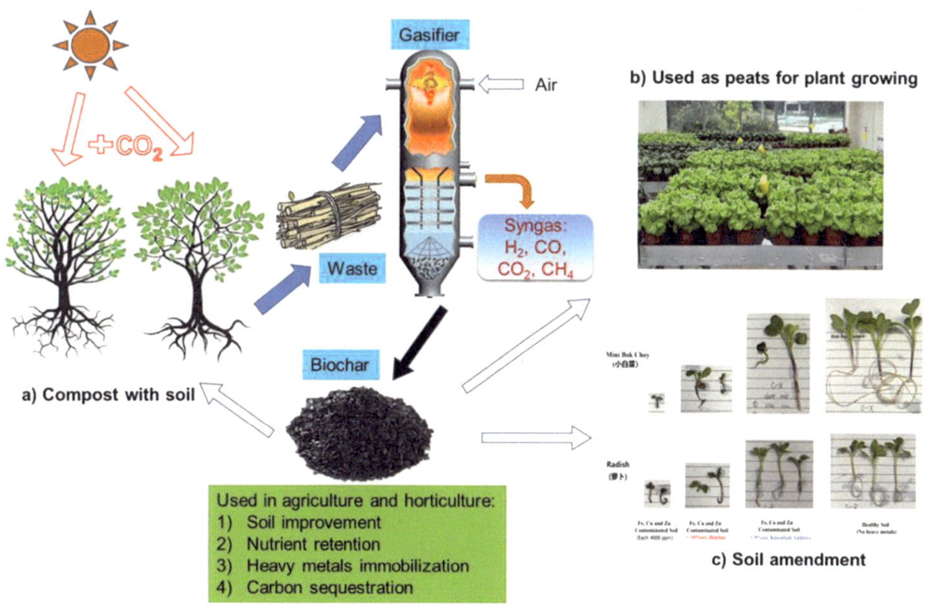

Figure 7.
Biochar uses in agriculture and horticulture and its contribution to the circular economy [78].

5.1.1 Biochar as a compost additive

Low soil organic carbon and fertility are challenges faced by many agricultural farmers around the globe. Biochar offers a solution to this challenge because it gives two options, i.e. returning nutrients and carbon to the soil while producing energy [90]. Also, the compositing rate can be increased by using biochar as an additive.

Process	Process temperature	Residence time	Solid product yield on a dry wood feedstock basis (mass %)	Carbon content of the solid product (mass %)	Carbon yield ($mass_{carbon, product}$/ $mass_{carbon feedstock}$
Slow pyrolysis	~ 400	Minute to days	≈ 30	95	≈ 0.58
Fast pyrolysis	~ 500	~ 1 s	12–26	74	0.2–0.26
Gasification	~ 800	~ 10–20 s	≈ 10	—	—
HTC	~ 180–250	1-12 h	< 66%	< 70%	≈ 0.88
Flash carbonization	~ 300–600	< 30 min	37	≈ 85	≈ 0.65
Torrefaction	~ 290	10–60 min	61–84	51–55	0.67–0.85

Table 2.
Comparison of typical operating conditions and product properties of various biochar production processes [81].

Zhang and Sun [91] have examined spent mushroom compost and biochar co-composting. Their results showed a great increase in nutrients content of the resultant compost product and an improved composed quality while reducing the composting time from 90 to 270 days to only 24 days. Also, the large porosity of biochar enables it to facilitate microbial growth in the compost pile, leading to accelerated nutrient recycling [92]. The addition of biochar to poultry manure has been found to increase the maximum temperature reached and shorten the thermophilic phase [93].

5.1.2 Biochar as an adsorbent

An issue of heavy metals/metalloids (HMS) and polycyclic aromatic hydrocarbons (PAHs) in soil and water poses detrimental environmental problems and poor quality of agriculture, affecting all forms of life [94, 95]. These pollutants are toxic, persistent, non-biodegradable and potentially bioaccumulate [96]. Among other bioremediation technologies used to solve the HMS and PAHs issue, biochar is one of the best solutions due to its advantages [97]. These advantages include sustainability, low costs, sequestration of carbon, etc. [94]. Various physical and chemical characteristics of biochar, such as pore structure, specific surface area and functional groups, have been used to adsorption different pollutants [98]. For instance, Mahmoud, et al. [99] have used modified Switchgrass biochar for efficient decolorization of reactive red 195 A dye from aqueous and wastewater samples. Other biomass materials such as rice husks and dairy manure have also been used for biochar production with varying adsorption capacities according to the biomass used upon other factors [100].

5.2 Biochar in construction

Biochar has been used in road construction and as a concrete admixture. Wang, et al. [77] assessed this where a novel production of fill material and pedestrian/vehicle paving blocks were done. In this study, biochar addition was found to be beneficial to cement hydration even though it was noticed that the studied particle sizes could incur microcracks and strength degradation. Also, biochar's incorporation resulted in enhanced immobilization of potentially organic contaminants and toxic elements in the sediment product, which is significant for moderately to heavily contaminated products. Therefore, biochar from wood can be used as a green combination for cement-based recycling procedures for highly contaminated waste. The use of biochar in construction material to trap atmospheric carbon dioxide in buildings also offers the potential to reduce greenhouse gasses by 25%. High pH and high water retention rate of biochar enable it to absorb some of the mixing water used in concrete mixing, thereby reducing the amount of free water in the concrete [101].

5.3 Future perspective

Since biochar's applications depend greatly on its properties, future research must elucidate the production process effects on biochar's properties. Biochar used in water treatment would differ from the one used in energy/agriculture. Likewise, there are diverse literature findings on the effects of biochar on agriculture, particularly on crop production caused by soils being different. For instance, crop yields may be increased or decreased by adding biochar depending on the soil type and fertilizer management [90, 102]. Also, the chemical behavior of biochar with heavy metal ions has been found to be inconsistent [103]. It is apparent that the interaction mechanisms between biochar, soil and plants are critical and yet not thoroughly known. Therefore, more efforts are still needed concerning biochar properties to soil and crop responses equally in the field and climate-controlled environment.

6. Conclusion

Biochar has been applied to remediate contaminated agricultural soil and improve soil fertility by reducing acidity and increasing the availability of nutrients. Thus, the addition of biochar to soils can be one of the best practices to overcome any biotic stress in soil and increase crop productivity, mainly in the agricultural sector. The properties of biochar have significantly been influenced by processes such as pyrolyscould, which have been discussed in this chapter. Thus, biochar appears as a highly promising option for pollutant removal. Economic impacts and recyclability should be considered in developing recoverable biochar for wide environmental applications. The relationship between various solutions for waste management and energy production differs in parameters and multiple techniques for its production and economic, social and ecological constraints. This review paper detailed the state-of-art information that would be helpful to find new opportunities in scientific innovation in the field of biochar research.

Acknowledgements

The authors are thankful to the Green Engineering and Sustainability research group in the Department of Chemical engineering at the Durban University of Technology, South Africa.

Conflict of interest

The authors declare no conflict of interest.

Author details

Edward Kwaku Armah[1,2*], Maggie Chetty[1], Jeremiah Adebisi Adedeji[3], Denzil Erwin Estrice[1], Boldwin Mutsvene[1], Nikita Singh[1] and Zikhona Tshemese[4]

1 Faculty of Engineering and the Built Environment, Department of Chemical Engineering, Durban University of Technology, Steve Biko Campus, Green Engineering and Sustainability Research Group, Durban, South Africa

2 Department of Applied Chemistry, C.K. Tedam University of Technology and Applied Sciences, School of Chemical and Biochemical Sciences, Navrongo, Ghana

3 University of KwaZulu-Natal, School of Engineering, Discipline of Chemical Engineering, Howard College Campus, Durban, South Africa

4 Faculty of Applied Sciences, Department of Chemistry, Durban University of Technology, Steve Biko Campus, Durban, South Africa

*Address all correspondence to: ekarmah@cktutas.edu.gh

References

[1] Amin FR, Huang Y, He Y, Zhang R, Liu G, Chen C. Biochar applications and modern techniques for characterization. Clean Technologies and Environmental Policy. 2016;**18**(5):1457-1473

[2] Borel LD, Lira TS, Ribeiro JA, Ataíde CH, Barrozo MA. Pyrolysis of brewer's spent grain: Kinetic study and products identification. Industrial Crops and Products. 2018;**121**:388-395

[3] Borel LD, Reis Filho AM, Xavier TP, Lira TS, Barrozo MA. An investigation on the pyrolysis of the main residue of the brewing industry. Biomass and Bioenergy. 2020;**140**:105698

[4] Zhou Y et al. Production and beneficial impact of biochar for environmental application: A comprehensive review. Bioresource Technology. 2021;**337**:125451

[5] Yaashikaa P, Kumar PS, Varjani S, Saravanan A. A critical review on the biochar production techniques, characterization, stability and applications for circular bioeconomy. Biotechnology Reports. 2020;**28**:e00570

[6] Al Arni S. Comparison of slow and fast pyrolysis for converting biomass into fuel. Renewable Energy. 2018;**124**:197-201

[7] Bridgwater A, Peacocke G. Fast pyrolysis processes for biomass. Renewable and Sustainable Energy Reviews. 2000;**4**(1):1-73

[8] Balogun AO, Sotoudehniakarani F, McDonald AG. Thermo-kinetic, spectroscopic study of brewer's spent grains and characterisation of their pyrolysis products. Journal of Analytical and Applied Pyrolysis. 2017;**127**:8-16

[9] Azargohar R, Nanda S, Dalai AK, Kozinski JA. Physico-chemistry of biochars produced through steam gasification and hydro-thermal gasification of canola hull and canola meal pellets. Biomass and Bioenergy. 2019;**120**:458-470

[10] Wei J et al. Assessing the effect of pyrolysis temperature on the molecular properties and copper sorption capacity of a halophyte biochar. Environmental Pollution. 2019;**251**:56-65

[11] Cantrell KB, Hunt PG, Uchimiya M, Novak JM, Ro KS. Impact of pyrolysis temperature and manure source on physicochemical characteristics of biochar. Bioresource Technology. 2012;**107**:419-428

[12] Gabhane JW, Bhange VP, Patil PD, Bankar ST, Kumar S. Recent trends in biochar production methods and its application as a soil health conditioner: A review. SN Applied Sciences. 2020;**2**(7):1-21

[13] Varma AK, Mondal P. Pyrolysis of pine needles: Effects of process parameters on products yield and analysis of products. Journal of Thermal Analysis and Calorimetry. 2018;**131**(3):2057-2072

[14] Park HC, Lee B-K, Yoo HS, Choi HS. Influence of operating conditions for fast pyrolysis and pyrolysis oil production in a conical spouted-bed reactor. Chemical Engineering & Technology. 2019;**42**(12):2493-2504

[15] Tripathi M, Sahu JN, Ganesan P. Effect of process parameters on production of biochar from biomass waste through pyrolysis: A review. Renewable and Sustainable Energy Reviews. 2016;**55**:467-481

[16] Liu W-J, Jiang H, Yu H-Q. Development of biochar-based functional materials: Toward a sustainable platform carbon material. Chemical Reviews. 2015;**115**(22):12251-12285

[17] Li W et al. Mechanistic insight into lignin slow pyrolysis by linking pyrolysis chemistry and carbon material properties. ACS Sustainable Chemistry & Engineering. 2020;**8**(42):15843-15854

[18] Xu W-C, Tomita A. The effects of temperature and residence time on the secondary reactions of volatiles from coal pyrolysis. Fuel Processing Technology. 1989;**21**(1):25-37

[19] Yogalakshmi K et al. Lignocellulosic biomass-based pyrolysis: A comprehensive review. Chemosphere. 2022;**286**:131824

[20] Chang C et al. Study on products characteristics from catalytic fast pyrolysis of biomass based on the effects of modified biochars. Energy. 2021;**229**:120818

[21] Dai L et al. Production of bio-oil and biochar from soapstock via microwave-assisted co-catalytic fast pyrolysis. Bioresource Technology. 2017;**225**:1-8

[22] Huang Y-F, Chiueh P-T, Kuan W-H, Lo S-L. Microwave pyrolysis of lignocellulosic biomass: Heating performance and reaction kinetics. Energy. 2016;**100**:137-144

[23] Bridgwater AV. Renewable fuels and chemicals by thermal processing of biomass. Chemical Engineering Journal. 2003;**91**(2-3):87-102

[24] Dabros TMH et al. Transportation fuels from biomass fast pyrolysis, catalytic hydrodeoxygenation, and catalytic fast hydropyrolysis. Progress in Energy and Combustion Science. 2018;**68**:268-309. DOI: 10.1016/j.pecs.2018.05.002

[25] Demirbas A, Arin G. An overview of biomass pyrolysis. Energy Sources. 2002;**24**(5):471-482

[26] Ghodake GS et al. Review on biomass feedstocks, pyrolysis mechanism and physicochemical properties of biochar: State-of-the-art framework to speed up vision of circular bioeconomy. Journal of Cleaner Production. 2021;**297**:126645

[27] Tripathi M, Sahu J, Ganesan P, Dey T. Effect of temperature on dielectric properties and penetration depth of oil palm shell (OPS) and OPS char synthesized by microwave pyrolysis of OPS. Fuel. 2015;**153**:257-266

[28] Tripathi M, Bhatnagar A, Mubarak NM, Sahu JN, Ganesan P. RSM optimization of microwave pyrolysis parameters to produce OPS char with high yield and large BET surface area. Fuel. 2020;**277**:118184

[29] Lam SS et al. Microwave vacuum pyrolysis conversion of waste mushroom substrate into biochar for use as growth medium in mushroom cultivation. Journal of Chemical Technology & Biotechnology. 2019;**94**(5):1406-1415

[30] Carrier M, Hugo T, Gorgens J, Knoetze H. Comparison of slow and vacuum pyrolysis of sugar cane bagasse. Journal of Analytical and Applied Pyrolysis. 2011;**90**(1):18-26

[31] Chen N, Ren J, Ye Z, Xu Q, Liu J, Sun S. Study on vacuum pyrolysis of coffee industrial residue for bio-oil production. In: IOP Conference Series: Earth and Environmental Science. Vol. 59, no. 1. United Kingdom: IOP Publishing; 2017. p. 012065

[32] Yazdani E, Hashemabadi SH, Taghizadeh A. Study of waste tire pyrolysis in a rotary kiln reactor in a wide range of pyrolysis temperature. Waste Management. 2019;**85**:195-201

[33] Kebelmann K, Hornung A, Karsten U, Griffiths G. Intermediate pyrolysis and product identification by TGA and Py-GC/MS of green microalgae and their extracted protein and lipid components. Biomass and Bioenergy. 2013;**49**:38-48

[34] Li G, Ji F, Bai X, Zhou Y, Dong R, Huang Z. Comparative study on thermal cracking characteristics and bio-oil production from different microalgae using Py-GC/MS. International Journal of Agricultural and Biological Engineering. 2019;**12**(1):208-213

[35] Stummann MZ, Høj M, Gabrielsen J, Clausen LR, Jensen PA, Jensen AD. A perspective on catalytic hydropyrolysis of biomass. Renewable and Sustainable Energy Reviews. 2021;**143**:110960

[36] Galiasso R, González Y, Lucena M. New inverted cyclone reactor for flash hydropyrolysis. Catalysis Today. 2014;**220**:186-197

[37] Thangalazhy-Gopakumar S, Adhikari S, Gupta RB, Tu M, Taylor S. Production of hydrocarbon fuels from biomass using catalytic pyrolysis under helium and hydrogen environments. Bioresource Technology. 2011;**102**(12):6742-6749

[38] Sekar M et al. A review on the pyrolysis of algal biomass for biochar and bio-oil–bottlenecks and scope. Fuel. 2021;**283**:119190

[39] Melligan F, Hayes M, Kwapinski W, Leahy J. A study of hydrogen pressure during hydropyrolysis of Miscanthus x giganteus and online catalytic vapour upgrading with Ni on ZSM-5. Journal of Analytical and Applied Pyrolysis. 2013;**103**:369-377

[40] Chen D, Gao A, Cen K, Zhang J, Cao X, Ma Z. Investigation of biomass torrefaction based on three major components: Hemicellulose, cellulose, and lignin. Energy Conversion and Management. 2018;**169**:228-237

[41] Yu H, Zhang Z, Li Z, Chen D. Characteristics of tar formation during cellulose, hemicellulose and lignin gasification. Fuel. 2014;**118**:250-256. DOI: 10.1016/j.fuel.2013.10.080

[42] Molton PM, Demmitt T. Reaction Mechanisms in Cellulose Pyrolysis: A Literature Review. United States. 1977. DOI: 10.2172/7298596

[43] Lindstrom JK, Proano-Aviles J, Johnston PA, Peterson CA, Stansell JS, Brown RC. Competing reactions limit levoglucosan yield during fast pyrolysis of cellulose. Green Chemistry. 2019;**21**(1):178-186

[44] Reza MS et al. Acacia holosericea: An invasive species for bio-char, bio-oil, and biogas production. Bioengineering. 2019;**6**(2):33

[45] Yang X, Zhao Y, Li W, Li R, Wu Y. Unveiling the pyrolysis mechanisms of hemicellulose: Experimental and theoretical studies. Energy & Fuels. 2019;**33**(5):4352-4360

[46] Huang J, Liu C, Tong H, Li W, Wu D. Theoretical studies on pyrolysis mechanism of xylopyranose. Computational and Theoretical Chemistry. 2012;**1001**:44-50

[47] Mu W, Ben H, Ragauskas A, Deng Y. Lignin pyrolysis components and upgrading—Technology

review. Bioenergy Research. 2013;**6**(4):1183-1204

[48] Santos RB, Hart P, Jameel H, Chang H-M. Wood based lignin reactions important to the biorefinery and pulp and paper industries. BioResources. 2013;**8**(1):1456-1477

[49] Brebu M, Vasile C. Thermal degradation of lignin—A review. Cellulose Chemistry & Technology. 2010;**44**(9):353

[50] Ma Z et al. In-depth comparison of the physicochemical characteristics of bio-char derived from biomass pseudo components: Hemicellulose, cellulose, and lignin. Journal of Analytical and Applied Pyrolysis. 2019;**140**:195-204

[51] Dasappa S, Shivapuji AM. Biomass gasification: Thermochemical route to energetic bio-chemicals. Advanced Biofuel Technologies: Elsevier. 2022;**12**: 305-332

[52] Galindo AL, Lora ES, Andrade RV, Giraldo SY, Jaén RL, Cobas VM. Biomass gasification in a downdraft gasifier with a two-stage air supply: Effect of operating conditions on gas quality. Biomass and Bioenergy. 2014;**61**:236-244

[53] Prabakar D, Manimudi VT, Sampath S, Mahapatra DM, Rajendran K, Pugazhendhi A. Advanced biohydrogen production using pretreated industrial waste: Outlook and prospects. Renewable and Sustainable Energy Reviews. 2018;**96**:306-324

[54] Maitlo G et al. Thermochemical conversion of biomass for syngas production: Current status and future trends. Sustainability. 2022;**14**(5):2596

[55] Prins M, Ptasinski K, Janssen F. More efficient gasification via torrefaction. Energy. 2006;**31**:3458-3470. DOI: 10.1016/j.energy.2006.03.008

[56] Sarker TR, Nanda S, Dalai AK, Meda V. A review of torrefaction technology for upgrading lignocellulosic biomass to solid biofuels. Bioenergy Research. 2021;**14**(2):645-669

[57] Ruiz JA, Juárez M, Morales M, Muñoz P, Mendívil M. Biomass gasification for electricity generation: Review of current technology barriers. Renewable and Sustainable Energy Reviews. 2013;**18**:174-183

[58] Bisht AS, Thakur N. Pine needles biomass gasification based electricity generation for Indian Himalayan region: Drivers and barriers. In: Green Buildings and Sustainable Engineering. New York, United States of America: Springer; 2020. pp. 47-59

[59] Kivisaari T, Björnbom P, Sylwan C, Jacquinot B, Jansen D, de Groot A. The feasibility of a coal gasifier combined with a high-temperature fuel cell. Chemical Engineering Journal. 2004;**100**(1-3):167-180. DOI: 10.1016/j.cej.2003.12.005

[60] Safarian S, Unnþórsson R, Richter C. A review of biomass gasification modelling. Renewable and Sustainable Energy Reviews. 2019;**110**:378-391. DOI: 10.1016/j.rser.2019.05.003

[61] Basu P. Combustion and Gasification in Fluidized Beds. Florida, United States of America: CRC press; 2006

[62] Puig-Arnavat M, Bruno JC, Coronas A. Review and analysis of biomass gasification models. Renewable and Sustainable Energy Reviews. 2010;**14**(9):2841-2851. DOI: 10.1016/j.rser.2010.07.030

[63] Shahbeig H, Shafizadeh A, Rosen M, Sels B. Exergy sustainability analysis of biomass gasification: A critical review. Biofuel Research Journal. 2022;**9**(1):1592-1607

[64] Islam MW. A review of dolomite catalyst for biomass gasification tar removal. Fuel. 2020;**267**:117095

[65] Obemah ND, Zhao B. Biochar Preparation, Characterization, and Adsorptive Capacityand its Effect on Bioavailability of Contaminants: An Overview. China: Hindawi Publishing Corporation; 2014

[66] Mary SG, Sugumaran P, Niveditha S, Ramalakshmi B, Ravichandran P, Seshadri S. Production, characterization and evaluation of biochar from pod (Pisum sativum), leaf (Brassica oleracea) and peel (Citrus sinensis) wastes. International Journal of Recyclying of Organic Waste Agriculture. 2017;**5**:43-53

[67] Sarfaraz Q et al. Characterization and carbon mineralization of biochars produced from different animal manures and plant residues. Scientific Reports. 2020;**10**(1):1-9

[68] Jindo K, Mizumoto H, Sawada Y, Sanchez-Monedero MA, Sonoki T. Physical and Chemical Characterization of Biochars Derived from Different Agricultural Residues. Spain: Biogeosciences; 2014

[69] Laird DA. The charcoal vision: A win–win–win scenario for simultaneously producing bioenergy, permanently sequestering carbon, while improving soil and water quality. Agronomy Journal. 2008;**100**(1):178-181

[70] Cha JS et al. Production and utilization of biochar: A review. Journal of Industrial and Engineering Chemistry. 2016;**40**:1-15

[71] Figueredo NAD, Costa LMD, Melo LCA, Siebeneichlerd EA, Tronto J. Characterization of biochars from different sources and evaluation of release of nutrients and contaminants. Artigo Científico. 2017;**48**(3):395-403

[72] Yang X, Ng W, Wong BSE, Baeg GH, Wang C-H, Ok YS. Characterization and ecotoxicological investigation of biochar produced via slow pyrolysis: Effect of feedstock composition and pyrolysis conditions. Journal of Hazardous Materials. 2019;**365**:178-185

[73] Wang D, Jiang P, Zhang H, Yuan W. Biochar production and applications in agro and forestry systems: A review. Science of the Total Environment. 2020;**723**:137775

[74] Woolf D, Amonette JE, Street-Perrott FA, Lehmann J, Joseph S. Sustainable biochar to mitigate global climate change. Nature Communications. 2010;**1**(1):1-9

[75] Kah M, Sigmund G, Xiao F, Hofmann T. Sorption of ionizable and ionic organic compounds to biochar, activated carbon and other carbonaceous materials. Water Research. 2017;**124**:673-692

[76] Pan J, Ma J, Zhai L, Luo T, Mei Z, Liu H. Achievements of biochar application for enhanced anaerobic digestion: A review. Bioresource Technology. 2019;**292**:122058

[77] Wang L et al. The roles of biochar as green admixture for sediment-based construction products. Cement and Concrete Composites. 2019;**104**:103348

[78] Hu Q et al. Biochar industry to circular economy. Science of the Total Environment. 2021;**757**:143820

[79] Laird DA, Brown RC, Amonette JE, Lehmann J. Review of the pyrolysis platform for coproducing bio-oil and biochar. Biofuels, Bioproducts and Biorefining. 2009;**3**(5):547-562

[80] Lehmann J. A handful of carbon. Nature. 2007;**447**(7141):143-144

[81] Meyer S, Glaser B, Quicker P. Technical, economical, and climate-related aspects of biochar production technologies: A literature review. Environmental Science & Technology. 2011;**45**(22):9473-9483

[82] Sohi SP, Krull E, Lopez-Capel E, Bol R. A review of biochar and its use and function in soil. Advances in Agronomy. 2010;**105**:47-82

[83] Glaser B, Lehmann J, Zech W. Ameliorating physical and chemical properties of highly weathered soils in the tropics with charcoal–a review. Biology and Fertility of Soils. 2002;**35**(4):219-230

[84] Pandit NR, Mulder J, Hale SE, Martinsen V, Schmidt HP, Cornelissen G. Biochar improves maize growth by alleviation of nutrient stress in a moderately acidic low-input Nepalese soil. Science of the Total Environment. 2018;**625**:1380-1389

[85] Purakayastha T et al. A review on biochar modulated soil condition improvements and nutrient dynamics concerning crop yields: Pathways to climate change mitigation and global food security. Chemosphere. 2019;**227**:345-365

[86] Mohamed BA, Ellis N, Kim CS, Bi X, Emam AE-R. Engineered biochar from microwave-assisted catalytic pyrolysis of switchgrass for increasing water-holding capacity and fertility of sandy soil. Science of the Total Environment. 2016;**566**:387-397

[87] Rizwan M et al. Cadmium phytoremediation potential of brassica crop species: A review. Science of the Total Environment. 2018;**631**:1175-1191

[88] Ruzickova J et al. A comprehensive assessment of potential hazard caused by organic compounds in biochar for agricultural use. Journal of Hazardous Materials. 2021;**403**:123644

[89] Tratsch MVM, Ceretta CA, Silva LSD, Ferreira PAA, Brunetto G. Composition and mineralization of organic compost derived from composting of fruit and vegetable waste. Revista Ceres. 2019;**66**:307-315

[90] Gaskin JW et al. Effect of peanut hull and pine chip biochar on soil nutrients, corn nutrient status, and yield. Agronomy Journal. 2010;**102**(2):623-633

[91] Zhang L, Sun X. Changes in physical, chemical, and microbiological properties during the two-stage co-composting of green waste with spent mushroom compost and biochar. Bioresource Technology. 2014;**171**:274-284

[92] Sanchez-Monedero M, Cayuela M, Roig A, Jindo K, Mondini C, Bolan N. Role of biochar as an additive in organic waste composting. Bioresource Technology. 2018;**247**:1155-1164

[93] Czekała W, Malińska K, Cáceres R, Janczak D, Dach J, Lewicki A. Co-composting of poultry manure mixtures amended with biochar–the effect of biochar on temperature and C-CO2 emission. Bioresource Technology. 2016;**200**:921-927

[94] Anae J et al. Recent advances in biochar engineering for soil contaminated with complex chemical mixtures: Remediation strategies and future perspectives. Science of the Total Environment. 2021;**767**:144351

[95] Pérez RM, Cabrera G, Gómez J, Abalos A, Cantero D. Combined strategy for the precipitation of heavy metals and biodegradation of petroleum in industrial

wastewaters. Journal of Hazardous Materials. 2010;**182**(1-3):896-902

[96] Hadia-e-Fatima AA. Heavy metal pollution–a mini review. Journal of Bacteriol Mycology Open Access. 2018;**6**(3):179-181

[97] Alaboudi KA, Ahmed B, Brodie G. Effect of biochar on Pb, Cd and Cr availability and maize growth in artificial contaminated soil. Annals of Agricultural Sciences. 2019;**64**(1):95-102

[98] Beesley L, Moreno-Jiménez E, Gomez-Eyles JL. Effects of biochar and greenwaste compost amendments on mobility, bioavailability and toxicity of inorganic and organic contaminants in a multi-element polluted soil. Environmental Pollution. 2010;**158**(6):2282-2287

[99] Mahmoud ME, Nabil GM, El-Mallah NM, Bassiouny HI, Kumar S, Abdel-Fattah TM. Kinetics, isotherm, and thermodynamic studies of the adsorption of reactive red 195 a dye from water by modified switchgrass biochar adsorbent. Journal of Industrial and Engineering Chemistry. 2016;**37**:156-167

[100] Xu X, Cao X, Zhao L. Comparison of rice husk-and dairy manure-derived biochars for simultaneously removing heavy metals from aqueous solutions: Role of mineral components in biochars. Chemosphere. 2013;**92**(8):955-961

[101] Gupta S, Kua HW. Factors determining the potential of biochar as a carbon capturing and sequestering construction material: Critical review. Journal of Materials in Civil Engineering. 2017;**29**(9):04017086

[102] Asai H et al. Biochar amendment techniques for upland rice production in northern Laos: 1. Soil physical properties, leaf SPAD and grain yield. Field Crops Research. 2009;**111**(1-2):81-84

[103] Abdelhafez AA, Li J, Abbas MH. "Feasibility of biochar manufactured from organic wastes on the stabilization of heavy metals in a metal smelter contaminated soil". Chemosphere. 2014;**117**:66-71

Chapter 3

Biochar from Cassava Waste: A Paradigm Shift from Waste to Wealth

Minister Obonukut, Sunday Alabi and Alexander Jock

Abstract

Waste is unwanted material left after useful parts have been removed and found to affect our environment and health adversely. Waste from agro-allied industries is massive and claims most land, which would have been used for agricultural purposes when used as a landfill, including other environmental and health issues. This chapter will assess wastes generated during the processing of cassava for variety of products and review their properties when characterized. In the course of characterizing the wastes, which emerged during processing, pre-processing, and post-processing depending on the products, various reports on the physical, chemical, and biological properties of cassava wastes will be presented. The properties of cassava waste when subjected to biochemical and thermochemical processes will be compared with those of conventional raw materials for biochar production. This chapter will showcase the potential of cassava wastes for efficient valorization, especially as adsorbents *via* biochar. It will be of great significance to engineers, farmers, and manufacturers in their quest to manage cassava wastes for the betterment of our environment and health.

Keywords: cassava waste, biochar, characterization, biochemical, thermochemical

1. Introduction

Manihot esculenta Crants (Manihot utilissima phol), commonly known as cassava, tapioca, mandioca, and manioc, is regarded as the bread of the tropics as it is mainly grown and consumed in the tropical world [1, 2]. Its primary attraction is its tuberous root, which serves as one of the highest yielding starchy staples [3]. The root has been processed into varieties of food, including garri and fufu, among others. Specifically, Abiagom [4] reported that 15% of cassava was consumed as fresh roots; 5% as garri; 10% as starch; and 10% as flour and others. Similarly, other parts of the crop are equally valuable. The stem of the cassava plant is mainly exploited for propagation, while the leaves—found to be nutritious—are equally consumed.

Besides domestic consumption, cassava has recently been processed into many products of high demand, including ethanol, glucose, starch, animal feed, baking flour, pulp, and paper [5, 6]. Its application beyond consumption has increased as cassava is presently regarded as an industrial/cash crop [7]. The economic importance of cassava transcends the tropics as a staple food to a global industrial raw material for the production of myriad products [8]. Despite being exploited in a variety of ways for domestic consumption and industrial products, waste generation is inevitable and this occurs at almost every stage of the production process [9]. However, the benefits of cassava wastes are yet to be assessed as they are generally discarded and disposed of due to their toxicity [6].

The adverse impact of cassava wastes on the environment and health has been one of the challenges confronting the tropics. Acknowledging cassava waste as a major source of pollution in the cassava processing areas, Okunade and Adakalu [10] reported that cassava waste has deleterious effects on the receiving soil and water source as well as the adjourning environment. Research shows that as far as cassava processing is concerned, waste would be inevitably generated [9]. Since human life generally revolves around the activities that result in the production of cassava wastes, it is necessary to examine previous research works on the characterization of cassava waste and they have been chronicled in one piece for accessibility and posterity. Due to the toxicity of the waste [11] and its adverse impact on the environment, wastes generated from cassava processing need to be handled bereft of levity.

This chapter presents the prospect and challenges confronting cassava processing, especially in Nigeria: the world's largest producer of cassava (Section 2). Waste is inevitable in most industrial processes and cassava processing is not an exemption. The estimated quantity of wastes generated during cassava processing as well as the nature of these wastes is considered next (Section 3). This review is necessary for better management, as the availability of the waste for valorization is paramount. Furthermore, a review on the characteristics of the wastes constituting Section 4 of this chapter showcased the biochemical and physicochemical properties of cassava processing wastes. The concluding section (Section 5) will be on biochar production using cassava waste and will focus on process parameters and the choice of feedstock. This information is a useful guide in our quest to manage cassava wastes for the betterment of our environment and health as well as reveals the potential of these wastes for efficient valorization.

2. Prospect and challenges of cassava processing in Nigeria

This section presents the prospect as well as challenges of processing cassava in Nigeria. Nigeria is the largest producer of cassava in the world for decades, and has a robust cassava industry with prospects and challenges. It is expected that other cassava-producing countries, such as Brazil and Thailand, among others, may have similar challenges. The progress in cassava production in Nigeria is reported next.

2.1 Progress in cassava production in Nigeria

Cassava has been identified as one of the most cost-effective and nutritionally vital native African tubers [12–14]. Its origin is traced back to South America, and the crop made its trans-Atlantics journey close to the beginning of the slave trade in the sixteenth century into Nigeria [15]. Cassava is a recurrent, vegetative bred shrub, and is cultivated through the plain tropics [16]. It is a dearth resilient crop grown mostly in temperate areas and adds appreciably to the nourishment and livelihood of many countries, including Nigeria. Cassava is one of the major staple foods in Nigeria, and its cultivation is a priority in almost every household, especially in the Southern part of the country [8, 15]. Hence, it is the most extensively farmed crop in Nigeria and it is largely cultivated by small-scale farmers that depend on seasonal rainfall [17].

Presently, cassava has been changed from a low-yielding dearth spare crop for consumption to a high-producing cash crop, with its many different uses in livestock feeds, source of raw materials for agro-industry, and beyond [18, 19]. Over half a billion people around the world depend on cassava as a major food source. It is the third largest source of calories after rice and corn [5, 6, 20]. The crop is known to thrive well on any soil and its ability to grow well in poor soils and withstand drought makes it an ideal crop to cultivate in places where other crops struggle [21, 22].

Fortunately, Nigerian soil is fertile and the crop thrives very well making Nigeria the world's largest producer of cassava for about a decade. The country went from harvesting 36 million tons of cassava in 2003 to 53 million in 2013, a 47% increase [15]. Growth was driven by a substantial increase in yields—which jumped by 44% over this period as Nigeria overtakes Brazil as the top producer of cassava. This increase is attributed to several interventions and initiatives, including the Institute of Tropical Agriculture (IITA), the presidential initiative on cassava, etc., by the Nigerian governments at all levels. It was reported that through these programs large hectares of land were dedicated to cassava cultivation in order to boost the production and exports of processed cassava products [23, 24]. With 59.5 million tons of cassava produced in 2019 (**Figure 1**), the country maintained its top spot in global cassava production since it outpaced Brazil in total production output in 1991 [26, 27].

The plant as a whole is useful as its roots and leaves are consumed, while the stems are mainly exploited for propagation. About 70% of Nigeria's cassava is processed into garri—a granular flour that is used mainly to make porridges and fufu—a type of mash [22]. Specifically, the roots are processed by several methods to form products that are used in diverse ways according to local preferences [28, 29]. However, when left unprocessed, the cassava roots perish quickly, often spoiling within 48 hours [30]. Processing offers not just the ability to produce higher-value, exportable cassava-derived items like garri and essential items, such as glucose, starch, and flour, but also to preserve the root (**Figure 2**).

Meanwhile, Thailand, the second-largest producer of cassava, considered the crop more of a cash crop than a staple food with the vast majority of the root being processed and exported [31–33]. Currently, Thailand is the world's leading country in the exportation of cassava-based products (**Figure 3**). It is reported that in Thailand, about 90% of all cassava is processed and exported [22]. Thailand accounted for approximately 76% of global trade and 84% of its cassava exports were sent to China,

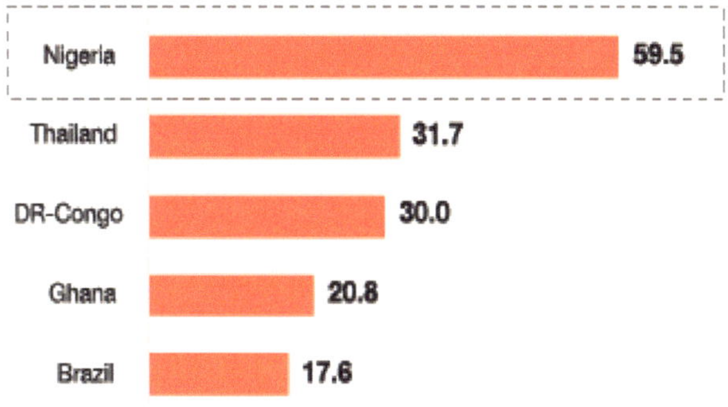

Figure 1.
Top cassava-producing countries (in million tons). Data are adapted from PwC [25].

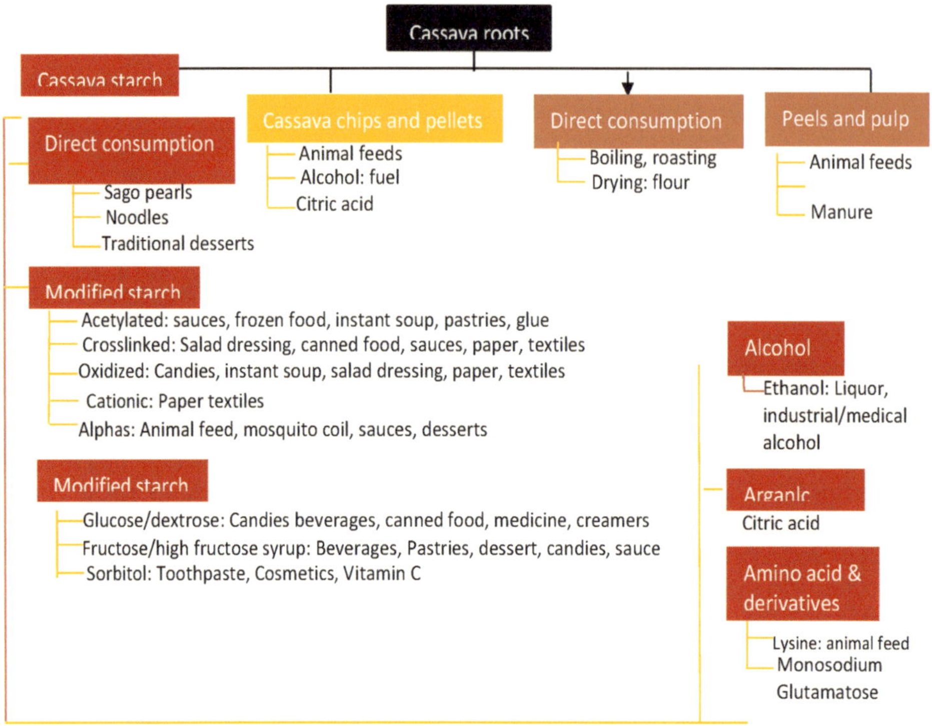

Figure 2.
Potentials of cassava root. Data are adapted from PwC [25].

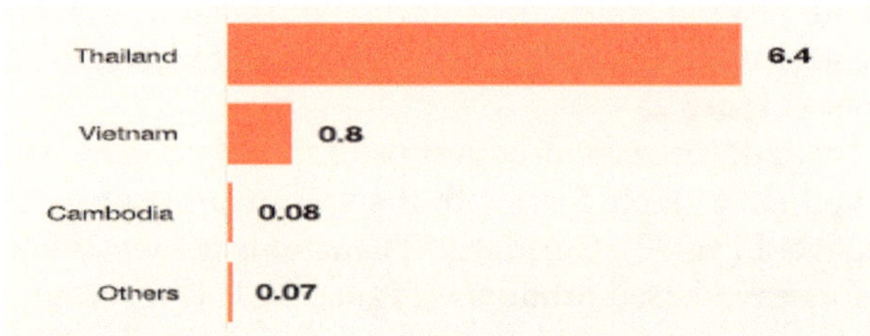

Figure 3.
Top cassava exporting countries (in million tons). Data are adapted from PwC [25].

where the products are largely used for ethanol production [34]. Specifically, over 6.4 million tons of cassava-based products were exported from Thailand accounting for about 50% of the total volume of exported products worldwide generating huge revenue for the country [22, 27].

2.2 Challenges of cassava processing in Nigeria

In Nigeria, about 80% of the cassava is processed into garri for local consumption and exportation [22]. Specifically, processing cassava into garri is relatively simple—it requires only that cassava roots are peeled, grated, and sieved, and then placed in a porous bag from which excess water can be squeezed out (**Figure 4**). The resulting

Figure 4.
Cassava processing facility in Nigeria.

dry flour can be stored for several months [29]. Approximately 70% of cassava processing occurs at small- and medium-size centers near villages.

In 2012, it was reported that there were 75,000 total small- and medium-processing centers that employed roughly 3 million people—most of which were small-scale farmers, and generated less than 5 tons of high-quality cassava flour per day [25]. However, medium- and large-scale processors struggle to stay afloat due to high transportation costs, mainly due to the poor condition of rural Nigerian roads [16]. The challenge of limited access to cassava processing facilities has not only hindered efficient large-scale processors, but also extends to storage facilities as unreliable transportation compounds the problem presented by the crop's perishability. Consequently, post-harvest losses for cassava are high as estimated by Deutsche Gesellschaft für Internationale Zusammenarbeit (GIZ) and found to be more than $600 million annually in Nigeria alone [22, 35].

The challenges confronting Nigeria's cassava processing capacity are responsible for the country's poor role in international cassava trade as large-quantity cassava needs to be processed in order to be exported. Thailand's position as the global leader in cassava processing and exportation is a clear indication that Nigeria—as the largest producer—is not doing well in the area of processing. Nigeria's emergence in the international cassava market coupled with achieving self-sufficiency in cassava-based products, such as starch, glucose, and flour production, is not out of reach. The country has already proven success in improving raw cassava production through its increase in yields and needs to extend these successes throughout the value chain. The economic potential of cassava is huge as seen in **Figure 5**.

Specifically, the country has the huge economic potential to generate USD 427.3 million from domestic value addition and derive an income of USD 2.98 billion in the exportation of cassava-based products [25, 36]. According to PwC [25], the local addition to cassava *via* local manufacturing and processing could potentially unlock about USD 16 million as revenue for Nigeria.

Realizing the incredible amount of untapped potential that lies waiting in this sector, Nigerian government has waded in to address the challenges confronting cassava processing. In an attempt to support the cassava processing industry, the government launched the Cassava Transformation Agenda in 2011 [22]. The initiative is working to expand the cassava value chain and Nigeria is making millions from cassava production and export. It was reported that Nigeria exported 509 tons of cassava products, half of which went to China, the world's top importer of the product in 2019 [8, 36]. Several medium- and large-scale cassava processing facilities are set up

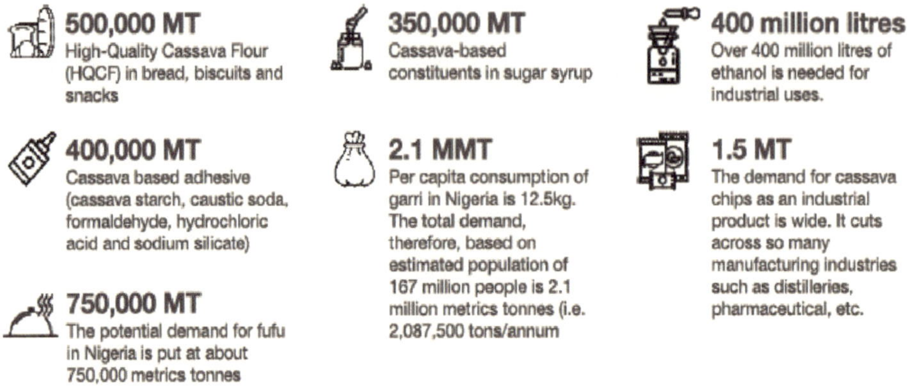

Figure 5.
Current demand for cassava-based products in Nigeria. Data are adapted from PwC [25].

Figure 6.
Cassava processing waste (cassava peel).

on a daily basis and processing clusters are domiciled in every community as seen in the Ojapata processing cluster, Kogi state, and Nigeria [8, 37].

Although the commercialization of cassava processing and subsequent exportation and expansion of its value chain generate huge revenue and job opportunities for Nigerians, this development is welcomed with mixed feelings due to poor waste management (**Figure 6**) and its adverse impact on the environment [38]. The cassava processing environment (**Figure 4**) is heavily polluted in all ramifications (air, water, and land) and the impact of environmental pollution attributed to cassava wastes is significant, especially with commercial processing of the cassava.

3. Cassava processing and quantity of wastes generated

Processing cassava into varieties of products inevitably result in the generation of wastes. Basically, according to Ubalua [37] and Zhang et al. [38], these wastes are categorized into (i) peels prior to crushing, (ii) sieved fibrous residue after crushing, (iii) bagasse and settling starch, and (iv) wastewater effluent. In view of this, cassava processing facilities continuously generate waste as the demands for cassava-based products soar. Chunk amount of cassava processing wastes and residues generated has been reported to be one of the major environmental threats, especially in rural regions of developing countries [39, 40]. These regions are mainly dominated by small-scale processing facilities managed mainly by rural dwellers bereft of standardized waste treatments and disposal strategies. The informed operators of these facilities considered the cost of treatment and disposal of these wastes a huge financial burden.

Zhang et al. [38] estimated that the processing of fresh cassava roots generates liquid waste between 8.85 and 10.62 MT per MT of fresh cassava processed, containing approximately 1% total solids (TS). In the case of dry cassava processed, the authors reported that between 0.93 and 1.12 MT of wet cassava bagasse and peels are produced per metric ton of dry cassava processed.

A breakdown of wastes generated from fresh cassava root during the production of high-quality cassava flour (HQCF), starch, garri, and fufu are presented in **Figures 7–10** respectively. Specifically, during the production of 150–200 kg HQCF from 1 MT of fresh cassava roots, 550–700 kg of wastes was generated constituting peels, fibrous waste, sifting juice, and wastewater (**Figure 7**).

A recent report stated that for every ton of cassava processed, 10–15% constituting 125 kg/tons are lost in form of wet peels, which are poorly utilized, dumped as waste, or burnt [42, 43]. These methods of disposing cassava wastes though easy and cheap

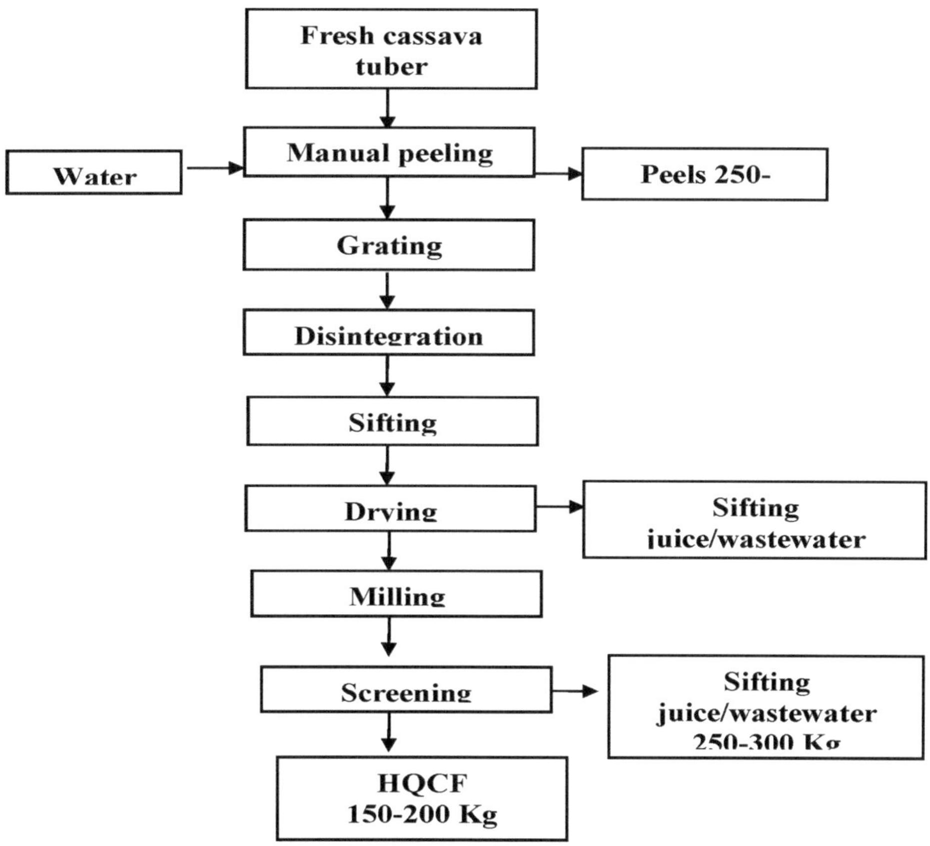

Figure 7.
Flow sheet of high-quality cassava flour production process. Data are adapted from Sanni and Jaji [7]; FAO [41].

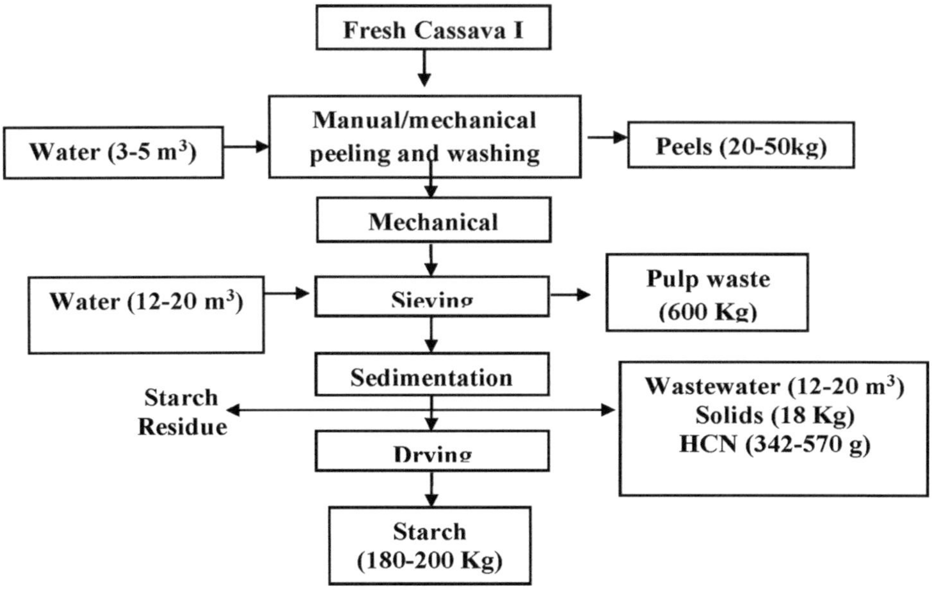

Figure 8.
Flow sheet of the starch production process and waste generated. Data are adapted from FAO [41].

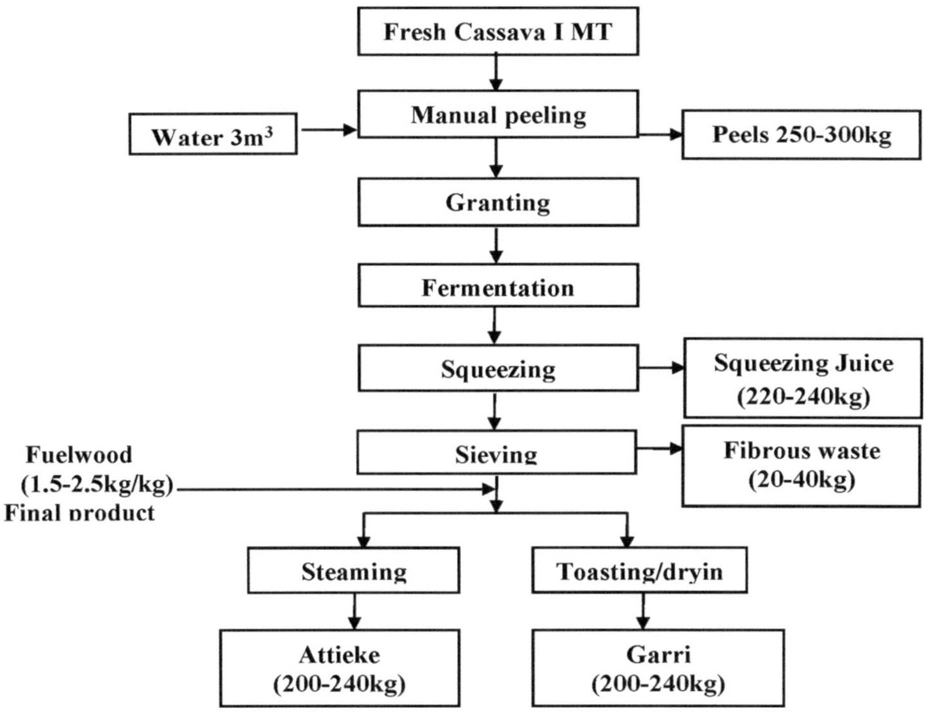

Figure 9.
Flow sheet of garri production process and waste generated. Data are adapted from Sanni and Jaji [7]; FAO [41].

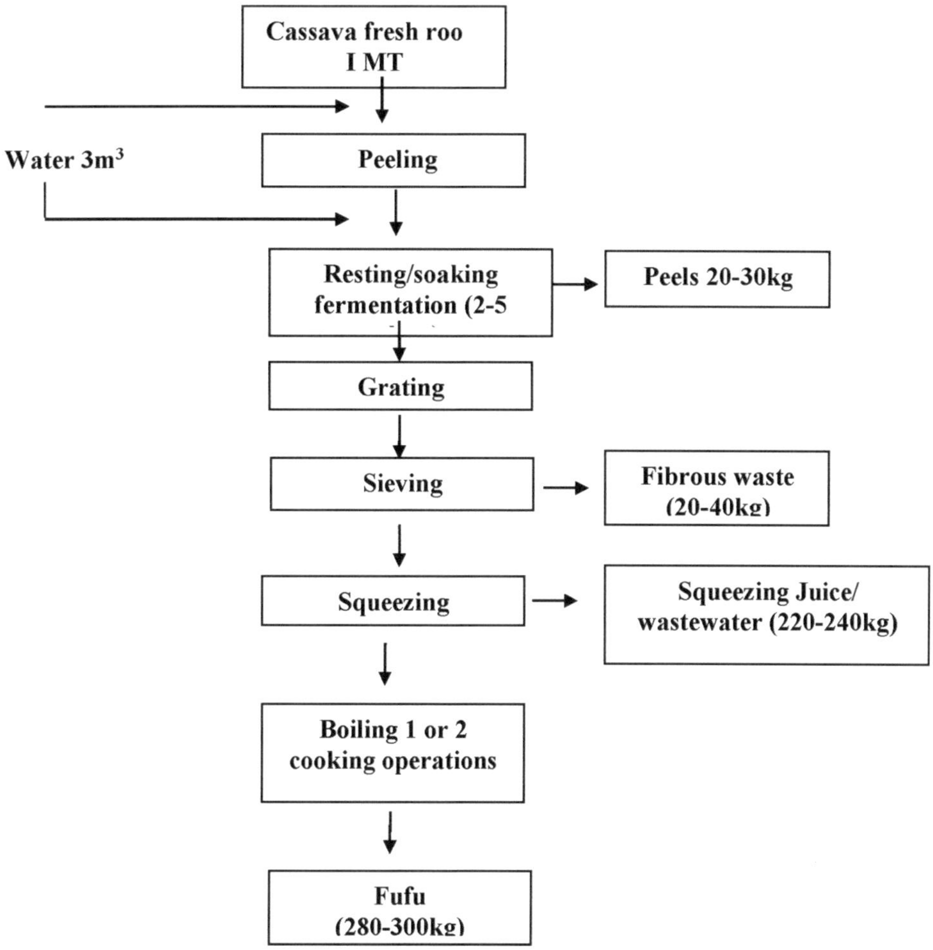

Figure 10.
Flow sheet of fufu production process and waste generated. Data are adapted from Sanni and Jaji [7]; FAO [41].

are not economically viable in terms of lands claimed and environmental-friendly due to pollution.

Starch is one of the cassava-based products with high demand in several industries, including laundry, and the flow sheet for its production process is presented in **Figure 8**. FAO [41] reported that 1 MT of fresh cassava roots when processed can produce between 180 and 200 kg starch with about 680 kg of waste generated.

Currently, garri (cassava flake) is the most preferred cassava-based product widely recognized as a staple food in the tropics. **Figure 9** presents the flow sheet of garri production process. However, it is reported that of the 200–240 kg of garri produced from 1 MT of fresh cassava roots, and 500–600 kg of waste was generated [7, 41].

Fufu is another cassava-based product known mostly in southern Nigeria. In terms of consumer preference, fufu was the most preferred in the early 1970s with more than 60% of cassava exploited for its production [21]. However, by the early 1980s, the consumption of fufu had declined to 14% of all cassava eaten, while consumption of garri rose to 65% according to a national consumption survey by the Federal Office of Statistics (FOS) [44]. It is considered that the consumer preference for fufu had reduced due to its inherent undesirable characteristics of poor odor, short shelf life, and tedious preparation [27, 29]. The flow sheet for the production of fufu is presented in **Figure 10**. It has been found that 1 MT of fresh cassava roots produced between 280 and 300 kg of fufu generating between 80 and 130 kg of waste [41].

4. Characteristics of cassava wastes

The properties of wastes generated from cassava processing constituting the peels prior to crushing, the sieved fibrous residue after crushing, the bagasse and settling starch, and the wastewater effluent are presented in this section. The physicochemical properties of the wastes are critical as they reveal the way in which the wastes interact with other substances physically and chemically when discharged. Section 4.1 presented the physicochemical characteristics of these wastes. The thermochemical properties of the wastes are presented in Section 4.2. The biochemical properties of the wastes are equally useful as the wastes when discharged are expected to interact with the fauna and flora content of the medium as well as the environment. This is discussed in Section 4.3.

4.1 Physicochemical characteristics of cassava wastes

These are the intrinsic physical and chemical characteristics of cassava wastes. These include appearance, boiling point, density, toxicity, volatility, water solubility, and flammability,. In the case of cassava wastes, several studies have been conducted by various researchers, and the outcome of some of the works is presented in this section. Zhang et al. [38] reported that cassava starch wastes (wastewater and solid waste) are weakly acidic liquids with high nitrogen and phosphorus contents of about 1300 and 780 mg/L, respectively. In addition, this category of cassava waste contains between 9.6 and 37.5 g/L of total carbohydrates and 2.3 total proteins. **Table 1**

Parameters/properties	Values (G/L)
pH	3.6–6.2
Total solid (TS)	4.5–38.2
Volatile solid (VS)	3.4–33.0
Total chemical oxygen demand (TCOD)	8.0–66.2
Soluble chemical oxygen demand (SCOD)	14.2–345
Biochemical oxygen demand (BOD)	—
Total carbohydrate	9.6–37.5
Solid carbohydrate	—
Oil and grease	0.6
Total protein	2.3
Total nitrogen	0.1–1.3
Total phosphorous	0.07–0.78

Data are adapted from Zhang et al. [38].

Table 1.
The physiochemical characteristics of cassava wastes.

presents the physicochemical properties of cassava starch wastes as well as their composition.

In the case of cassava bagasse, Zhang et al. [38] further reported that a typical solid residue of cassava processing contains between 40.1 and 75.1% starch (dry weight) and between 14.9 and 50.6% fiber. **Table 2** shows the composition of the cassava residue.

Wastewater is inevitably generated during cassava processing either as a byproduct of the initial production process or arises when the cassava tubers are indiscriminately discharged to a nearby water body. Okunade and Adekalu [10] reported on the organic components of cassava wastewater (**Table 3**) as they

Composition	% by dry weight
Starch	40.1–75.1
Crude fiber	14.9–50.6
Cellulose	4.1–11.4
Hemicellulose	4.2–8.3
Lignin	1.2
Crude fat (lipids)	0.5–1.1
Crude protein	0.3–1.6
Total ash	0.7–11.9
Total solid	—
Volatile solid	—
Total nitrogen	—

Data are adapted from Zhang et al. [38].

Table 2.
Cassava residue composition % by dry weight.

Component	Concentration in ppb
3-penten-2-ol	276.007
1-butanol	259.561
3-hexanol	95.897
Octadecanoic acid	495.085
Oleic acid	135.546
n-hexadecanoic acid	71.417
Acetoin	362.956
Dibutyl phthalate	140.801
Squalene	76.9188
Bis (2-ethylhexyl) phthalate	73.686

Data are adapted from Young and Markmanuel [35].

Table 3.
Concentration of organic compounds in cassava wastewater.

found that cassava wastewater, which is five times denser than water contains alcohols, acids, and others (3-penten-2-ol, 1-butanol, 3-hexanol, octadecanoic acid, oleic acid, n-hexadecanoic acid, acetoin, dibutyl phthalate, squalene, and bis (2-ethylhexyl) phthalate). The rust-removing properties of the wastewater from metallic substances, such as nails, are attributed to the presence of these organic compounds [10].

In the related development, Aripin et al. [45] have recognized that without proper waste management, and the organic wastes like cassava peels could result in an increased amount of solid waste dumped into landfills. This eventually leads to less soil available for agricultural purposes as most of these soils are rendered infertile, in an attempt to utilize these organic wastes as pulp for paper-making industries and to promote the concept of “from waste to wealth and recyclable material.” The authors exploited Kurscher-Hoffner and Chlorite methods to determine the chemical properties of the wastes in accordance with the relevant Technical Association of the Pulp and Paper Industry (TAPPI) test. It was found that the cassava waste was rich in holocellulose, cellulose, hemicellulose, lignin, and ash content with 1% of sodium hydroxide and hot water solubility (**Table 4**).

In order to determine the suitability of cassava peel as an alternative fiber resource in pulp and paper making, its properties were compared to other published literature, especially from wood sources. Aripin et al. [45] reported that the amount of holocellulose contents in cassava peels (66%) is within the limit suitability to produce paper although it is the least when compared with that of the wood (70–80.5%) and canola straw (77.5%). Similarly, the lignin content (7.52%) is the lowest than those of all wood species (19.9–26.22%). However, the morphological properties of the cassava peel are promising as the authors went further to subject the peels to scanning electron microscopy (SEM) under different magnifications (**Figures 11** and **12**).

Aripin et al. [45] observed that under different levels of magnifications (27 and 300), there exist differences in fiber morphology of the peels (**Figures 11** and **12**). The surface morphology of cassava peels depicted in **Figure 11** shows that it is dominated by a low (micro) pore size structure, while that of **Figure 12** differs. The differences in fiber morphology of the peel indicate variation in the major character of the fiber's physical structure. This variation had been reported to be attributed to differences in the physical properties of the cassava peels [46]. The porous structure of the peel can be exploited for several industrial applications.

Constituents	W/W oven-dried materials
Holocellulose	66.0
Cellulose	37.9
Hemicellulose	37.0
Lignin	7.5
1% NaOH	27.5
Hot water	7.6
Ash	4.5

Data are adapted from Aripin et al. [45].

Table 4.
Chemical composition of oven-dried cassava peels.

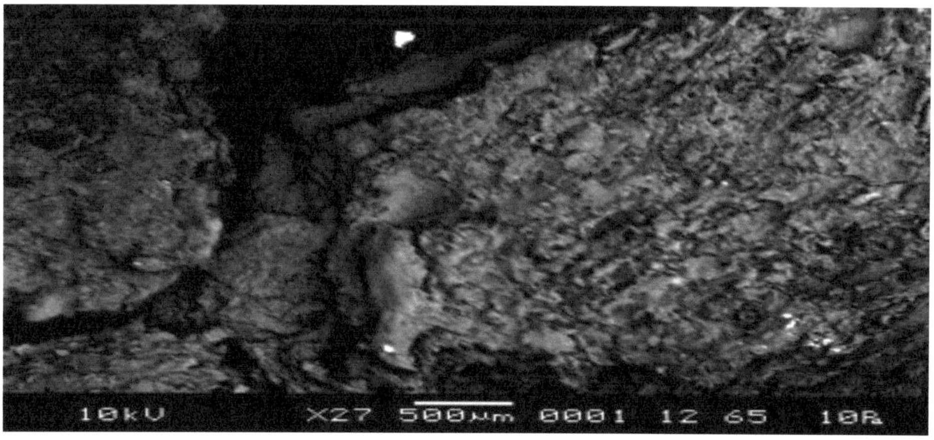

Figure 11.
Surface morphology (SEM) of cassava peel at ×27. Data are adapted from Aripin et al. [45].

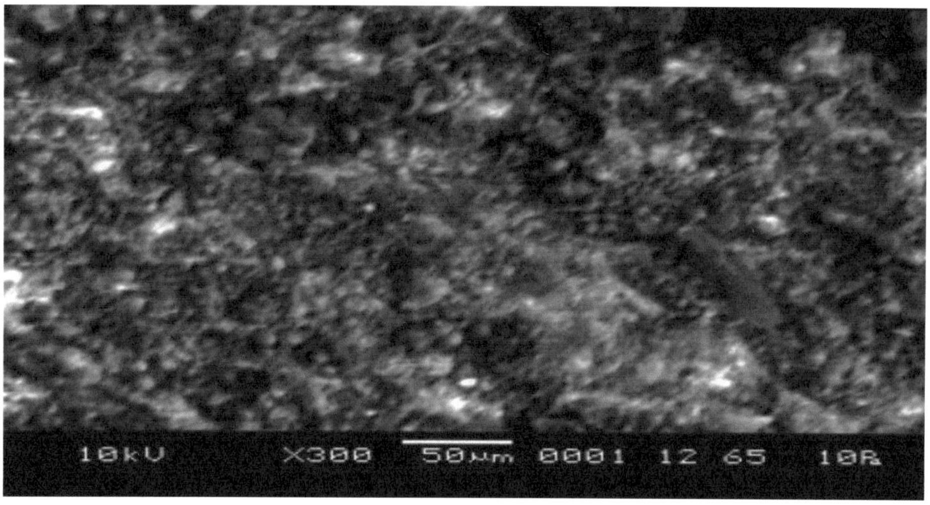

Figure 12.
Surface morphology (SEM) of cassava peel ×300. Data are adapted from Aripin et al. [45].

4.2 Thermochemical characteristics of cassava wastes

The thermochemical properties of agricultural wastes have been a subject of research interest recently. One of these was reported by Pattiya [47] on cassava wastes. The study includes proximate, ultimate, structural, inorganic matter, heating value, and thermogravimetric analyses. Cassava waste was found as shown in **Table 5** to have high volatile contents (78–80%, dry basis) and contains 51% carbon, 7% hydrogen, 41% oxygen, 0.7–1.3% nitrogen, and <0.1% sulfur. Structural analysis reveals that cassava residues are composed of about 36% cellulose, 44% hemicellulose, and 24% lignin. The main inorganic elements found are potassium, phosphorus, and calcium. The lower heating values (LHV) of the biomass are approximately 18 MJ kg^{-1}.

Similarly, Aro et al. [48] carried out a similar study on cassava tuber wastes (CTW) produced by a cassava starch-processing factory in the Ondo State of Nigeria. They investigated the properties of five different types of CTW wastes: cassava starch

Composition	Value (% dry basis)
Volatile organic contents	78.0–80.0
Carbon	51.0
Hydrogen	7.0
Oxygen	41.0
Nitrogen	0.7–1.3
Sulphur	<0.1
Cellulose	36.0
Hemicellulose	44.0
Lignin	24.0
Heating Values	18 MJ kg^{-1}

Data are adapted from Pattiya [47].

Table 5.
Proximate, ultimate, structural, inorganic matter, heating value, and thermo-gravimetric analyses of cassava wastes.

Parameters	CAP	CAE	CAW	CSR	CAS
Dry matter	17.9	8.63	3.34	15.8	35.9
Crude protein	4.20	2.92	2.46	1.12	1.71
Crude fiber	29.6	6.69	ND	19.3	12.9
Fat	3.26	1.75	ND	2.37	5.35
Ash	7.47	3.16	1.88	2.84	3.39
Moisture	82.1	91.4	96.7	84.2	64.1
N.F.E.*	55.5	85.5	95.7	74.4	72.3

**N.F.E. = nitrogen free extractives, CAP = cassava peels, CAE = cassava effluent, CAW = cassava whey, CSR = cassava starch residues, CAS = cassava stumps, ND = not detected, and D.M. = dry matter. Data are adapted from Aro et al. [48].*

Table 6.
Proximate composition (g/100g D.M.) of different types of fresh cassava tuber wastes (CTW) collected from the factory.

residues (CSR) or pomace cassava peels (CAP), cassava effluent (CAE), cassava stumps (CAS), and cassava whey (CAW). The proximate composition of samples collected in respect of these five types of wastes (**Tables 6–8**) showed that moisture was the highest in CAW (96.7%) and the lowest in CAS (64.1%). Crude fiber was highest in CAP (29.6%) but was not detected in the whey (CAW). The CAS had the highest content of fat (5.35%), while it was not detected in CAW. Protein was the highest in CAP (4.20%) and lowest in CSR (1.12%). Ash content was the highest in CAP (7.47%) and lowest in CAW (1.88%). The nitrogen-free extractives (NFE) were the highest in CAW (95.7%) and lowest in CAP (55.5%).

The proximate analysis of cassava waste conducted by Janz and Uluwaduge [49] is presented in **Table 9**, while that of Obadina et al. [50] on cassava peel is presented in **Table 10**.

Parameters	CAP	CAE	CAW	CSR	CAS
Potassium	269	67.7	42.7	138	117
Calcium	122	50.2	5.17	60.0	15.2
Magnesium	236	33.2	22.7	129	117
Iron	14.7	16.2	2.21	5.66	2.46
Manganese	0.43	ND	ND	ND	ND
Copper	0.24	0.13	ND	10.8	0.19
Molybdenum	ND	ND	ND	ND	ND
Cobalt	ND	ND	ND	ND	ND
Arsenic	ND	ND	ND	ND	ND
Selenium	1.16	0.42	ND	1.27	0.46
Sodium	261	24.0	7.54	93.4	46.9
Phosphorus	3233	2517	251	2251	1663

Data are adapted from Aro et al. [48].

Table 7.
Mineral composition (mg/kg D.M.) of fresh cassava tuber wastes (CTWs) collected from the factory site.

Parameters	CAP	CAE	CAW	CSR	CAS
Cyanide, mg/kg	32.9	54.4	61.1	15.5	34.8
Phytate, mg/kg	8238	4264	3947	15930	9276
Oxalate, mg/kg	330	674	520	270	610
Tannins, %	3.90	2.16	0.98	270	3.44
Saponin, %	0.06	0.08	0.02	2.53	0.15
Total alkaloids, %	0.48	0.40	0.16	0.10	0.52

Data are adapted from Aro et al. [48].

Table 8.
Antinutrient composition of cassava wastes (dry matter basis).

Parameters	Values (g)
Moisture	59.4
Total carbohydrate	38.1
Protein	0.7
Lipid	0.2

Data are adapted from Janz and Uluwaduge [49].

Table 9.
Proximate analysis of cassava waste on the basis of 100 g.

Composition	Value (% dry basis)
Carbohydrate	42.6
Protein	1.6
Ether extract	12.1
Total ash	5.0
Crude fiber	22.5
Data are adapted from Obadina et al. [50].	

Table 10.
Proximate analysis of cassava peel.

4.3 Biochemical characteristics of cassava wastes

Apart from the nutritional value of cassava, the chemical oxygen demand (COD), biochemical oxygen demand (BOD), and cyanide content are the parameters of interest because of their effect on the flora and fauna as well as the environment. Based on cyanide contents, cassava can be classified as sweet and bitter and it wastes are often laden with suspended solids, high COD, and BOD making them toxic. Some of the works carried out in this direction show that one liter of cassava wastewater has 23.9 g of COD, 23.1 g of volatile solids (VS), and 22.9 g of total solids (TS) [51, 52].

In the case of nutritional value, some of the works presented earlier (vide supra) indicated the presence of carbohydrates, proteins, and other nutritional components that support life. Dresden [11] conducted research on the nutritional profile of cassava waste and reported its nutritional composition as presented in **Table 11**.

Glanpracha et al. [53] reported on the cyanide content of cassava waste as presented in **Table 12**. The cyanide content includes hydrocyanic acid (HCN) and cyanogenic glucoside (linamarin).

Nutritional composition	Value
Calories	330 kcal
Protein	2.8 g
Carbohydrate	78.4 g
Fiber	3.7 g
Calcium	33.0 mg
Magnesium	43.0 mg
Potassium	558.0 mg
Vitamin C	42.4 mg
Thiamine	0.087 mg
Riboflavin	0.048 mg
Niacin	0.854 mg
Data are adapted from Dresden [11].	

Table 11.
The nutritional profile of cassava waste.

Properties/parameters	Values
Cyanide (cyanohydrins), hydrocyanic acid (HCN), and cyanogenic glucoside (linamarin)	45–154 mg
Fibers	—
Protein	—
C/N	17.64 ± 3.1
Total nitrogen (% dry weight)	0.3 ± 0.1
Total organic carbon (% dry weight)	52.1 ± 3.2
Volatile solids (% wet weight)	19.6 ± 0.8
Total solids (% wet weight)	20.9 ± 0.9
Moisture content (% wet weight)	78.9 ± 0.7

Data adapted from Glanpracha et al. [53].

Table 12.
Biochemical composition of cassava waste.

Parameter	Mean ± SD
Cyanide (mg/100 g)	10.15 ± 4.36
Dry matter (g/100 g)	33.70 ± 5.85
Moisture (g/100 g)	66.31 ± 5.83
Carbohydrate (g/100 g)	30.53 ± 6.16
Starch (g/100 g)	12.40 ± 8.357
Energy value (kcal/100 g)	142.48 ± 22.60
Reducing sugar (g/100 g)	1.09 ± 0.48
Ash (g/100 g)	0.44 ± 0.18
Acidity (meq/100)	82.19 ± 21.33
Total sugar (g/100 g)	16.75 ± 6.08
Fat (g/100 g)	1.55 ± 1.11
Ph	5.95 ± 0.12
Carotenoid (mg/100 g)	0.11 ± 0.06
Vitamin A (mg/100 g)	17.89 ± 15.42
Protein (mg/100 g)	1.61 ± 0.70
Vitamin C (g/100 g)	6.89 ± 2.17

Data adapted from Mégnanou et al. [54].

Table 13.
Physicochemical and biochemical characteristics of nine cassava varieties.

Mégnanou et al. [54] conducted a study on the physicochemical and biochemical characteristics of nine varieties of cassava roots (V4, V23, V60, V61, V62, V63, V64, V65, and V66). The mean values of their physicochemical and biochemical characteristics are presented in **Table 13**.

In a similar development, Izah et al. [55] investigated the heavy metal content of cassava mill effluents (cassava processing wastes) collected from a cassava processing mill at Ndemili in Ndokwa west local government area of delta state, Nigeria. It was found that the effluent contains 1.46 mg/l of copper, which is comparable to the value 1.83 mg/l reported by Orhue et al. [56], 1.91 mg/l reported by Adejumo and Ola [57], and lower than the value of 2.50 mg/l as reported by Patrick et al. [58] as well as 2.60 mg/l by Olorunfemi and Lolodi [59] and higher than the value of 0.00 mg/l reported by Omomowo et al. [60]. Another heavy metal present was zinc with a concentration of 4.35 mg/l. This is comparable to the value of 4.1 mg/l reported by Patrick et al. [58] and lower than the value of 5.90 mg/l reported by Olorunfemi and Lolodi [59] and higher than the value of 1.07 mg/l reported by Orhue et al. [56] as well as 0.00 mg/l as reported by Adejumo and Ola [57].

Manganese was equally found with a concentration of 4.64 mg/l, which is lower than the value of 0.71 mg/l reported by Adejumo and Ola [57], as well as 0.00 mg/l by Omomowo et al. [60] and lower than the value of 7.10 mg/l reported by Olorunfemi and Lolodi [59]. About 28.27 mg/l of iron was reported to be found in the effluent, which is far higher than the value of 2.35 mg/l reported by Adejumo and Ola [57], as well as 2.30 mg/l reported by Omomowo et al. [61] and 2.00 mg/l by Orhue et al. [56] and lower than the value of 30.9 mg/l reported by Olorunfemi and Lolodi [59]. The study further revealed the presence of 0.18 mg/l of chromium, which was lower than the value of 1.14 mg/l reported by Olorunfemi and Lolodi [59].

Generally, the arbitrary variation (with no trend) in the heavy metal concentration (**Table 14**) could be attributed to the age of the cassava prior to processing, activities leading to individual heavy metals disposition in the plantation where the cassava was cultivated, and possible leaching of metals from the processing equipment.

From the foregoing, it is obvious that a huge quantity of waste from cassava processing would unavoidably be generated irrespective of the cassava-based products of interest. The waste constituents are not all toxic as researchers found that in it (the waste) contains 11% of the crop energy coupled with valuable mineral nutrients [61–64]. These valuable contents of the waste can be exploited to boost the economic potential of cassava processing. Thus, the characteristics of cassava wastes as reviewed are within the range, presented in **Table 15**.

Heavy metals	Izah et al. [55]	Omomowo et al. [60]	Orhue et al. [56]	Adejumo and Ola [57]	Patrick et al. [58]	Olorunfemi and Lolodi [59]
Cu, mg/l	1.460 ± 0.460	0.00	1.83	1.91	2.50	2.60
Zn, mg/l	4.353 ± 0.365	NA	1.07	0.0	4.10	5.90
Mn, mg/l	4.637 ± 0.195	0.0	NA	0.71	NA	7.10
Fe, mg/l	28.270 ± 1.130	2.30	2.00	2.35	NA	30.9
Cr, mg/l	0.180 ± 0.020	NA	NA	NA	NA	NA
Ni, mg/l	1.810 ± 0.110	NA	NA	NA	NA	NA

NA: not available.

Table 14.
Heavy metal contents of various effluents from cassava processing mill.

Parameters	Range of value
C/N	17.64–30.0
Total solid (TS)	4.5–38.2 mg
Volatile solid (VS)	3.4–33.0 mg
pH	3.6–6.2
Total chemical oxygen demand (TCOD)	8.0–66.2
Soluble chemical oxygen demand (SCOD)	14.2–345
Total carbohydrate	9.6–37.5 mg
Cellulose	36.0–43.2 mg
Hemicellulose	44.0–64.4 mg
Lignin	24.0–46.2 mg
Cyanide	45–154 mg

Table 15.
Parameters of interest.

5. Biochar from cassava wastes: parameters and choice of feedstock

Characterizations of cassava wastes are necessary for efficient valorization. Through characterization, it is obvious that embedded in the wastes are fractions of the crop energy and minerals. The huge quantity of waste generated during cassava processing translates to a huge quantity of energy and minerals that need to be recovered. The cassava wastes can be converted into biogas (energy recovery) as well as digestate filtrate and residue for biofertilizer, bio-oil, and biochar (mineral recovery) [65]. Several studies have been carried out on bio-oil production using cassava wastes but much has not been reported on biochar from cassava wastes [66, 67]. A brief description of biochar properties would be beneficial to identify its applications.

Biochar is a carbon-rich product obtained when biomass (cassava waste) is heated in a closed system with restricted oxygen. Structurally, it is similar to charcoal but with different properties. However, biochar has a high surface area (highly porous) and negative surface charge, and charge density [68]. Due to its superlative adsorption properties, biochar has been extensively used as an adsorbent [69]. Biochar as adsorbent finds application in removing “emerging contaminants” from flue gas and wastewater. It is equally applied to soil to improve soil properties as biochar can hold nutrients and become more stable than most fertilizer or other organic matter in soil [69].

Biochar (pyrochar) is a solid product from the pyrolysis process. Of all the thermochemical conversion processes (combustion, incineration, etc.), pyrolysis offers a great opportunity of transforming wastes into wealth. Varieties of biomass, including cassava wastes, are exploited as feedstock to produce valuable gas, liquid, or solid products, including biochar. Research has shown that pyrolysis is relatively environmentally friendly when compared with its counterparts as it produces low emissions [70]. During the production process, it was reported that biochar is able to scrub carbon dioxide, nitrous oxides, and sulfur dioxide from the flue gas constituting greenhouse gases (GHG) [68]. These gases (GHG) contribute immensely to global warming leading to climate change with heat waves, flooding, and typhoons [71–73].

Scrubbing GHG during biochar production can be harnessed as a potential tool to slow global warming [74].

Pyrolysis process is versatile—fast or slow depending on residence time—and can be optimized to enhance the production of desired product. Fast pyrolysis with residence time in seconds generates more liquid products (bio-oil), while slow pyrolysis with residence time in hours favors more solid products (biochar) [75–77]. In addition, the properties of biochar produced are varied by the pyrolysis parameters and choice of feedstock.

Besides residence time, temperature and heating rates equally influence the yield and composition of the pyrolysis products. The temperature and heating rates are two of the pyrolysis parameters that affect the yield and composition of the pyrolysis [70, 78]. Noor et al. [79] reported that temperature has a more significant influence than the heating rate during biochar production. Although the nature of feedstock determines the fixed carbon content in the biochar produced, it was found that a higher pyrolysis temperature increased more fixed carbon in the biochar than a higher heating rate [79, 80]. Thus, the effect of production parameters is significant. However, the choice of feedstock on the quality of biochar is equally important.

We extensively reviewed the complementary role of thermochemical conversion process after subjecting the wastes to a biochemical (anaerobic) process [81]. Anaerobic digestate of cassava waste is a valuable pyrolysis feedstock for biochar production due to its high volatile matter content, low ash, and sulfur content [81]. Consequently, thermally treated digestate is more suitable than any other materials subjected to pyrolysis for biochar production. Meanwhile, cassava plantation residues: cassava stem and cassava rhizome have been exploited for biochar production [79, 80]. However, the biochar produced from cassava wastes contains a high percentage of fixed carbon, which is about five to eight times higher than that from cassava plantation residues.

Cassava irrespective of its components can be exploited for biochar production *via* pyrolysis. Three categories of cassava waste can be exploited as pyrolysis feedstock for biochar production (cassava plantation residues, cassava processing waste, and digestate from anaerobic digestion of cassava processing waste). The quality of biochar produced depends on the process parameters and choice of feedstock. Slow pyrolysis when optimized produces high-quality biochar suitable for several applications [79, 80]. Nevertheless, digestate from anaerobic digestion of cassava waste is the most preferred pyrolysis feedstock. This is followed by cassava processing wastes and the least is the cassava plantation residues, especially the stem, this is subject to further investigations. Besides the production of high-quality biochar, the digestate from cassava processing waste has been effectively exploited for biogas and biofertilizer production.

6. Conclusion

This review has established that waste generation during cassava processing is inevitable irrespective of the cassava-based products. In the course of characterizing the wastes, which emerged during processing, pre-processing, and post-processing depending on the products, various researchers reported that the physical, chemical, and biological properties are within the range as paired: carbon/nitrogen (17.64–30.0), total solid (4.5–38.2 mg), volatile solid (3.4–33.0 mg), pH (3.6–6.2), total chemical oxygen demand (8.0–66.2), soluble chemical oxygen demand (14.2–34.5), total carbohydrate (9.6–37.5 mg), cellulose (36.0–43.2 mg), hemicellulose (44.0–64.4 mg), lignin (24.0–46.2 mg), and cyanide (45–154 mg).

It can be confirmed that the cassava waste through toxicity contains a valuable component of interest. This was proven as its energy content is about 11% of the crop energy. This can be harnessed and added to our energy mix. The microporous structure of cassava residue, especially the peel is equally promising for adsorbent formulation. This review pinpoints the potential of these wastes for biochar production. The quality of biochar produced depends on the process parameters and choice of feedstock.

Three categories of cassava waste can be exploited as pyrolysis feedstock for biochar production (cassava plantation residues, cassava processing waste, and digestate from anaerobic digestion of cassava processing waste). Digestate from anaerobic digestion of cassava waste is the most preferred pyrolysis feedstock. Slow pyrolysis when optimized produces high-quality biochar suitable for several applications. Anaerobic digestion of cassava processing wastes generates much more than high-quality biochar. It is an effective waste to wealth strategy.

Conflicts of interest

The authors declare that they have no conflicts of interest regarding the publication of this paper.

Data availability

No data were used to support this study.

Author details

Minister Obonukut[1*], Sunday Alabi[1] and Alexander Jock[2]

1 Department of Chemical Engineering, Topfaith University, Mkpatak, Nigeria

2 Department of Chemical and Petroleum Engineering, University of Uyo, Uyo, Nigeria

*Address all correspondence to: obonukutminister@gmail.com

References

[1] Blanshard AFJ. Quality of processed cassava food in Sierra Leone [PhD thesis]. Loughborough, UK: University of Nothingham; 1994

[2] Hahn SK. An overview of traditional cassava processing and utilization in Africa. In: IITA/ILCA/University of Ibadan Workshop on the Potential Utilization of Cassava as Livestock Feed in Africa, Ibadan. Ibadan, Nigeria: International Institute of Tropical Agriculture (IITA); 14-18 November, 1988. 1988

[3] Coker AO, Achi CG, Sridhar MK. The utilization of cassava processing waste as a viable and sustainable strategy for meeting cassava processing needs. Case study from Ibadan city, Nigeria. In: The 13th International Conference on Solid Waste Technology. Vol. 76. Ibadan, Nigeria: International Institute of Tropical Agriculture (IITA): 2014. pp. 473-482

[4] Abiagom JD. Report on production/utilisation/processing accounts of food crops in Nigeria for the purpose of preparing food balance sheet for Nigeria. In: Prepared for the National Agricultural Development Committee of Nigeria, 1961-69. Lagos (Mimeographed): Federal Office of Statistic (FOS); 1971

[5] Eke-Okoro ON, Njoku DN. A review of cassava development in Nigeria from 1940-2010. Journal of Agricultural and Biological Science. 2012;7(1):59-65

[6] Sehkar CSC, Roy D, Bhatt Y. Food inflation and food price volatility in India: Trends and determinants, markets, trade and institutions division. In: IFPRI Discussion Paper 01640. Bangladesh: International Food Policy Research Institute; 2017

[7] Sanni LO, Jaji FF. Effect of drying and roasting on the quality attributes of fufu powder. International Journal of Food Properties. 2003;**6**(2):229-238

[8] Uba G. Making millions from cassava production and export. Business Day. 2019;**18**:7-8

[9] Ekop IE, Simonyan KJ, Evwierhoma ET. Utilization of cassava wastes for value added products: An overview. International Journal of Scientific Engineering and Science. 2019; **3**(1):31-39

[10] Okunade D, Adekalu K. Characterization of cassava-waste effluents contaminated soils in Ile-Ife, Nigeria. European International Journal of Science and Technology. 2014;**3**(4): 173-182

[11] Dresden D. What to know about cassava: Nutrition and toxicity. Medical News Today. 2021. Article 323756. Available from: www.medicalnewstoday.com [Accessed: February 02, 2022]

[12] Dipeolu AO, Adebayo K, Ayinde IA, Oyewole OB, Sanni LO, Pearce D, et al. Fufu Marketing Systems in South-West Nigeria. In: NRI Report Natural Resources Institute (NRI) Report Number R2626. Chatham Maritime, Kent ME4 4TB, UK: University of Greenwich Central Avenue; 2001. pp. 55-78

[13] Collins RC, Dent B, Bonney LB. A Guide to Value-Chain Analysis and Development for Overseas Development Assistance Projects, Publication of Australian Center of International Agricultural Research (Publication Code: MN178). 2015. pp. 21-28

[14] Bamidele OP, Fasogbon MB, Oladiran DA, Akande EO. Nutritional

composition of fufu analog flour produced from Cassava root (*Manihot esculenta*) and Cocoyam (*Colocasia esculenta*) tuber. Food Science and Nutrition. 2015;**3**(6):597-603

[15] Akpata TG. Analysis of cassava value chain in Nigeria: Pro-poor approach and gender perspective. International Journal of Value Chain Management. 2019;**10**(3): 219-237

[16] Ganeshkumar C, Pachayappan M, Madanmohan G. Agri-food supply chain management: Literature review. Intelligent Information Management. 2017;**9**:68-96

[17] Manvong M, Oyewole B, Olaniyan GO. Report of Field Trip to Southern Guinea Savannah of Nigeria. Nigeria: IITA Ibadan; 1995. pp. 79-88

[18] Tell J, Hoveskog M, Ulvenblad P, Ulvenblad PO, Barth H, Ståhl J. Business model innovation in the agri-food sector: A literature review. British Food Journal. 2016;**118**:1462-1476

[19] Sanni LO. Dehydration characteristics of cassava chips in a solar cabinet dryer [MSc thesis]. Ibadan, Nigeria: University of Ibadan; 1992

[20] Asante-Pok A. Analysis of incentives and disincentives for cassava in Nigeria. In: Technical Notes Series. Rome: MAFAP, FAO; 2013. pp. 57-63

[21] Ikuemonisan ES, Mafimisebi TE, Ajibefun I, Adenegan K. Cassava production in Nigeria: Trends, instability and decomposition analysis (1970-2018). Heliyon. 2020;**6**(10):56-78

[22] Adebowale AA, Sanni LO, Kuye A. Effect of roasting methods onsorption isotherm of tapioca grits. Electronic Journal of Environmental, Agricultural and Food Chemistry. 2006;**5**(6):1649-1653

[23] Daniels A, Udah A, Elechi N, Oriuwa C, Tijani G, Sanni A. Report on Cassava Value Chain Analysis in the Niger Delta, Foundation for Partnership Initiatives in the Niger Delta (PIND). Abuja, Nigeria; 2011. pp. 74-79

[24] Dufour D, O'Brein GM, Best R. Cassava Flour and Starch: Progress in Research and Development. Apartado Aereo 6713, Cali, Colombia: International Center for Tropical Agriculture (CIAT); 2002

[25] PwC. Harnessing Economic Potential of Cassava Production in Nigeria. 2020 Cassava Production in Nigeria Report. 2020. pp. 1-12. Available from: www.pwc.com/ng [Accessed: October 21, 2021]

[26] FAO. International Fund for Agricultural Development (IFAD)/Food and Agriculture Organization (FAO): A review of cassava in Africa with country case studies on Nigeria, Ghana, Tanzania, Uganda and Benin. Rome: Proc. valid.forum Glob. cassava Dev. Stra.; 2005

[27] Oyewole OB, Sanni LO. Constraints in traditional cassava processing—The case of 'fufu' production. In: Agbor Egbe T, Brauman A, Griffon D, Treche S, editors. Cassava Food Processing. France: ORSTOM; 1995. pp. 523-529

[28] Falade SB, Akingbala BD. Productivity analysis of cassava based production systems in guinea savannah. American-Eurasian Journal of Scientific Research. 2010;**3**(1):33-39

[29] Okpokiri AO, Ljioma BC, Alozie SO, MAN E. Production of improved fufu. Nigerian Food Journal. 1985;**3**:145-148

[30] Sanni LO, Maziya-Dixon B, Akanya CI, Alaya Y, Egwuonwu CV, Okechukwu RU, et al. Standards for

Cassava Products and Guidelines for Export. Ibadan, Nigeria: International Institute of Tropical Agriculture; 2005. pp. 11-39

[31] Umeh SO, Odibo FJC. Amylase activity as affected by different retting methods of cassava tubers. International Journal of Agricultural Policy and Research. 2014;**2**(7):267-272

[32] Maziya-Dixon B, Dixon AGO, Adebowale AA. Targeting different end uses of cassava: Genotypic variations for cyanogenic potentials and pasting properties. In: A Paper Presented at ISTRC-AB Symposium. Mombassa, Kenya: International Society for Tropical Root Crops-African Branch (ISTRC-AB). 31 October-5 November 2004

[33] Awoyale W, Asiedu R, Kawalawu WKC, Maziya-Dixon B, Abass A, Edet M, et al. Assessment of the heavy metals and microbial contamination of garri from Liberia. Food Science and Nutrition. 2018;**6**:62-66

[34] Zainuddin IM, Fathoni A, Sudarmonowati E, Beeching JR, Gruissem W, Vanderschuren H. Cassava post-harvest physiological deterioration: From triggers to symptoms. Postharvest Biology and Technology. 2018;**142**: 115-123

[35] Young E, Markmanuel D. Characterisation of Cassava (*Manihots calentacrantz*) waste-water. International Journal of Scientific Engineering and Science. 2019;**3**(6):30-33

[36] Uche C. Cost and returns structure in garri and fufu processing in Rivers State, Nigeria. Nigerian Agricultural Policy Research Journal. 2016;**1**(1): 131-138

[37] Ubalua AO. Cassava wastes: Treatment options and value addition alternatives. African Journal of Biotechnology. 2007;**6**:2065-2073

[38] Zhang M, Xie L, Xin Z, Khanal SK, Zhou Q. Biorefinery approach for cassava-based industrial wastes: Current status and opportunities. Bioresource Technology. 2016;**215**:50-62

[39] Vivekanandan S, Suresh R. Two stage anaerobic digestion over single stage on biogas yields from edible and non-edible de-oiled cakes under the effect of single and co-digestion system. International Journal of Civil Engineering and Technology. 2017;**8**(3): 565-574

[40] Kumar V, AdelAl-Gheethi S, Asharuddin M, Othman N. Potential of cassava peels as a sustainable coagulant aid for institutional wastewater treatment: Characterisation, optimisation and techno-economic analysis. Chemical Engineering Journal. 2021;**420**(2):123-142

[41] FAO. Food Agriculture Organisation of the United Nations, FAO Statistical Database (1990-2011, 2012). 2012. Available from: www.fao.org [Accessed: February 30, 2022]

[42] Berk Z. Food Process Engineering and Technology. London: Academic Print, Elsevier; 2013

[43] Nweke FI, Okorji EC, Njoku JE, King DJ. Elasticities of demand for major food items in a root and tuber based food system: Emphasis on yam and cassava in southeast Nigeria. In: RCMD Research Monograph No.11. Ibadan, Nigeria: Resource and Crops Management Division, International Institute of Tropical Agriculture; 1997. pp. 27-45

[44] FOS. Nigeria Federal Office of Statistics. National Consumer Survey, 1980 (8); Lagos. 1981

[45] Aripin AM, Mohd-Kassim AS, Daud Z, Mohd-Hatta M. Cassava peels for alternative fibre in pulp and paper industry: Chemical properties and morphology characterization. International Journal of Integrated Engineering. 2013;**5**(1):30-33

[46] Rowell RM, Han JS, Rowell JS. Characterization and factors effecting fiber properties. Natural Polymers and Agrofibers Composites. 2000;**1**:115-134

[47] Pattiya A. Thermochemical characterization of agricultural wastes from thai cassava plantations. Energy Sources, Part A: Recovery, Utilization, and Environmental Effects. 2011;**33**: 691-701

[48] Aro SO, Aletor VA, Tewe OO, Agbede JO. Nutritional potentials of cassava tuber wastes: A case study of a cassava starch processing factory in south-western Nigeria. Livestock Research for Rural Development. 2010;**22**(11):34-44

[49] Janz ER, Uluwaduge DI. Biochemical aspect of cassava with special emphasis on cynogenic glucosides: A review. Journal of the National Science Council of Sri Lanka. 1997;**25**(1):1-24

[50] Obadina A, Oyewole O, Sanni L, Abiola S. Fungal enrichment of cassava peels proteins. African Journal of Biotechnology. 2006;**5**(3):302-304

[51] Plevin R, Donnelly D. Converting waste to energy and profit; tapioca starch powerin Thailand. Renewable Energy World. 2004;**1**:74-81

[52] Barros F, Dionísio A, Silva J, Pastore G. Potential uses of cassava wastewater in biotechnological processes. In: Pace CM, editor. Agriculture Issues and Policies. Cassava Farming, Uses and Economic Impact. New York: NOVA Science Publishers Inc.; 2012

[53] Glanpracha N, Basnayake B, Rene E, Lens P, Annachhatre A. Cyanide degradation kinetics during anaerobic co-digestion of cassava pulp with pig manure. Water Science and Technology Bonus Issue. 2017;**3**:650-660

[54] Mégnanou R, Kouassi S, Akpa E, Djedji C, Bony N, Lamine S. Physico-chemical and biochemical characteristics of improved cassava varieties in Cote d'Ivoire. Journal of Animal and Plant Sciences. 2009;**5**(2):507-514

[55] Izah S, Bassey S, Ohimain E. Removal of heavy metals in cassava mill effluents by saccharomyces cerevisiae isolated from palm wine. Medcrave Online Journal of Toxicology. 2017;**3**(4):83-87

[56] Orhue E, Imasuen E, Okunima D. Effect of Cassava mill effluent on some soil chemical properties and the growth of fluted pumpkin (Telfairiaoccidentalis Hook F.). Journal of Applied Natural Science. 2014;**6**(2):320-325

[57] Adejumo B, Ola F. The effect of cassava effluent on the chemical composition of agricultural soil. Journal of Environmental Monitoring Assessment. 2014;**187**(7):418

[58] Patrick A, Egwuonwn N, Augustine O. Distribution of cyanide in a cassava–mill–effluent polluted eutric tropofluvent soils of Ohaji Area, South–eastern Nigeria. Journal of Soil Science and Environmental Management. 2011; **2**(2):49-57

[59] Olorunfemi D, Lolodi O. Effect of cassava processing effluents on antioxidant enzyme activities in *Allium cepa* L. Biokemistri. 2011;**23**(2):49-61

[60] Omomowo O, Omomowo A, Adeeyo O. Bacteriological screening and pathogenic potential of soil receiving cassava mill effluents. International

Journal of Basic Applied Science. 2015; **3**(4):26-36

[61] Kainthola J, Ajay S, Kalamdhad V, Goud V. A review on enhanced biogas production from anaerobic digestion of lignocellulosic biomass by different enhancement techniques. Process Biochemistry. 2019;**84**:81-90

[62] de Carvalholvo J, Alberto B, Lorenci W, Vanete T, Carlos F, Soccol R. Biorefinery integration of microalgae production into cassava processing industry: Potential and perspectives. Bioresource Technology. 2017;**247**: 1165-1172

[63] de Amorim M, Silva P, Barbosa, Montefusco N. Anaerobic biodegradation of cassava wastewater under different temperatures and inoculums. Comunicata Scientiae. 2019;**10**(1):65-76

[64] Cruz I, Larissa R, Andrade R, Bharagava R, Ashok K, Nadda M, et al. An overview of process monitoring for anaerobic digestion. Biosystems Engineering. 2021;**207**:106-121

[65] Obonukut M, Inyang U. Advances and challenges of anaerobic biodigestion technology. London Journal of Engineering Research. 2022;**2022**(1):23-40

[66] Pattiya A, Titiloye J, Bridgewater A. Fast pyrolysis of agricultural residues from cassava plantations for bio-oil production. Asian Journal on Energy and Environment. 2007;**8**(2):496-502

[67] Pattiya A. Bio-oil production via fast pyrolysis of biomass residues from cassava plants in a fluidised-bed reactor. Bioresource Technology. 2011;**102**: 1959-1967

[68] Lehmann J. Bio-energy in the black. Frontiers in Ecology and the Environment. 2006;**5**(7):381-387

[69] Lehmann J, Joseph S. Biochar for environment management–An introduction. In: Lehmann J, Joseph S, editors. Biochar for Environmental Management: Science and Technology. UK and USA: Earthscan Publisher; 2009. pp. 1-12

[70] Zajec L. Slow pyrolysis in a rotary kiln reactor: Optimization and experiments [M. S. thesis]. Akureyri, Iceland: School for Renewable Energy Sciences in Affiliation with University of Iceland and University of Akureyri; 2009

[71] Marais EA, Jacob DJ, Wecht K, Lerot C, Zhang L, Yu K, et al. Anthropogenic emissions in Nigeria and implications for atmospheric ozone pollution. A view of space. Atmospheric Environment. 2014;**99**:32-40

[72] Lou S, Liao H, Yang Y, Mu Q. Simulation of the inter-annal variations of tropospheric ozone over China: Roles of variations in meteorological parameters and anthropogenic emissions. Atmospheric Environment. 2015;**122**:839-851

[73] Shang-Shyng Y, I-Chu C, Ching-Pao L, Li-Yun L, Cheng-Hsiung C. Carbon dioxide and methane emission from Tanswei River in Northen Taiwan. Atmospheric Pollution Research. 2015;**1**: 52-61

[74] Kwapinski W, Byrne C, Kryachko E, Wolfram P, Adley C, Leahy E, et al. Biochar from biomass and waste. Waste Biomass Valor. 2010;**1**:177-189

[75] Woolf D, Amonette J, Street-Perrott F, Lehmann J, Joseph S. Sustainable biochar to mitigate global climate change. Nature Communication. 2010; **1**(56):1-9

[76] Isaac F, Adetayo O. Small scale biochar production technologies—A

review. Journal of Emerging trends in Engineering and Applied Sciences. 2010; **1**(2):151-156

[77] Goyal H, Seal D, Saxena R. Biofuels from thermochemical conversion of renewable resources—A review. Renewable and Sustainable Energy Reviews. 2008;**21**:504-517

[78] Williams P, Besler S. The influence of temperature and heating rate on the slow pyrolysis of biomass. Renewable Energy. 1996;**7**(3):233-250

[79] Noor N, Shariff A, Abdullah N. Slow pyrolysis of cassava wastes for biochar production and characterization. Iranica Journal of Energy & Environment. 2012; **3**(Special Issue on Environmental Technology):60-65

[80] Obonukut M, Alabi S, Alexander J, Egemba K. Digestate valorization: A value addition to anaerobic biodigestion technology. Chinese Journal of Chemical Engineering. 2022 (in press)

[81] Bulmău C(G), Mărculescu C, Badea A, Apostal T. Pyrolysis parameters influencing the bio-char generation from wooden biomass. UPB Scientific Bulletin. 2010;**72**(1):30-38

Section 2

Environmental Applications

Chapter 4

Biochar for Environmental Remediation

Dinesh Chandola and Smita Rana

Abstract

The environment is deteriorating rapidly, and it is essential to restore it as soon as possible. Biochar is a carbon-rich pyrolysis result of various organic waste feedstocks that has generated widespread attention due to its wide range of applications for removing pollutants and restoring the environment. Biochar is a recalcitrant, stable organic carbon molecule formed when biomass is heated to temperatures ranging from 300°C to 1000°C under low (ideally zero) oxygen concentrations. The raw organic feedstocks include agricultural waste, forestry waste, sewage sludge, wood chips, manure, and municipal solid waste, etc. Pyrolysis, gasification, and hydrothermal carbonization are the most frequent processes for producing biochar due to their moderate operating conditions. Slow pyrolysis is the most often used method among them. Biochar has been utilised for soil remediation and enhancement, carbon sequestration, organic solid waste composting, water and wastewater decontamination, catalyst and activator, electrode materials, and electrode modification and has significant potential in a range of engineering applications, some of which are still unclear and under investigation due to its highly varied and adjustable surface chemistry. The goal of this chapter is to look into the prospective applications of biochar as a material for environmental remediation.

Keywords: biochar, biochar properties, biochar reactivity, environmental remediation

1. Introduction

Biochar (biomass-derived char) is a versatile renewable source and is gaining popularity due to its diverse raw material sources, high porosity, large surface area, surface functional groups, and high treatment efficacy for a variety of contaminants [1]. Biochar is produced from three types of materials (plant residue, sewage sludge, and animal litter) that are pyrolyzed with little or no oxygen (typically below 1000°C) [2]. Biochar production not only deals with waste, but also benefit from waste, for example, pyrolysis of sewage sludge can reduce pollutants and turn it into a valuable resource [3]. Therefore, it is a great way to make biochar out of solid waste. Because of its unique properties, biochar has sparked widespread concern about its potential for use in the environment [4]. As indicated by the increase in the number of published publications regarding biochar in the last 10 years, it has gotten a lot of attention (**Figure 1**). Biochar's main technique for removing contaminants and remediating the environment is sorption. And, biochar's sorption capacity is directly related to its physiochemical features, such as surface area, pore size distribution, functional groups,

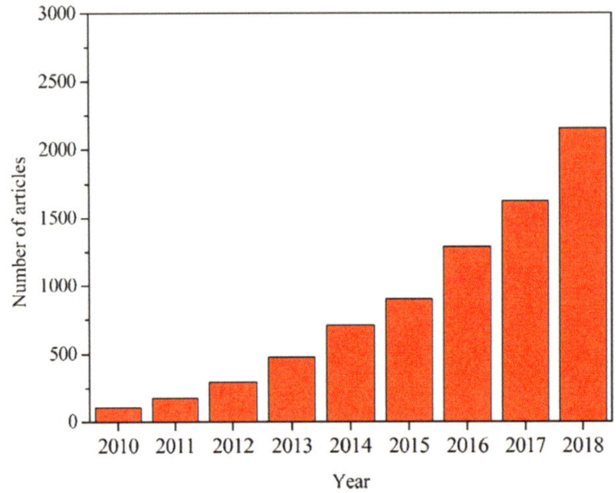

Figure 1.
The number of articles published in recent 10 years. (Source: [5].)

and cation exchange capacity, which vary depending on the preparation conditions [4]. Like, biochar produced at high temperatures has a larger surface area and carbon content than biochar produced at lower temperatures, due to the rising micro-pore volume caused by the elimination of volatile organic molecules at high temperatures [4]. The yields of biochar, on the other hand, decreases as the temperature goes up [6]. Therefore, in terms of biochar yields and adsorption capacity, an ideal synthesis method is required. To increase its physiochemical characteristics, biochar can further be modified with different chemicals like acids, alkalis, oxidizing agents, and ions for various environmental processes [7]. Due to its own properties such as large surface area, recalcitrance, and catalysis, biochar has been widely used in environmental applications such as soil remediation, carbon sequestration, water treatment, and wastewater treatment. In addition, biochar's application for energy and as an agricultural amendment is not a new concept. Biochar has also found its application in climate change mitigation and as a renewable energy source [8]. Biochar's use in engineering applications has received far less attention, despite the fact that economic estimates for biochar production for direct agricultural use have been poor for some time [9]. To that aim, a summary of our current understanding of biochar's potential for use in a variety of environmental remediation applications, as well as emerging obstacles and prospects for biochar usage in environmental remediation, is discussed below.

2. Biochar mechanisms for contaminants removal

Biochar's function mostly refers to its ability to uptake (e.g., sorption) other substances. The sorption of biochar can be divided into two categories, chemical sorption and physical sorption. Moreover, in term of biochar's interaction with other substances, there are three types of interactions: sorption, catalysis, and redox as shown in **Figure 2**. Sorption is a major environmental process that has a major impact on pollutant biogeochemistry. In sorption, the surface properties of biochar, which includes surface functional groups (carboxyl, carbonyl, phenolic–OH, ester, aliphatic, aromatic, hydroxyl, amino, and azyl groups), surface charges, and free radicals,

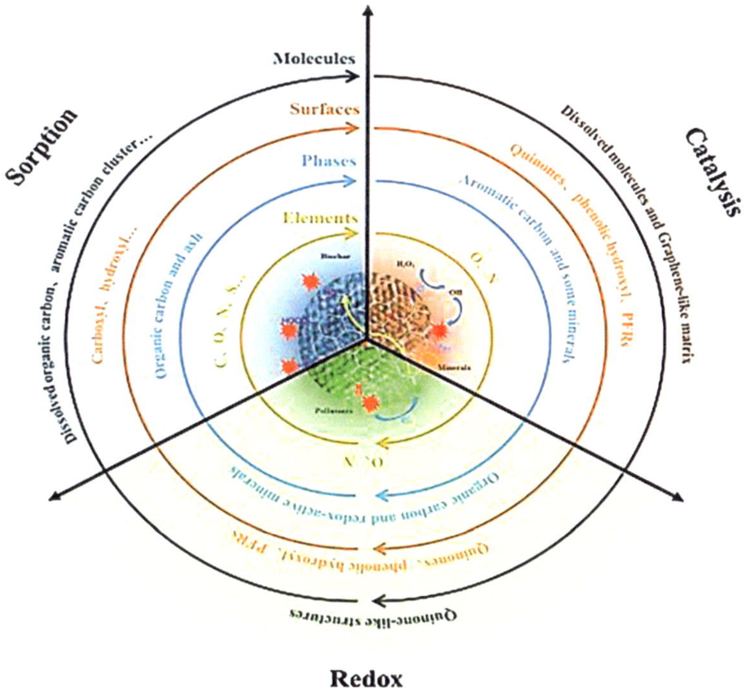

Figure 2.
Biochar remediation mechanisms. (Source: [10]).

are important for the behaviour of the interface between biochar and organic and inorganic pollutants, as it provides important sites for sorption and catalytic degradation of pollutants. These functional groups can form hydrogen bonds with other substances, As a result, Biochar can adsorb a variety of pollutants, including organic compounds, metals, nutrients, gases, and microbes [11, 12]. Moreover, the removal of some contaminants are also achieved by partitioning, electrostatic interaction, and pore-filling between biochar and pollutants and depends largely on biochar and pollutant characteristics [5]. Biochar also aids in the transformation of abiotic contaminants through various methods such as free radicals mediated transformation. Free radicals on the surface of biochar can react with chemicals like hydrogen peroxide and persulfate and promote the breakdown of organic pollutants [13]. Apart from that, biochar surfaces contain a variety of catalytic sites, such as quinone and phenolic functional groups, as well as persistent free radicals (PFRs), they enable biochar-mediated pollutants transformation [14]. For example, surface functional groups like quinones, convert sulphide into polysulfides, which accelerates the breakdown of azo dyes by increasing electron transport [14]. PFRs on the surface of biochar have a high reactivity and act as a catalyst in pollutant breakdown [13]. Also, the dissolved fractions in biochar, which are primarily composed of aliphatic and aromatic with quinone-like structures, has been tested and found to enhance the photochemical transformation of many organic pollutants by generating reactive intermediates or reactive oxygen species (ROS) [15]. Surface redox active moieties are the main contributors to the redox of biochar even though there are only a handful of relevant reports in publication so far. The surface redox-active moieties in biochar can directly react with pollutants via non-radical pathways, as well as activate some oxidants to form reactive radicals like OH and SO_4. For example, OH generated from the activation of H_2O_2 in biochar reduces about 20% of *p*-nitrophenol (PNP); however, about 80% of PNP is degraded by directly interacting with reactive sites, most likely the hydroquinone in biochar. Therefore, biochar not only enhances the degradation or transformation of pollutants by facilitating the transfer of electrons as a catalyst, but it can also directly react with pollutants, which will have a significant influence on the environmental behaviour of contaminant [16]. Apart from that, In terms of element composition, the major elements that make up the matrix of biochar are C, H, O, and N, while other elements like Si, P, and S have varying mass percentages in different biochars and play a special or even major role in sorption of various other specific pollutants. For example the sorption of Pb and Al on biochar is attributed to coprecipitation with P and Si in the biochar as $Pb_5\ (PO_4)_3\ (OH)$ and $KAlSi_3O_8$, respectively. An overview of metal ion precipitation and coprecipitation is shown in **Figure 2**. Ion exchange is another crucial phenomenon in the sorption of some heavy metals by biochar [11]. Furthermore, in biochar, there are two different phases: organic and inorganic. By raising pyrolysis temperatures, which results in increased surface area, pore volume, and aromaticity, sorption mechanisms evolved from partitioning-dominant to adsorption-dominant, and sorption components developed from polar-selective to porosity-selective [4, 17]. Furthermore, due to the movement of the organic components from aliphatic to aromatic, the sorption rate shows a transitional process: from fast to slow, then back to fast. In terms of inorganic components, it was discovered that ash has a catalytic effect on the formation of biochar with more orderly graphitic structures during the pyrolysis process; additionally, deashing after pyrolysis increases hydrophobic sorption sites, favouring the sorption of hydrophobic organic contaminants [10]. Therefore, the surface structure, functional groups and surface area and mechanisms of these functional groups are observed in the removal of pollutants.

3. Environmental remediation by biochar

3.1 Soil remediation and amelioration

Biochar can be used to clean up soil pollution caused by organic contaminants and heavy metals. Soil remediation using biochar is mostly accomplished by sorption and the mechanisms involved are surface complexation, hydrogen binding, electrostatic attractions, acid-base interactions, and π–π interactions as shown in **Figure 3**. For example, biochar produced from *Carya tomentosa* (a tree in the Juglandaceae or walnut family) and Pecan (*Carya illinoinensis*) (the tree is cultivated for its seed in the southern United States) can adsorb Clomazone and Bispyribac sodium (herbicides used in agriculture) in soil, and effectively reduce the leaching of clomazone and bispyribac sodium. Similarly, sawdust-derived biochar and wheat straw-derived biochar, on adding to the soil, significant reduces the polycyclic aromatic hydrocarbons (PAHs) [13]. **Table 1** shows how adding biochar to soil can help remove several forms of organic contaminants. However, there are several factors such as the types of feedstock, the applied dose, the targeted pollutants, and their concentrations all affect the removal of organic pollutants in soil by biochar. Biochar has the potential to absorb heavy metal ions as well in soil. The heavy metal adsorption mechanism on biochar includes surface complexation, precipitation, cation exchange, chemical reduction, and electrostatic attraction [29]. For example, the adsorption of Pb, Cd, Cr, Cu,

DOI: http://dx.doi.org/10.5772/intechopen.105430

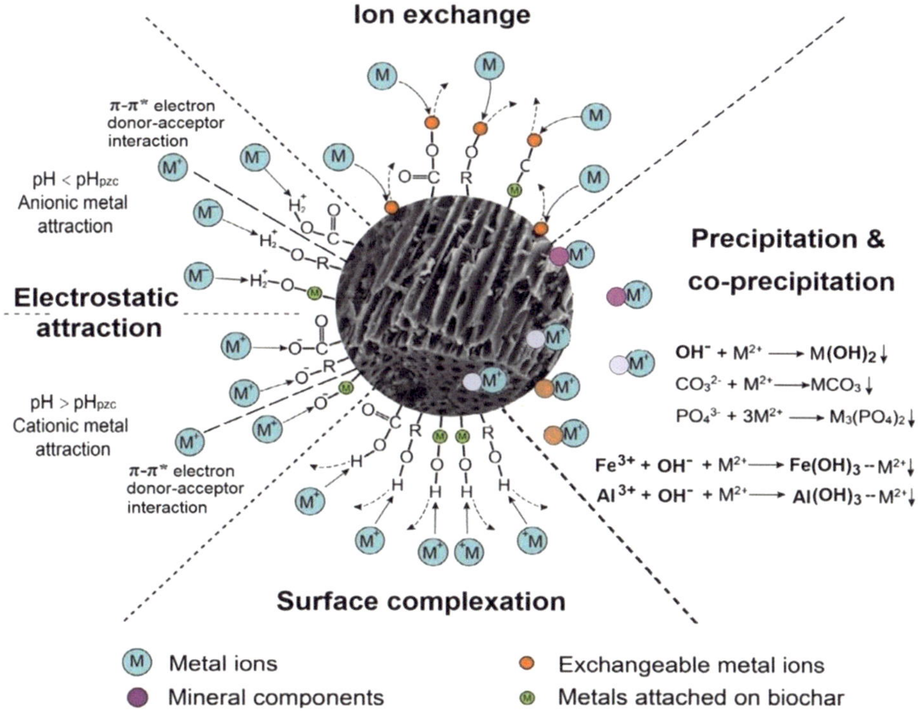

Figure 3.
Biochar mechanisms in soil for contaminants removal. (Source: [18]).

Reference	Organic pollutants	Removal efficiency	Feedstock
[19]	Dibutyl phthalate	87.5%	Bamboo
[20]	Phenanthrene	100%	Conifer
[21]	Imidacloprid	—	Rice-straw
[3]	Diethyl phthalate	90%	Bamboo
[22]	Carbaryl	71.8%	Pig manure
[23]	Tylosin	66%	Hardwood
[24]	Acetamiprid	52.3%	*Eucalyptus* spp.
[25]	Atrazine	>66%	Dairy Manure 450
[26]	Pentachlorophenol	96.2%	Rice-straw
[27]	Chlorpyrifos	34%	*Gossypium* spp.
[28]	Terbuthylazine	>88%	Sawdust

Table 1.
Adsorption of organic pollutants in soil by biochar.

and Zn by sesame straw-derived biochar demonstrates varied adsorption capacities for each among them. Pb absorption is the highest in biochar among the metals. Furthermore, when the metals are present together, Cd adsorbed on by sesame biochar is easily replaced by other metal ions. And water hyacinth-derived biochar can adsorb around 90% of As (V) whereas rice straw-derived biochar is able to adsorb

Zn^{2+} [30, 31]. Adsorption of antibiotics like sulfamethazine on biochar increases and subsequently decreases with pH, which affects the surface charge of both biochar and sulfamethazine, and the sorption processes evolve from electron donor–acceptor interaction to negative charge-assisted H-bond. And, metal ion adsorption occurs on the biochar surface's proton-active carboxyl and phenolic hydroxyl functional groups, and adsorption increased with pH in the range of pH 7. Apart from that, ion exchange and cation bonding are also found responsible for the sorption of K^+ and Cd^{2+} by [32]. The types of feedstock and experimental conditions have a big impact on the removal efficiencies. A number of parameters affect the adsorption capacity of biochar, including pH, surface functional groups, porosity, surface charge, and mineral composition. Therefore, when biochar is used as a remediation method, optimization of various parameters should be done based on the targeted organic contaminants. **Table 2** summarizes the removal of heavy metals from soil by biochar. Tables show how different biochars remove organic pollutants and heavy metals at varying rates. As shown in the **Table 2**, the types of biochar used and heavy metals are so different, it is difficult to compare them [46, 47]. Because different biochars have distinct physiochemical properties, they have varying adsorption capacities for inorganic and organic contaminants. As a result, selecting the right feedstock is more significant

Reference	Heavy metal	Removal efficiency	Feedstock
[33]	Cd^{2+}	80%	Eucalyptus wood
	Pb^{2+}	93.7%	
	Zn^{2+}	97.1%	
	Cu^{2+}	99.8%	
	Cd^{2+}	90%	Poultry litter
	Pb^{2+}	99.8%	
	Zn^{2+}	99.3%	
	Cu^{2+}	99.9%	
[34]	Pb^{2+}	55.9%	Sewage sludge
	Zn^{2+}	51.2%	
[35]	Cd^{2+}	56%	Bamboo
[36]	Pb^{2+}	—	Pine cone
[37]	Cd^{2+}	97.1%	Rice straw
[38]	Cd^{2+}	>99%	Tree bark
[39]	Cu^{2+}	>99%	Pine bark
[40]	Ni^{2+}	93%	Woody biomass, Gliricidia sepium
[41]	Zn^{2+}	54%	Sugar cane straw
[42]	Ni^{2+}	99.5%	Deinking paper sludge
[43]	Pb^{2+}	90%	Soybean stover
[44]	Pb^{2+}	93.5%	Chicken manure
[45]	Cd^{2+}	93.6%	Wheat straw

Table 2.
Heavy metal stabilization in soil by biochar.

for removing impurities than adjusting the pyrolysis temperature or changing the surface characteristics of biochar [19]. Additionally, modification of biochar is another option for increasing the removal capability of heavy metals. Apart from the removal of organic contaminants and heavy metal from soil, biochar can neutralize acidic soil, boost cation exchange capacity, and improve soil fertility, for example, the acidity of soil can be enhanced by 2 units after 1 month of treatment with soy bean stover-derived biochar and oak-derived biochar. Moreover, the cation exchange capacity can be increased significantly with 5% biochar. As a result, it aided maize growth and with 3% biochar [13]. The addition of biochar made from bamboo also enhances maize production and growth [8]. The addition of biochar to soil improves soil fertility due to the following reasons: (1) increased water retention capacity (2) increased soil aggregate stability; (3) reduction of soil compaction; and (4) decreased soil bulk density and increased porosity. The aforementioned factors may encourage root growth, boosting crop growth and yields even more. However, based on varied soil and feedstock, the most important reason for improving soil fertility needs to be investigated further.

3.2 Carbon sequestration

The process of storing carbon in soil organic matter and thereby removing carbon dioxide from the atmosphere is known as carbon sequestration. As part of attempts to establish climate resilient agriculture practices, the idea of using biochar to trap carbon in the soil has gotten a lot of attention in recent years. Biochar (biological charcoal) is a carbon sink that absorbs carbon from the atmosphere and stores it on agricultural grounds. Biochar is biologically inert, allowing it to retain fixed carbon in the soil for years to millennia while also absorbing net carbon from the atmosphere [20]. In addition, agriculture fixes 30 gigatons of carbon per year, but 30 gigatons of carbon return to the atmosphere as the plants die, resulting in no net change. When Biochar is combined with compost, soil, and plants, it recovers and stores a significant amount of carbon in the ground, resulting in a continuous and significant reduction in atmospheric greenhouse gas (GHG) levels. In recent years, climate change has sparked an increased interest in lowering carbon dioxide emissions into the atmosphere. Soil, being a major carbon sink, plays a critical role in the global carbon cycle, which has a direct impact on climate change. Carbon sequestration has offered as a strategy to reduce carbon dioxide emissions. Biochar has a great resistance to biodegradation due to its extremely condensed aromatic structure. As a result, biochar is thought to have a positive impact on soil carbon sequestration. Many investigations have been carried out to determine the impact of biochar on soil for carbon sequestration. However, due to the variability in carbon dioxide emissions, no consistent result can be presented. For example, adding carbon from fire to soil increased soil organic carbon turnover. However, adding biochar made of wood sawdust to soil inhibited carbon mineralization, resulting in more carbon sequestration. The mineralization of soil organic matter after the addition of biochar is shown to be higher in low-fertility soils than in high-fertility soils [21]. Carbon mineralization is also higher in soils with low organic carbon concentration than in soils with high organic carbon content. Also, the application of biochar to soil has found an increase in the rate of organic matter decomposition. This so-called “priming effect” affects carbon sequestration efforts since increased microbial activity might lead to breakdown rates exceeding carbon input rates. While the exact mechanism causing this impact has yet to be determined, it could be due to the increase of microbial activity as bacteria consume the carbon and nitrogen in biochar. However, the carbon in biochar can be separated into two types: liable and recalcitrant carbon. When biochar is introduced to the soil, soil microbes may quickly consume available carbon, resulting in an increase in carbon mineralization at first. This explains why adding biochar to soil accelerates carbon mineralization. Moreover, recalcitrant carbon content in biochar is significantly higher than labile carbon concentration. In soil, recalcitrant carbon can persist for a long time. As a result, the carbon input generated by biochar is more than the carbon outflow induced by relevant carbon mineralization. And, shorter pyrolysis times and higher pyrolysis temperatures, according to recent research [4], result in more recalcitrant biochar (i.e., it persists for longer periods in the soil). However, these pyrolysis conditions yields less biochar per unit feedstock, there are trade-offs. The effect of biochar addition on carbon sequestration is largely unknown in general. The priming impact varies depending on the feedstock and pyrolysis conditions, suggesting that the relationship between biochar's effect and feedstock type must be investigated further. The inherent properties of biochar, as determined by feedstock and pyrolysis conditions, interact with environmental factors like precipitation and temperature to determine how long biochar carbon is held in the soil. Soil texture, as is typically the case, plays an important influence in the stability of biochar carbon. Biochar interacts with soil particles to stabilize itself in the soil.

However, numerous uncertainties remain about the efficiency of biochar in carbon sequestration. It is also crucial to investigate the link between pyrolysis conditions and biochar's carbon sequestration ability. While biochar contains a lot of carbon, it is unclear how long that carbon will stay in the soil after it has been applied. In terms of boosting soil carbon reserves and combating climate change, biochar remains a hot topic. Many uncertainties remain, however, before definitive conclusions can be drawn about what conditions allow biochar to contribute positively to soil carbon sequestration.

3.3 In organic solid waste composting

The constant increase in solid waste seems to have a negative impact on human society's long-term development, which has raised numerous concerns. Organic waste accounts for around half of all solid waste generated. The ability to effectively treat organic solid waste is critical for successful solid waste disposal. Composting has received a lot of attention as a waste treatment method because of its benefits, such as low cost. Composting is a biological process that takes place. Organic matter from raw materials is exposed to biological breakdown during the process. Biochar has a direct influence on microbes, which has an impact on composting. Many researches have been carried out to see how biochar affects the composting of organic waste. The following are the effects of biochar on microorganisms during the composition of organic solid waste: (1) providing a habitat for microorganisms; (2) providing ideal growing conditions for microorganisms; (3) enriching the microbial diversity. It is documented that biochar addition accelerated the decomposition of organic solid waste due to the favorable effect of biochar addition on composting. **Table 3** shows the impact of adding biochar to the composting process. In general, adding biochar to compost has a good impact on the process. The priming effect, on the other hand, can be overlooked in low-fertility, alkaline, temperate soil. The type of soil affects the performance of biochar in compositing [22]. Furthermore, the types and doses of biochar, as well as the soil types, have a significant impact on the composting of organic solid waste. As a result, a biochar application strategy should be developed depending

Reference	Feedstock	Applied dose	Performance
[48]	Peanut shell	0.75% biochar and 0.75% compost (w %)	Increase the growth of sesbania, seashore mallow, and overall biomass.
[49]	Rice husk	24 g compost + 16 g biochar in 400 g soil	Reduce the availability of Cd and Zn and enhance the availability of Cu by increasing total organic carbon and water-extract organic carbon.
[50]	Acacia	2 t ha^{-1} biochar, 10 t ha^{-1} compost and 92 kg N ha^{-1}	Improve the grain yields and N uptake
[51]	Acacia green waste	47 t ha^{-1} biochar and 10 t ha^{-1} compost	Enhance macroporosity and bioturbation; increase microbial abundance; modify microbial structure
[52]	Logs	2.5 t ha^{-1} biochar and 25 t ha^{-1} compost	Increase soil organic carbon, nutritional status, and water content, as well as maize output.
[53]	Hardwood, coniferous wood	8 t ha^{-1} biochar and 55 t ha^{-1} compost	Vine growing on low-fertility, alkaline, temperate soil has no immediate commercial value.
[54]	Wood	0.3 kg compost and 0.27 kg biochar	Increase the oxygen intake by accelerating the humification of sludge organics.
[55]	Beech wood	100 mg/kg biochar and 100 mg/kg compost	Increase plant height, total organic carbon, and total nitrogen content; decrease ammonium content
[56]	Quercus serrate	10% biochar and 90% compost	Change the microbial community structure Increase
[57]	Hardwood coniferous wood	8 t ha^{-1} biochar and 63 t ha^{-1} compost	Increase microbial number and activity while having no influence on the amount of copper available.

Table 3.
Impact of adding biochar to the composting process.

on the characteristics of organic solid waste composting and soil. Furthermore, it was discovered that bacterial consortiums combined with biochar can stimulate microbial activity to accelerate degradation, increase bacterial community richness, and change the specific selection of bacteria, providing a method for effectively improving microbial activity and enhancing organic solid waste degradation.

3.4 Decontamination of water and wastewater

Many studies have demonstrated that biochar may adsorb contaminants from water and wastewater, including both organic and inorganic pollutants. Antibiotics, for example, are becoming common organic contaminants in the environment. Sludge-derived biochar has been shown to be a cost-effective and reusable adsorbent for the elimination of antibacterial drugs. **Table 4** shows how biochar can remove organic pollutants from water via adsorption [68, 69].

The adsorption of pollutants by biochar in water depends on the physiochemical characteristics of targeted pollutants and the types of biochar. For example, the sawdust-derived biochar can remove entirely 20.3 mg/l of sulfamethoxazole while

Reference	Feedstock	Removal efficiency	Organic pollutant
[58]	Chicken manure	100%	Microcystin-LR
[59]	Sewage sludge	26%-60%	Tetracycline
[60]	Corn stalks	97.62%	Norfloxacin
[61]	Pinus radiata sawdust	100%	Sulfamethoxazole
[58]	Mangosteen peel	80%	Methylene blue
[3]	Cool Planet LLC	<6%	Ibuprofen
	Organic Farms LLC	<10%	Sulfamethoxazole
	Corncob	—	Bisphenol A
[62]	Waste Douglas fir	100%	salicylic acid
[63]	Corn straw	100%	Atrazine
[64]	Wood	20%-30%	Sulfamethoxazole
[65]	Rice-husk	~90%	Tetracycline
[66]	Buffalo-weed	88.47%	Trichloroethylene
[67]	Soybean Stalk	99.5%	Phenanthrene

Table 4.
Organic pollutant removal by biochar in waste.

wood-derived biochar demonstrates substantially lower removal effectiveness of sulfamethoxazole (20–30%). For biochar obtained from organic farm, it demonstrates the lowest removal effectiveness of sulfamethoxazole (<6%) [23]. Varying pyrolysis temperatures led in different tetracycline removal efficiencies for biochar generated with rice husk [24]. The removal efficiency of tetracycline ranged from 26% to 60% when the pyrolysis temperature was 800°C and the initial concentration of tetracycline was 200 mg/l. When the pyrolysis temperature was 500°C and the initial tetracycline concentration was 5 mg/l, the removal efficiency was around 90%. It is therefore, established that pyrolysis temperature had important effect on the adsorption capacity of biochar. Other parameters such as pyrolysis time, in addition to pyrolysis temperature, can influence the physiochemical characteristics of biochar, which in turn affects the adsorption capacity of biochar. Heavy metal contamination is a major problem that requires immediate attention. Heavy metals can be removed from the aquatic environment using adsorption as well. Biochar's ability to remove heavy metal ions is listed in **Table 5** [80]. The removal of heavy metals by biochar is dependent on the types of heavy metals and the types of feedstock, similar to the removal of organic pollutants by biochar. Biochar has a lower removal capacity for Cd^{2+} and As^{5+} than other heavy metals like Pb^{2+} and Zn^{2+} among the major heavy metals [25]. Biochar produced from corn straw, for example, had a different Cu^{2+} adsorption capability like 0.1 g/l of biochar can remove 1 mM of Cu^{2+} when the pyrolysis temperature is set at 800°C. And, when the pyrolysis temperature is set to 400°C, 20 g/l biochar can remove 20 mg/l Cu^{2+} [26]· Similarly, biochar produced from water hyacinths shows different adsorption capacities for Cd^{2+} and Pb^{2+}, demonstrating that biochar adsorption capability varies depending on the targeted heavy metals. Zhang et al. [27] discovered that biochar prepared at high temperatures was effective in removing Cr (VI). A recent study found that sludge-derived biochar may successfully remove ammonium by monolayer chemical adsorption [59], implying

Reference	Feedstock	Removal efficiency	Heavy metal
[70]	Corn straws	97.7%	Cu^{2+}
[33]	Rape straw	100%	Cd^{2+}
[49]	Sawdust and swine manure	100%	Pb^{2+}
[71]	Mangosteen peel	80%	Cd^{2+}
[72]	Corn straw	99.24%	Cd^{2+}
[73]	Celery	97.7%	Pb^{2+}
[74]	Scots pine	~23%	Cd^{2+}
[75]	Water hyacinths	~60%	Cd^{2+}
[75]	Sugar cane bagasse	~80%	Pb^{2+}
[76]	Macroalga	~80%	Cu^{2+}
[77]	Wheat straw	100%	Cd^{2+}
[78]	Hickory wood	95.9%	Cd^{2+}
[79]	Pinewood	~35%	As^{5+}
[3]	Rice husk	~100%	Cr^{6+}
[3]	Anaerobic digested sludge	26%	Ni^{2+}

Table 5.
Heavy metal uptake by biochar in water.

that competition adsorption occurred when biochar was utilised as adsorbents for the removal of heavy metals and organic pollutants in the presence of ammonium. It should be highlighted that the adsorption capacity of the functional groups-modified biochar is clearly improved by the functional groups. The amino-modified biochar, for example, significantly increases the adsorption of Cu (II) due to strong complexation [60]. Moreover, biochar can enrich microorganisms, which can aid in the removal of organic matter, in addition to adsorption. Luo et al. [48] discovered that the proportion of Archaea was significantly greater in the presence of fruitwood-derived biochar, which relieved the stress of ammonia and acids on the microbes, raising microbial activity even more. Lu et al. [35] discovered a similar phenomenon as well. When using biochar for water and wastewater treatment, it's important to keep in mind that it can be recycled and reused. Based on the foregoing findings, biochar performs well in batch experiments in removing the contaminants of concern. However, various contaminants coexist in water and wastewater. Competitive adsorption may occur, resulting in results that differ from those obtained in the laboratory. In addition, the adsorption of contaminants by biochar may be affected by actual flow conditions. As a result, more research should be done in the lab to imitate the real-world condition and study the efficacy of biochar in the removal of contaminants.

3.5 Building sector

Biochar is a good building material for insulating buildings and managing humidity because of its low thermal conductivity and capacity to absorb water. Biochar, together with cement mortar clay and lime, can be used with sand in a 1: 1 ratio. As a result, the plaster made using this technology has excellent insulation and breathing capabilities, allowing it to sustain humidity levels of 45–70% in both summer and winter. This prevents dry air, which can cause respiratory problems and allergies, as well as moisture caused by air condensing on the outer walls, which can lead to mould growth [27].

4. Future research

The capacity to carefully adjust the structure and chemistry of biochar at nanoscale (nm) scales allows certain aspects of the biochar to be altered to target certain environmental engineering solutions, comparable to the proposed “designer biochar” for agricultural uses. It is crucial to remember, however, that once in the field; biochar characteristics do not remain constant over time. Even at ambient temperatures, ageing, oxidation, and microbial degradation can modify surface functional groups and chemistry, affecting sorption characteristics. The list of biochar’s potential engineering applications is continually growing. Due to its unique magnetic properties, magnetic biochar opens the door to facilitating removal of various contaminants from soil or other media. This broadens the scope of biochar’s possible use in environmental remediation.

5. Environmental concern of biochar

Along with the widespread use of biochar, it may have some disadvantages which may lead to harmful impact on the environment. When using bio-char in the environment, one of the most crucial aspects to consider is stability. The carbon structure makes up the majority of biochar. Biochar stability refers to the stability of the carbon structure in general. Aromaticity and the degree of aromatic condensation in biochar are markers of its carbon structure. Biochar stability must be considered because different biochars have varying physiochemical properties. Due to the instability of biochar, Huang et al. [28] observed the potential dissolution of organic matter from biochar in the complexation of heavy metals, implying that dissolved organic matter from biochar can be discovered in solution. Furthermore, the aromaticity, stability, and resistivity of the dissolved organic matter may be high. When biochar is used in the treatment of water and wastewater, the carbon content of the water body may rise due to the release of carbon from the biochar. Furthermore, biochar, particularly sludge-derived biochar, includes heavy metals, which may leach out during the water and wastewater treatment process, resulting in heavy metal contamination. When biochar is used as a catalyst support, the catalyst’s stability tends to deteriorate after a few uses. One reason for the lower catalyst stability could be charcoal structural degradation. As a result, biochar stability is also linked to water and wastewater treatment quality. In conclusion, the stability of biochar has a significant impact on its environmental applicability. As a result, more research is needed in the future to determine the stability of biochar. Because pyrolysis conditions can change carbon content and structure however, research into the relationship between biochar stability and pyrolysis conditions is important. Biochar’s possible toxicity on microorganisms should be considered in addition to its stability. Biochar increases the enzymatic activities of soil microorganisms at low doses, according to Gong et al. [75], demonstrating that low doses of biochar had no toxicity on the bacteria. Dong et al. [79] shown that Fe_3O_4-modified bamboo biochar has a low cytotoxicity potential. In contrast, high doses of tobacco stem-derived biochar exhibited cytotoxic and genotoxic effects in epithelial cells through promoting ROS production. As previously stated, biochar has a wide range of physical and chemical properties. More research into the potential toxicity of biochar to the environment is needed to support its effective application. Fish, algae, water fleas, and luminous bacteria can all be used to conduct toxicity tests.

6. Conclusions and remarks

This chapter provided an overview of biochar application and its interaction with other substances, focusing on its use in environmental remediation. Firstly, the raw material especially waste materials used for biochar production offers a treatment option for wastes that contributes to environmental sustainability. Furthermore, biochar's practical applicability is aided by its low-cost feedstock and simple preparation technique. Biochar has the ability to remediate, improve soil, and mitigate climate change, all of which contribute to environmental sustainability. However, the primary explanation for the increase in soil fertility remained unknown, and the work on the impact of biochar on carbon sequestration needs to be conducted and understood. Composting organic waste using biochar can help promote biological decomposition of organic waste. However, different doses of biochar were required for various organic wastes and biochar kinds. As a result, a biochar application strategy should be developed depending on the characteristics of organic solid waste composting and soil. Biochar can be employed as absorbents in the decontamination of water and wastewater, but its adsorption capacity and stability must be improved. Biochar can activate persulfate, which can be used to remove hazardous organic pollutants from water and wastewater, however the relationship between biochar structure and persulfate activation needs to be studied further to figure out how it works. In conclusion, biochar has a bright future in improving environmental sustainability. The majority of bio-char research is currently being done in laboratories. Biochar's environmental impact has yet to be fully understood. Furthermore, the real world is more complex than the laboratory, resulting in ambiguity about biochar's environmental impact. More in situ tests are needed to determine the true impact of biochar on the environment, such as environmental microorganisms, before it is used on a broad basis. Furthermore, the preparation conditions of biochar for industrial use must be enhanced depending on the various environmental reasons.

Author details

Dinesh Chandola* and Smita Rana
Centre for Land and Water Resource Management (CLWRM), GB Pant National Institute of Himalayan Environment (GBPNIHE), India

*Address all correspondence to: chandola.dinesh@gmail.com

References

[1] Hashem MA, Hasan M, Momen MA, Payel S, Nur-A-Tomal MS. Water hyacinth biochar for trivalent chromium adsorption from tannery wastewater. Environmental and Sustainability Indicators. 2020;**5**:100022. DOI: 10.1016/J.INDIC.2020.100022

[2] Hagemann N, Spokas K, Schmidt HP, Kägi R, Böhler MA, Bucheli TD. Activated carbon, biochar and charcoal: Linkages and synergies across pyrogenic carbon's ABCs. Water (Switzerland). 2018;**10**(2): 1-19. DOI: 10.3390/w10020182

[3] Fathianpour A, Taheriyoun M, Soleimani M. Lead and zinc stabilization of soil using sewage sludge biochar: optimization through response surface methodology. CLEAN—Soil, Air, Water. 2018;**46**(5):1-10. DOI: 10.1002/clen.201700429

[4] Jindo K, Mizumoto H, Sawada Y, Sanchez-Monedero MA, Sonoki T. Physical and chemical characterization of biochars derived from different agricultural residues. Biogeosciences. 2014;**11**(23):6613-6621. DOI: 10.5194/bg-11-6613-2014

[5] Mohan D, Sarswat A, Ok YS, Pittman CU. Organic and inorganic contaminants removal from water with biochar, a renewable, low cost and sustainable adsorbent - A critical review. Bioresource Technology. 2014;**160**:191-202. DOI: 10.1016/j.biortech.2014.01.120

[6] Chen C, Zhou W, Lin D. Sorption characteristics of *N*-nitrosodimethylamine onto biochar from aqueous solution. Bioresource Technology. 2015;**179**:359-366. DOI: 10.1016/j.biortech.2014.12.059

[7] Ma Y, Liu WJ, Zhang N, Li YS, Jiang H, Sheng GP. Polyethylenimine modified biochar adsorbent for hexavalent chromium removal from the aqueous solution. Bioresource Technology. 2014;**169**:403-408. DOI: 10.1016/j.biortech.2014.07.014

[8] Luo L et al. The characterization of biochars derived from rice straw and swine manure, and their potential and risk in N and P removal from water. Journal of Environmental Management. 2019;**245**:1-7. DOI: 10.1016/J.JENVMAN.2019.05.072

[9] Yang X, Zhang S, Ju M, Liu L. Preparation and modification of biochar materials and their application in soil remediation. Applied Sciences. 2019;**9**(7). DOI: 10.3390/app9071365

[10] Yi Y et al. Magnetic biochar for environmental remediation: A review. Bioresource Technology. 2020;**298**:122468. DOI: 10.1016/J.BIORTECH.2019.122468

[11] Fan J et al. Remediation of cadmium and lead polluted soil using thiol-modified biochar. Journal of Hazardous Materials. 2020;**388**:122037. DOI: 10.1016/J.JHAZMAT.2020.122037

[12] Rani P, Singh AP, Rai S. Effect of rice husk biochar and lime treated sludge on NPK concentration and uptake by rice crop. Environment and Ecology. 2015;**33**(3A):1218-1224

[13] Wang J, Wang S. Preparation, modification and environmental application of biochar: A review. Journal of Cleaner Production. 2019;**227**:1002-1022. DOI: 10.1016/J.JCLEPRO.2019.04.282

[14] Lütke SF, Igansi AV, Pegoraro L, Dotto GL, Pinto LAA, Cadaval TRS. Preparation of activated carbon from black wattle bark waste and its application for phenol adsorption.

Journal of Environmental Chemical Engineering. 2019;7(5):103396. DOI: 10.1016/j.jece.2019.103396

[15] Li H, Zhang T, Tsang DCW, Li G. Effects of external additives: Biochar, bentonite, phosphate, on co-composting for swine manure and corn straw. Chemosphere. 2020;**248**:125927. DOI: 10.1016/J.CHEMOSPHERE.2020.125927

[16] Çeçen F, Aktaş Ö. Water and wastewater treatment: Historical perspective of activated carbon adsorption and its integration with biological processes. In: Cecen F, Aktas O, editors. Activated Carbon for Water and Wastewater Treatment. 1st ed. Weinheim: WILEY-VCH Verlag GmbH & Co. KGaA; 2011. pp. 1-11. DOI: 10.1002/9783527639441.ch1

[17] Chen B, Zhou D, Zhu L. Transitional adsorption and partition of nonpolar and polar aromatic contaminants by biochars of pine needles with different pyrolytic temperatures. Environmental Science & Technology. 2008;**42**(14):5137-5143. DOI: 10.1021/es8002684

[18] Guo M, Song W, Tian J. Biochar-facilitated soil remediation: Mechanisms and efficacy variations. Frontiers in Environmental Science. 2020;**8**:521512. DOI: 10.3389/fenvs.2020.521512

[19] Luo M, Lin H, Li B, Dong Y, He Y, Wang L. A novel modification of lignin on corncob-based biochar to enhance removal of cadmium from water. Bioresource Technology. 2018;**259**:312-318. DOI: 10.1016/j.biortech.2018.03.075

[20] Mehmood K et al. Biochar research activities and their relation to development and environmental quality. A meta-analysis. Agronomy for Sustainable Development. 2017;**37**(3):1-22. DOI: 10.1007/s13593-017-0430-1

[21] Marschner B, Werner S, Alfes K, Lubken M. Potential dual use of biochar for wastewater treatment and soil amelioration. In: EGU General Assembly. Geophysical Research Abstracts. 2013;**15**:EGU2013-11260

[22] Awasthi MK et al. Emerging applications of biochar: Improving pig manure composting and attenuation of heavy metal mobility in mature compost. Journal of Hazardous Materials. 2020;**389**:122116. DOI: 10.1016/J.JHAZMAT.2020.122116

[23] Shimabuku KK, Kearns JP, Martinez JE, Mahoney RB, Moreno-Vasquez L, Summers RS. Biochar sorbents for sulfamethoxazole removal from surface water, stormwater, and wastewater effluent. Water Research. 2016;**96**:236-245. DOI: 10.1016/j.watres.2016.03.049

[24] Jing XR, Wang YY, Liu WJ, Wang YK, Jiang H. Enhanced adsorption performance of tetracycline in aqueous solutions by methanol-modified biochar. Chemical Engineering Journal. 2014;**248**:168-174. DOI: 10.1016/j.cej.2014.03.006

[25] Rasheed A, Sana S, Kashif S, Umer Z, Khatoon M. To evaluate the efficiency of char and biochar for waste water treatment. Journal of Waste Recycling. 2017;**2**(2):7. Available from: http://www.imedpub.com/resources-recycling-and-waste-management/

[26] Chen X et al. Adsorption of copper and zinc by biochars produced from pyrolysis of hardwood and corn straw in aqueous solution. Bioresource Technology. 2011;**102**(19):8877-8884. DOI: 10.1016/j.biortech.2011.06.078

[27] Schmidt H. Novel uses of biochar. In: USBI North American Biochar Symposium. 2013

[28] Wang H, Lin K, Hou Z, Richardson B, Gan J. Sorption of the herbicide terbuthylazine in two New Zealand forest soils amended with biosolids and biochars. Journal of Soils and Sediments. 2010;**10**(2):283-289. DOI: 10.1007/s11368-009-0111-z

[29] Zhang P, Sun H, Yu L, Sun T. Adsorption and catalytic hydrolysis of carbaryl and atrazine on pig manure-derived biochars: Impact of structural properties of biochars. Journal of Hazardous Materials. 2013;**244-245**:217-224. DOI: 10.1016/j.jhazmat.2012.11.046

[30] Xiao R et al. Recent developments in biochar utilization as an additive in organic solid waste composting: A review. Bioresource Technology. 2017;**246**:203-213. DOI: 10.1016/J.BIORTECH.2017.07.090

[31] Chen J, Liao Z, Lu S, Hu G, Liu Y, Tang C. Study on a stepped eco-filter for treating greywater from single farm household. Applied Water Science. 2017;7(7):3849-3857. DOI: 10.1007/s13201-017-0536-2

[32] González-García CM, González-Martín ML, Denoyel R, Gallardo-MorenoAM,Labajos-BroncanoL, Bruque JM. Ionic surfactant adsorption onto activated carbons. Journal of Colloid and Interface Science. 2004;**278**(2):257-264. DOI: 10.1016/j.jcis.2004.06.012

[33] Li B et al. Adsorption of Cd(II) from aqueous solutions by rape straw biochar derived from different modification processes. Chemosphere. 2017;**175**:332-340. DOI: 10.1016/j.chemosphere.2017.02.061

[34] Fathianpour A, Taheriyoun M, Soleimani M. Lead and zinc stabilization of soil using sewage sludge biochar: Optimization through response surface methodology. CLEAN – Soil, Air, Water. 2018;**46**:1700429

[35] Lu K et al. Effect of bamboo and rice straw biochars on the mobility and redistribution of heavy metals (Cd, Cu, Pb and Zn) in contaminated soil. Journal of Environmental Management. 2017;**186**:285-292. DOI: 10.1016/j.jenvman.2016.05.068

[36] Igalavithana AD et al. Heavy metal immobilization and microbial community abundance by vegetable waste and pine cone biochar of agricultural soils. Chemosphere. 2017;**174**:593-603. DOI: 10.1016/j.chemosphere.2017.01.148

[37] Zhang RH et al. Immobilization and bioavailability of heavy metals in greenhouse soils amended with rice straw-derived biochar. Ecological Engineering. 2017;**98**:183-188. DOI: 10.1016/j.ecoleng.2016.10.057

[38] Venegas A, Rigol A, Vidal M. Changes in heavy metal extractability from contaminated soils remediated with organic waste or biochar. Geoderma. 2016;**279**:132-140. DOI: 10.1016/j.geoderma.2016.06.010

[39] Oustriere N et al. Influence of biochars, compost and iron grit, alone and in combination, on copper solubility and phytotoxicity in a Cu-contaminated soil from a wood preservation site. Science of the Total Environment. 2016;**566-567**:816-825. DOI: 10.1016/j.scitotenv.2016.05.091

[40] Herath I, Kumarathilaka P, Navaratne A, Rajakaruna N, Vithanage M. Immobilization and phytotoxicity reduction of heavy metals in serpentine soil using biochar. Journal of Soils and Sediments. 2015;**15**(1):126-138. DOI: 10.1007/s11368-014-0967-4

[41] Puga AP, Abreu CA, Melo LCA, Beesley L. Biochar application to a contaminated soil reduces the availability and plant uptake of zinc, lead and cadmium. Journal of Environmental Management. 2015;**159**:86-93. DOI: 10.1016/j.jenvman.2015.05.036

[42] Méndez A, Paz-Ferreiro J, Araujo F, Gascó G. Biochar from pyrolysis of deinking paper sludge and its use in the treatment of a nickel polluted soil. Journal of Analytical and Applied Pyrolysis. 2014;**107**:46-52. DOI: 10.1016/j.jaap.2014.02.001

[43] Moon DH et al. Immobilization of lead in contaminated firing range soil using biochar. Environmental Science and Pollution Research. 2013;**20**(12):8464-8471. DOI: 10.1007/s11356-013-1964-7

[44] Park JH, Choppala G, Lee SJ, Bolan N, Chung JW, Edraki M. Comparative sorption of Pb and Cd by biochars and its implication for metal immobilization in soils. Water, Air, & Soil Pollution. 2013;**224**(12):1711. DOI: 10.1007/s11270-013-1711-1

[45] Bian R et al. Biochar soil amendment as a solution to prevent Cd-tainted rice from China: Results from a cross-site field experiment. Ecological Engineering. 2013;**58**:378-383. DOI: 10.1016/j.ecoleng.2013.07.031

[46] Marschner B, Werner S, Alfes K. Potential dual use of biochar for wastewater treatment and soil amelioration. In: EGU General Assembly 2013. 2016

[47] M. K. Awasthi et al., "Emerging applications of biochar: Improving pig manure composting and attenuation of heavy metal mobility in mature compost," Journal of Hazardous Materials, vol. 389, p. 122116, May 2020, DOI: 10.1016/J.JHAZMAT.2020.122116.

[48] Luo X, Liu G, Xia Y, Chen L, Jiang Z. Use of biochar-compost to improve properties and productivity of the degraded coastal soil in the Yellow River Delta, China. Journal of Soils and Sediments. 2017;**17**:780-789. DOI: 10.1007/s11368-016-1361-1

[49] Liang J et al. Changes in heavy metal mobility and availability from contaminated wetland soil remediated with combined biochar-compost. Chemosphere. 2017;**181**:281-288. DOI: 10.1016/j.chemosphere.2017.04.081

[50] Agegnehu G, Nelson PN, Bird MI. The effects of biochar, compost and their mixture and nitrogen fertilizer on yield and nitrogen use ef fi ciency of barley grown on a Nitisol in the highlands of Ethiopia. Science of the Total Environment. 2016;**569-570**:869-879. DOI: 10.1016/j.scitotenv.2016.05.033

[51] Abujabhah IS, Bound SA, Doyle R, Bowman JP. Effects of biochar and compost amendments on soil physico-chemical properties and the total community within a temperate agricultural soil. Applied Soil Ecology. 2016;**98**:243-253. DOI: 10.1016/j.apsoil.2015.10.021

[52] Agegnehu G, Bass AM, Nelson PN, Bird MI. Benefits of biochar, compost and biochar-compost for soil quality, maize yield and greenhouse gas emissions in a tropical agricultural soil. Science of the Total Environment. 2016;**543**:295-306. DOI: 10.1016/j.scitotenv.2015.11.054

[53] Schmidt HP, Kammann C, Niggli C, Evangelou MWH, Mackie KA, Abiven S. Biochar and biochar-compost as soil amendments to a vineyard soil: Influences on plant growth, nutrient uptake, plant health and grape quality. Agriculture, Ecosystems & Environment. 2014;**191**:117-123. DOI: 10.1016/j.agee.2014.04.001

[54] Zhang J, Lü F, Shao L, He P. The use of biochar-amended composting to improve the humification and degradation of sewage sludge. Bioresource Technology. 2014;**168**:252-258. DOI: 10.1016/j.biortech.2014.02.080

[55] Schulz H, Dunst G, Glaser B. Positive effects of composted biochar on plant growth and soil fertility. Agronomy for Sustainable Development. 2013;**33**(4):817-827. DOI: 10.1007/s13593-013-0150-0

[56] Jindo K et al. Biochar influences the microbial community structure during manure composting with agricultural wastes. Science of the Total Environment. 2012;**416**:476-481. DOI: 10.1016/j.scitotenv.2011.12.009

[57] Mackie KA, Marhan S, Ditterich F, Schmidt HP, Kandeler E. The effects of biochar and compost amendments on copper immobilization and soil microorganisms in a temperate vineyard. Agriculture, Ecosystems & Environment. 2015;**201**:58-69. DOI: 10.1016/j.agee.2014.12.001

[58] Li J, Cao L, Yuan Y, Wang R, Wen Y, Man J. Comparative study for microcystin-LR sorption onto biochars produced from various plant- and animal-wastes at different pyrolysis temperatures: Influencing mechanisms of biochar properties. Bioresource Technology. 2018;**247**:794-803. DOI: 10.1016/j.biortech.2017.09.120

[59] Tang L et al. Sustainable efficient adsorbent: Alkali-acid modified magnetic biochar derived from sewage sludge for aqueous organic contaminant removal. Chemical Engineering Journal. 2018;**336**:160-169

[60] Wang B, Jiang YS, Li FY, Yang DY. Preparation of biochar by simultaneous carbonization, magnetization and activation for norfloxacin removal in water. Bioresource Technology. 2017;**233**:159-165. DOI: 10.1016/j.biortech.2017.02.103

[61] Reguyal F, Sarmah AK, Gao W. Synthesis of magnetic biochar from pine sawdust via oxidative hydrolysis of FeCl2 for the removal sulfamethoxazole from aqueous solution. Journal of Hazardous Materials. 2017;**321**:868-878. DOI: 10.1016/j.jhazmat.2016.10.006

[62] A. G. Karunanayake et al., "Rapid removal of salicylic acid, 4-nitroaniline, benzoic acid and phthalic acid from wastewater using magnetized fast pyrolysis biochar from waste Douglas fir," Chemical Engineering Journal, Jul. 2017;**319**:75-88. DOI: 10.1016/J.CEJ.2017.02.116.

[63] Tan G, Sun W, Xu Y, Wang H, Xu N. Sorption of mercury (II) and atrazine by biochar, modified biochars and biochar based activated carbon in aqueous solution. Bioresource Technology. 2016;**211**:727-735. DOI: 10.1016/j.biortech.2016.03.147

[64] Shimabuku KK, Kearns JP, Martinez JE, Mahoney RB, Moreno-Vasquez L, Summers RS. Biochar sorbents for sulfamethoxazole removal from surface water, stormwater, and wastewater effluent. Water Research. 2016;**96**:236-245. DOI: 10.1016/j.watres.2016.03.049

[65] Jing XR, Wang YY, Liu WJ, Wang YK, Jiang H. Enhanced adsorption performance of tetracycline in aqueous solutions by methanol-modified biochar. Chemical Engineering Journal. 2014;**248**:168-174. DOI: 10.1016/j.cej.2014.03.006

[66] Ahmad M et al. Production and use of biochar from buffalo-weed (*Ambrosia trifida* L.) for trichloroethylene removal from water. Journal of Chemical Technology & Biotechnology. 2014;**89**(1):150-157. DOI: 10.1002/jctb.4157

[67] HuoLiang K, Jiao H, YanZheng G, HuiFang W, XueZhu Z. Cosorption of phenanthrene and mercury(II) from aqueous solution by soybean stalk-based biochar. Journal of Agricultural and Food Chemistry. 2011;**59**(22):12116-12123

[68] A. Rasheed, S. Sana, S. Kashif, Z. Umer, and M. Khatoon, "To evaluate the efficiency of char and biochar for waste water treatment," Journal of Waste Recycling, 2017;**2**(2):7. [Online]. Available: http://www.imedpub.com/resources-recycling-and-waste-management/.

[69] Chen X et al. Adsorption of copper and zinc by biochars produced from pyrolysis of hardwood and corn straw in aqueous solution. Bioresource Technology. 2011;**102**(19):8877-8884. DOI: 10.1016/j.biortech.2011.06.078

[70] Yu W, Lian F, Cui G, Liu Z. N-doping effectively enhances the adsorption capacity of biochar for heavy metal ions from aqueous solution. Chemosphere. 2018;**193**:8-16. DOI: 10.1016/j.chemosphere.2017.10.134

[71] Ruthiraan M, Abdullah EC, Mubarak NM, Noraini MN. A promising route of magnetic based materials for removal of cadmium and methylene blue from waste water. Journal of Environmental Chemical Engineering. 2017;**5**(2):1447-1455. DOI: 10.1016/j.jece.2017.02.038

[72] Chi T, Zuo J, Liu F. Performance and mechanism for cadmium and lead adsorption from water and soil by corn straw biochar. Frontiers of Environmental Science & Engineering. 2017;**11**(2):1-8. DOI: 10.1007/s11783-017-0921-y

[73] Zhang T et al. Efficient removal of lead from solution by celery-derived biochars rich in alkaline minerals. Bioresource Technology. 2017;**235**:185-192. DOI: 10.1016/j.biortech.2017.03.109

[74] Baltrenaite JKE. Biochar as adsorbent for removal of heavy metal ions [cadmium (II), copper (II), lead (II), zinc (II)] from aqueous phase. International Journal of Environmental Science and Technology. 2016;**13**:471-482. DOI: 10.1007/s13762-015-0873-3

[75] Online VA et al. Competitive removal of Cd (II) and Pb (II) by biochars produced from water hyacinths: Performance and mechanism. RSC Advances. 2018;**6**:5223-5232. DOI: 10.1039/C5RA26248H

[76] Hoon S, Jung H, Ryu C, Park Y. Removal of copper (II) in aqueous solution using pyrolytic biochars derived from red macroalga *Porphyra tenera*. Journal of Industrial and Engineering Chemistry. 2016;**36**:314-319. DOI: 10.1016/j.jiec.2016.02.021

[77] Bogusz A, Oleszczuk P, Dobrowolski R. Application of laboratory prepared and commercially available biochars to adsorption of cadmium, copper and zinc ions from water. Bioresource Technology. 2015;**196**:540-549. DOI: 10.1016/j.biortech.2015.08.006

[78] Wang W, Li M, Zeng Q. Adsorption of chromium (VI) by strong alkaline anion exchange fiber in a fixed-bed column: Experiments and models fitting and evaluating. Separation and Purification Technology. 2015;**149**(5):16-23. DOI: 10.1016/j.seppur.2015.05.022

[79] Wang S et al. Removal of arsenic by magnetic biochar prepared from pinewood and natural hematite. Bioresource Technology. 2015;**175**:391-395. DOI: 10.1016/j.biortech.2014.10.104

[80] Schmidt H. Novel uses of biochar. In: USBI North American Biochar Symposium. 2013

Chapter 5

The Potential Roles of Biochar in Restoring Heavy-Metal-Polluted Tropical Soils and Plant Growth

Abdul Kadir Salam

Abstract

Biochar shows interesting and environmentally useful properties, among which is its relatively high cation exchange capacity (CEC). High CEC may lower the easily plant-available heavy metals in soils due to the increase in the soil adsorption capacity resulted from biochar application. Quite a lot of current researches reveal that the extracted heavy metals in tropical soils particularly Cu and Zn were significantly lowered in the presence of biochar at 5–10 Mg ha^{-1}. Heavy metal–contaminated tropical soils planted with corn plants (*Zea mays* L.) show significant decreases in Cu and Zn concentrations at moderate- and high-level addition of heavy metal–containing waste. The growth and dry masses of roots and shoot of corn plant improved immediately as a result of biochar amendment. Planting heavy metal–polluted soils treated with biochar with thorny amaranth (*Amaranthus spinosus*) also demonstrated a similar phenomenon.

Keywords: biochar, heavy metals, tropical soils

1. Introduction

Heavy metal contamination and pollution in soils and environment are still of a serious concern since the presence of heavy metal may directly and indirectly endanger living things [1–12]. Reports on the occurrence of soil contamination and pollution come intensively from all over the world related to modern industries [1–4, 7, 10, 13–31]. The negative effects of heavy metals on plants, animals and human beings are also documented in the current literature [5, 6, 8, 9, 25, 26, 30, 32–35]. One important case of the negative effects currently documented was the occurrence of Minamata and Itai-itai diseases in Japan [2]. These suggest that the problem related to heavy metals in the soil environment must be more extensively studied.

Among the various chemical methods available to cope with heavy metal contamination and pollution in soils is the use of organic materials [13, 36–43]. Organic materials such as plant compost may enhance the capability of soil materials to immobilize soil mobile heavy metals. Composted organic matters may effectively lower the soil mobile heavy metals to lower their concentrations to the levels that are not harmful to plants and animals. Organic matters may consist of various functional groups such as phenolic, carboxylic and hydroxyl that may increase the soil cation

IntechOpen

adsorption capacity [2]. Therefore, the addition of organic matter compost into heavy-metal polluted soils was reported to significantly decrease the soil mobile heavy metals [41, 42]. For example, the addition of cassava (*Manihot utilissima*) leaf compost into tropical soils amended with heavy metals containing waste significantly lowers the soil DTPA extractable Cu and Zn [41]. This phenomenon was observed in the laboratory and greenhouse experiment employing some tropical soils of Alfisols, Ultisols and Oxisols from Lampung, Indonesia. A recent report also showed that the residual Cu and Zn in industrial waste amended soils were lower in soils treated also with cassava-leaf compost [41, 42]. The effect was more significant at sampling time < 10 years amendment [42].

Some researchers [41, 42, 44] reported that the effect of organic matter compost was more significant when added simultaneously with other potential materials. The addition of organic matter compost and lime was shown to better decrease the soil mobile heavy metals [37, 41, 42, 44]. The results of research in [41, 42] showed that the lowering effect on soil heavy metals of cassava-leaf compost and $CaCO_3$ was significantly greater than addition of organic matter or lime alone. The DTPA extracted Cd from Ultisols, Oxisols and Alfisols was significantly lowered by additions of cassava leaf compost and lime [41, 42]. The residual Cu and Zn were also lower in soils amended with cassava-leaf compost and $CaCO_3$ than with organic compost or $CaCO_3$ alone [42]. The presence of increasing OH^- ion by the increase in soil pH [45] may have stimulated the H releases from the organic functional groups and thus widened the capability of the soil materials in adsorbing the heavy metal ions from the soil solution. The adsorption of heavy metal free ions by soil materials may stimulate the releases of heavy metals held as chelates and complexes and also soil heavy metal precipitates and thus finally lower the soil extracted heavy metals.

As shown by numerous data, organic matter compost may significantly affect the soil concentrations of heavy metals. Most reports show that various organic matter may significantly decrease the soil concentrations of heavy metals. However, several reports demonstrated that organic matter may relatively quickly decay in soil system [13, 42, 43, 46]. These observations suggest that the use of organic matters to lower the concentrations of heavy metals in soils is limited for a short duration. Their effectiveness is lower for long-time uses. The problem will be more significant in tropical regions where the soil average temperature and moisture content are relatively high. Therefore, other materials with high durability to organic decomposition are needed. Current literature suggests that biochar will be the best candidate for this purpose [38, 44, 45, 47–62]. As reported by [45, 57], biochar is produced through pyrolysis or charring, causing their structure and composition to be more stable and durable in soil system. In addition, biochar also possesses chemical properties better than ordinary organic materials in terms of cation exchange capacity, pH, specific surface area and nutrient contents.

This chapter was to evaluate the properties and effects of biochar in restoring heavy metal–contaminated or contaminated soils and their effect on the concentration of heavy metals in soils affected by heavy metal–containing materials like industrial wastes.

2. Effects of high concentrations of heavy metals on plant growth

Heavy metals are detrimental to living things, particularly at high concentrations [2]. As mentioned previously, their negative effects are reported from various sites in the world. Research report in [63] shows the negative effect of heavy metal–containing waste on the growth of water spinach, caisim and lettuce in 23 years old heavy metal–containing waste amended tropical soils. Clearly found that the growth of these plants was depressed at high heavy metals and the growth in control soil was the best (**Figure 1**). Lettuce was not survived at high heavy metal contents only until 2 weeks after planting (WAP). It is also obvious that water spinach grew better than the other two plants at any level of soil-heavy metals.

The data above demonstrated that high concentrations of heavy metals (in this case Cu and Zn) were detrimental to plants (**Figure 1**). Their effects are dependent on their concentrations and plant species. Higher concentration of heavy metals gave more significant effects. Water spinach was more adaptable to high concentrations of heavy metal and therefore it grew much better. It is possible to employ plants like water spinach in phytoremediation. Biomass analysis showed also that the plant uptake of Cu and Zn of water spinach was much higher than were other two plants [63].

A similar phenomenon was demonstrated by thorny amaranth. The growth of thorny amaranth was significantly retarded in 24 years old waste amended soils with high heavy metals (treated with 60 Mg waste ha^{-1}) (**Figure 2**). The retardation occurred along the growing time from 0 to 6 WAP. Low heavy metals (treated with 15 Mg waste ha^{-1}) only slightly lowered the growth of this plant.

The effect of heavy metals was more clearly shown by the growth of plant roots. In general, the growth of plant roots may adjust to the high concentrations of Cu and Zn and probably of other heavy metals. This environmental stress by heavy metals may stimulate plant roots to work harder and cause plant biomass to distribute more to plant roots (**Figure 3**). The root/shoot was shown to positively and linearly correlate with the soil-heavy metal concentration. The writer in [64] stated that higher root weight may cause higher root cation exchange capacity (CEC) that may retain

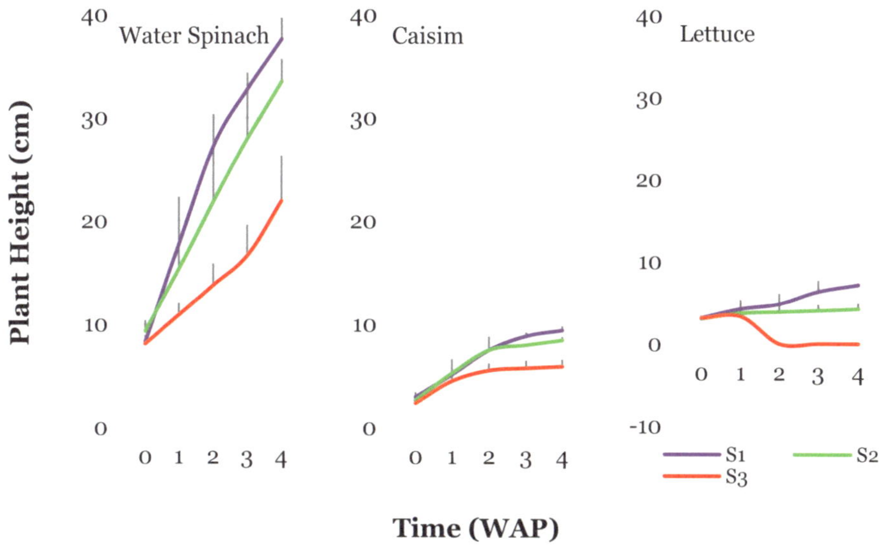

Figure 1.
The growth of several plants in heavy metal contaminated soil (S1 control, S2 low heavy metals, S3 high heavy metals; lettuce dead in S3, WAP weeks after planting) (after [63] with permission).

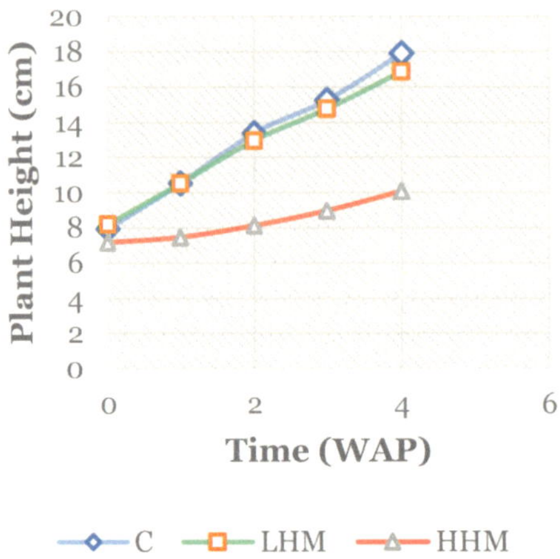

Figure 2.
The growth of thorny amaranth in heavy-metal polluted soils (C control, LHM low heavy metal, HHM high heavy metal, WAP weeks after planting).

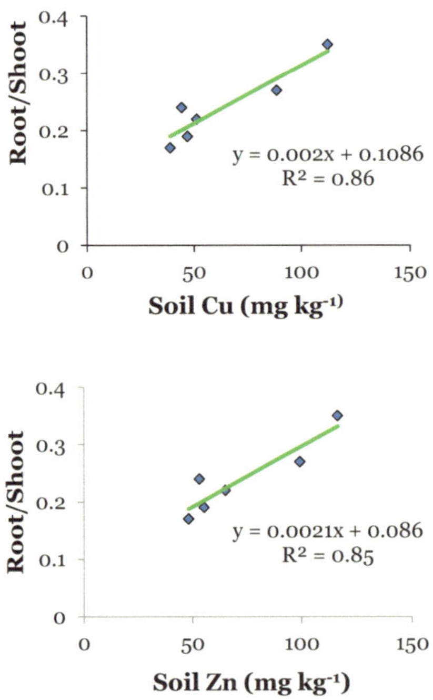

Figure 3.
The relationship between the root/shoot and the soil DTPA extracted Cu and Zn (after [64] with permission).

more heavy metal cations on the surface of plants' roots so that less heavy metals may move to plant shoots. Higher soil CEC may then lower the stimulation of the growth of plant roots. High concentrations of heavy metals in soils caused more biomass distribution to plant roots (**Figure 3**). Higher CEC can be attained by increasing soil pH [2, 65]. Plant roots also produce some exudates such as low molecular organic acids that may chelate heavy metal cations in soil solution and lower heavy metal effects on plants [66, 67].

3. Some physical and chemical properties of biochar

Organic compost is significantly different from biochar both in the process of production and in its properties. Organic compost was produced by a complete decomposition of plant materials in the presence of microorganisms in a well-regulated condition of O_2, heat and water moisture. Urea N is usually added to accelerate the decomposition process while the soil pH is maintained high by lime addition. Microorganism is introduced through cow dung addition. Low C/N ratio is used as a measure of compost maturity. Biochar is produced by incomplete thermo-decomposition of some feedstocks like woods, leaves, feces, straws, husks and manure in a limited or no oxygen supply called pyrolysis or charring [45, 57]. Therefore, biochar consists of much higher C content and consequently, it is more stable with high durability in soils. Reports of [45, 57] show that biochar also showed several better physical and chemical properties. Some of feedstocks abundantly available in Indonesia are woods, straws of corn and rice, bagasse and dairy manure. Therefore, application of biochar may provide a low-cost method of coping with environmental problems. One example of biochar is shown in **Figure 4**, which shows the production of biochar from rice husk and the physical appearance of the rice husk biochar.

Biochar shows porous surfaces so that in the soil system it may physically absorb pollutants like heavy metals. Combined with the increase in the soil adsorption capacity the biochar porosity may significantly enhance the soil retainment on heavy metal cations in biochar-treated soils. In addition to the better physical properties, biochar also shows better, interesting and useful chemical properties [45, 57]. Like organic matters in general, biochar possesses some functional groups like hydroxyl and carboxyl that may bear great amounts of negative charges. It shows a high CEC

Figure 4.
The production of rice husk biochar in the University of Lampung experimental farm (courtesy of Sri Yusnaini with permission).

of 28.8–327 mmol kg^{-1} and high pH depending on the charring temperature, higher at higher charring temperature. The pH of biochar ranges from 5.81–10.1. Biochar also shows high specific surface area (SSA) ranging from 40.99 to 189.8 $m^2 g^{-1}$.

The potential of biochar at increasing the soil pH may raise the soil adsorption capacity. The increase in OH-ions by biochar treatment may dehydrogenase the biochar functional groups of hydroxyl and carboxyl raising the soil adsorption capacity. Finally, through the synergic works of its high porosity, abundant functional groups and potential to increase the soil pH, biochar may significantly immobilize heavy metal cations in soils.

Therefore, the most important properties of biochar useful in the management of heavy metals in soils is its high SSA, abundant functional groups, high cation exchange capacity and potential to increase the soil pH [45, 57]. Therefore, its presence in heavy metal contaminated or polluted soils may significantly lower heavy metal contaminants. Several mechanisms may involve in the immobilization of heavy metals in soil-biochar mixtures that include physical sorption, ion exchange, chemisorption, complexation and precipitation. Biochar may eventually reduce heavy metal mobility and bioavailability [45]. Wastewater treatment with biochar is reported to immobilize up to 99% of Cd, Pb and Zn in an optimum condition [57]. The effectiveness of biochar is dependent on biomass and soil types and also on heavy metals [60].

4. Improvement of soil chemical properties by biochar

There are several forms of heavy metals in the soil environment [2]. Of which, heavy metal cation is the most directly affected by the active negative charges of soils through adsorption and desorption processes [68–72]. The adsorption of heavy metals that decrease the concentration of heavy metal cations in soil solution may, of course, stimulate the release of heavy metals of other forms such as chelates through de-chelation, complexes through decomplexation, precipitates through dissolution, and other soil chemical reactions that may altogether lower the total concentration of total soil heavy metals as shown in **Figure 5** [2].

The above interrelationship shows the importance of heavy metal cation form in the soil environment and therefore the effort to cope with the problem of heavy metals in soils must be first focused on lowering the concentration of heavy metal cations. The increase in the soil's negative surfaces was repeatedly suggested to suffice this relationship [2]. The presence of soil solid negative surfaces may electrostatically decrease the mobility of heavy metals cations through immobilization process. Heavy metal cations are strongly held by the soil materials and finally decreased the total soil heavy metals in soils as shown in **Figure 6**.

The quantity of heavy metals held by soil materials is negatively charged surface-dependent. High amounts of negative charges are attainable by enrichment with high quantity of negatively charged materials and/or negative charge stimulating materials. Previous observation shows that this condition can be attained by the addition of cassava leaf compost and/or lime materials that were reported to lower the soil concentration of Cd [41]. The cassava leaf compost may provide high amounts of negative charges to its various functional groups. The lime materials may raise the soil pH that may then stimulate the release of H ions from organic matter functional groups. The addition of organic materials and lime material may then finally widen the total negative charges and may increase the immobilization of heavy metal cations in soils.

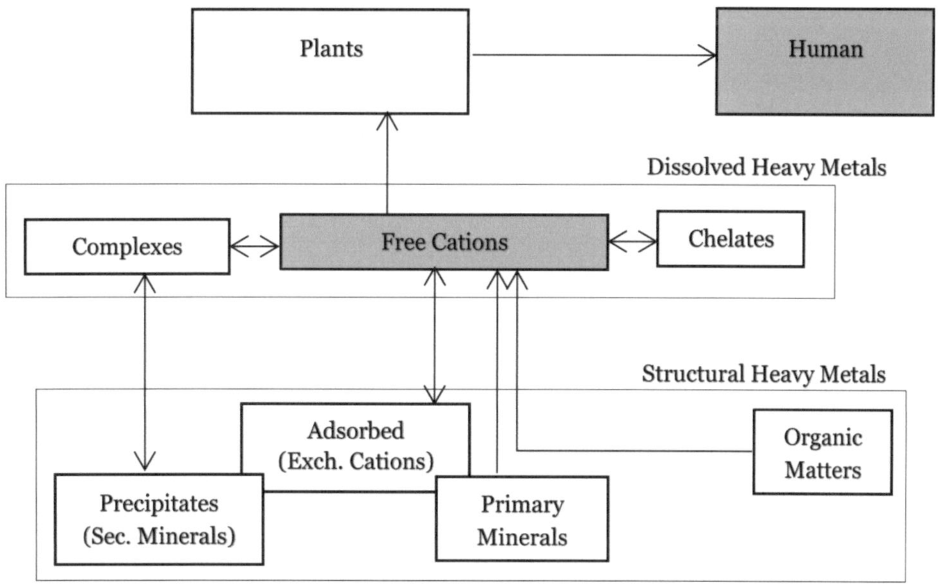

Figure 5.
The interrelationships between various forms of dissolved and structural heavy metals in soils, plants and human (after [2] with permission).

The improvement of the soil negative charges by biochar application may give more significant effect on the amount of the soil negative charges since as stated previously the biochar possesses high amounts of negative charges [57, 59]. The CEC of biochar ranges from 28.8 to 327 mmol kg^{-1} [45, 57]. The increase in soil pH caused by biochar addition may increase the significance of biochar application. Consequently, biochar application may enhance the retainment of soluble heavy metals in soils and finally lower the total extractable heavy metals in soils. This process will provide suitable soluble heavy metal levels in soils and enable plants to grow better.

5. Restoration of heavy metal–polluted soils and plant growth

The relationship between the biochar application, the increase in the soil negative charges, and the improvement of plant growth stated in Section 4 is exemplified in **Figure 6**. The improvement of plant growth by this process is expected in soil contaminated or polluted by heavy metals. Better growth of plants may absorb heavy metals at safe levels and may lower the soil heavy metals from immobilized forms like soil precipitates or soil adsorbed heavy metals much faster. The danger of heavy metals to plants may also be alleviated since plants may absorb heavy metals at lower levels of solubility in the presence of biochar. By this means, the soil's heavy metals are lowered by plants that grow better at safe levels of heavy metals. Thereby plants may also grow better in heavy metal polluted soils.

The decrease in soil Cu and Zn levels in the presence of biochar was currently reported from 23-years old polluted tropical soils planted with corn (*Zea mays* L.) as shown in **Figure 7**. The lowering effect of biochar on the soil extracted Cu and Zn is clearly depicted. The soil concentrations of Cu and Zn decreased in the order of soil treatment with $10 < 5 < 0$ Mg biochar ha^{-1}, indicating that the presence of biochar

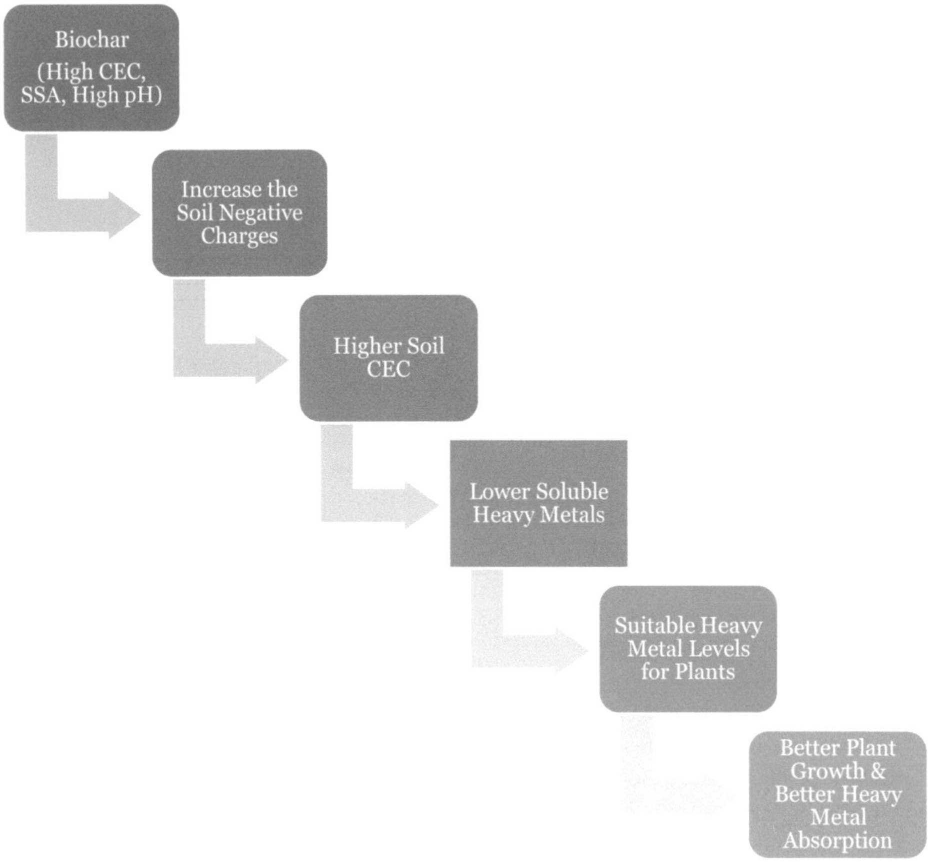

Figure 6.
The effect of biochar application on the soil heavy metal levels and plant growth.

lowered the soil extracted Cu and Zn. The most possible reason for this phenomenon is that the soil adsorption sites for heavy metals were enlarged by the presence of biochar. The enhancement in the soil adsorption capacity towards heavy metals was also probably associated with the significant increase in soil pH by biochar application. This synergic effect of biochar presence in soils may have finally lowered the soil concentrations of Cu and Zn in soils (**Figure 7**).

As the consequence (**Figures 5** and **6**), the growth of corn plants was significantly altered by biochar application, which was indicated by plant height (**Figure 8**) and plant biomasses (**Figure 9**). The trend in the corn plant height was clearly associated with the significant increase in the soil Cu and Zn concentration and the significant decrease in the soil Cu and Zn in the presence of biochar (**Figure 7**). The decrease in plant height was associated with the increase in the levels of amended soils that increase the soil Cu and Zn while the increase in plant height was associated with the decrease in heavy metal concentrations stimulated by the presence of biochar. A similar trend was also indicated by the changes in the plant biomasses as affected by the levels of amended waste and biochar application (**Figure 9**). The corn plant biomasses including corn roots and corn shoots were lowered by soil concentrations of heavy metals and increased in the presence of biochar associated with the decrease in the soil heavy metals (**Figure 9**).

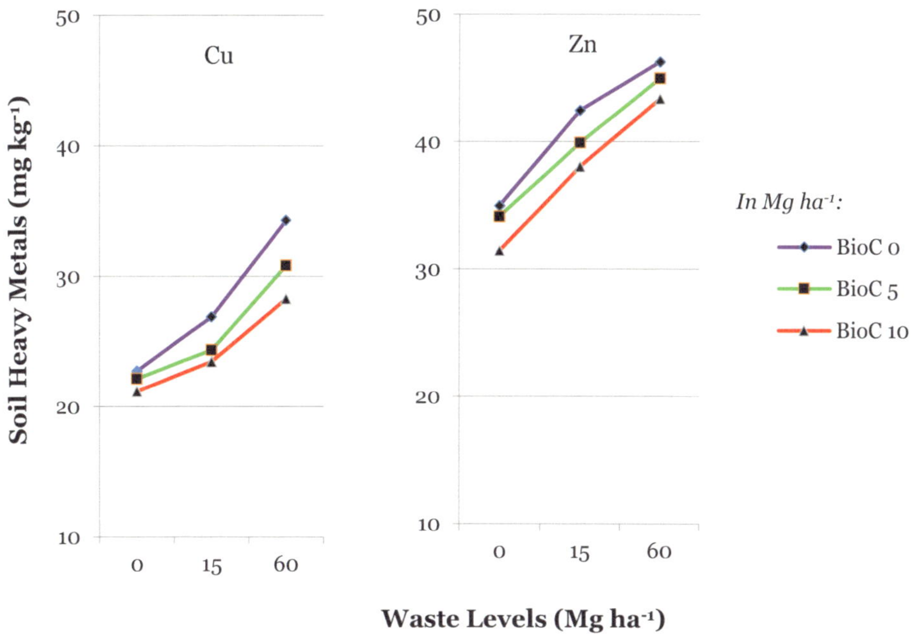

Figure 7.
The effect of biochar on Cu and Zn concentrations in waste-amended soil extracted by N HNO_3 (after [73] with permission).

The research result in [73] showed that the related analysis of variance (ANOVA) also indicated that the amended waste levels significantly enhanced the soil concentrations of heavy metals particularly Cu and Zn and significantly depressed the plant height and plant biomasses (roots, shoots, and the whole plant). Several previous research also showed that the waste-borne Cu and Zn in the soils depressed the growth of several other plants including caisim, corn plant, lettuce, Napier grass, and water spinach [63, 64, 73]. Elevated concentrations of heavy metals in soil system are detrimental to plants. Biochar at 5–10 Mg ha^{-1} was generally effective in changing plant characteristics in heavy metal–containing waste-amended tropical soils. Biochar significantly affected the soil heavy metals, organic C and pH, and also Cu accumulated in corn plant shoots as well as plant height and biomass dry-weight.

The effect of biochar in alleviating the high concentration of heavy metals particularly Cu and Zn was also reported for thorny amaranth [74]. Thorny amaranth was demonstrated to absorb quite high heavy metals from polluted soils and shown to be one of the heavy-metal bio-accumulators and therefore significantly decreased the Cu and Zn concentrations in the 23 years old waste amended tropical soils (**Figure 10**). The presence of thorny amaranth was shown to significantly lower the soil Cu from 79.3 to 60.0 mg kg^{-1} (24.3% decrease) and the soil Zn from 69.2 to 57.4 mg kg^{-1} (17.1% decrease) at the waste level of 60 Mg ha^{-1}. The decreases were much higher or 46.0% for Cu and 24.3% for Zn at lower waste level of 15 Mg ha^{-1}. Copper and Zn showed similar behavior in response to planting but the per cent decrease of Cu was higher than that of Zn, demonstrating that Zn was less mobile and less easily absorbed by plant roots than was Cu. It is stated in [74] that not all lost Cu and Zn was absorbed by plant roots. Some of these heavy metals may have also

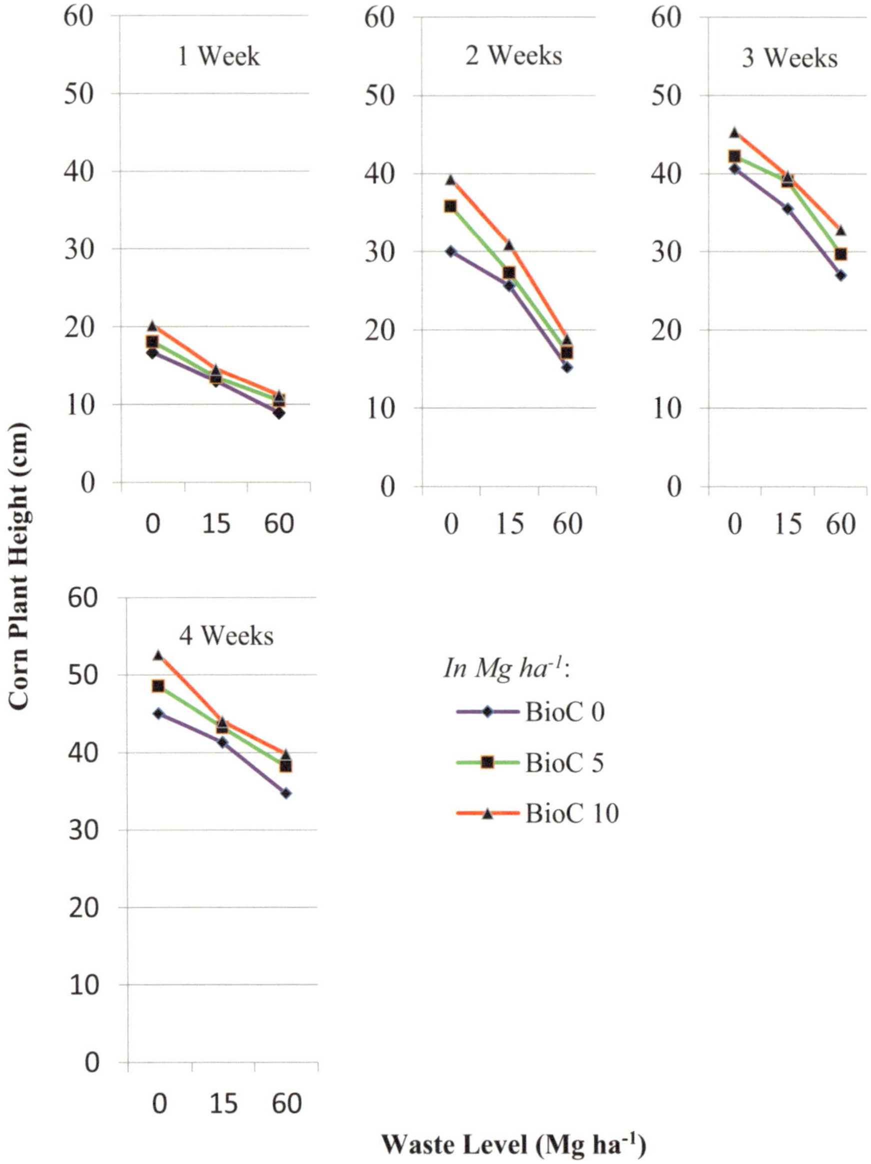

Figure 8.
The improvement of corn plant height in waste-amended soil by biochar (after [73] with permission).

shifted to more strongly adsorbed heavy metals due to the increase in soil pH caused by planting. Copper was probably more easily and strongly adsorbed by soil colloids or precipitated than was Zn.

The lowering of total heavy metals was also expected in phytoremediation. As stated in [75], at suitable levels, the absorption of heavy metals by plant roots may proceed fast enough since the presence of lower levels of heavy metals will not disturb the physics and works of plant roots during phytoremediation. The amount

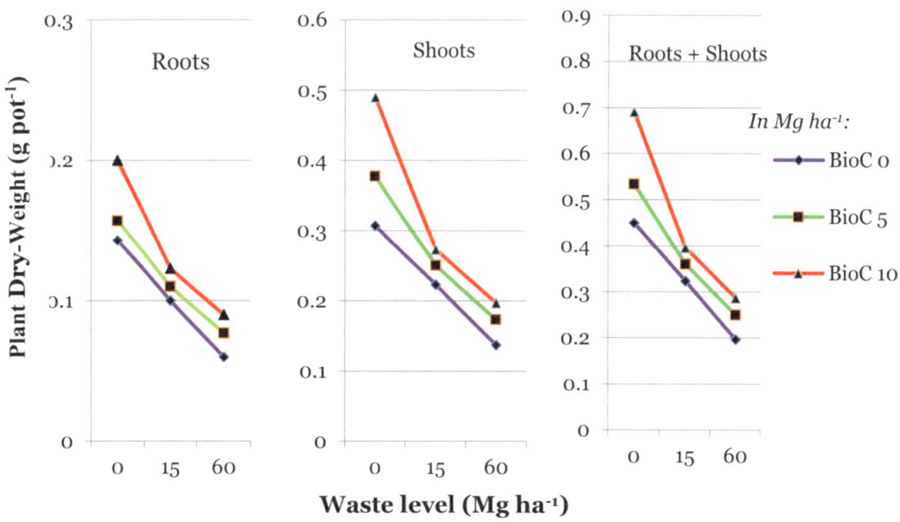

Figure 9.
The improvement of corn plant biomasses in waste-amended soil by biochar (after [73] with permission).

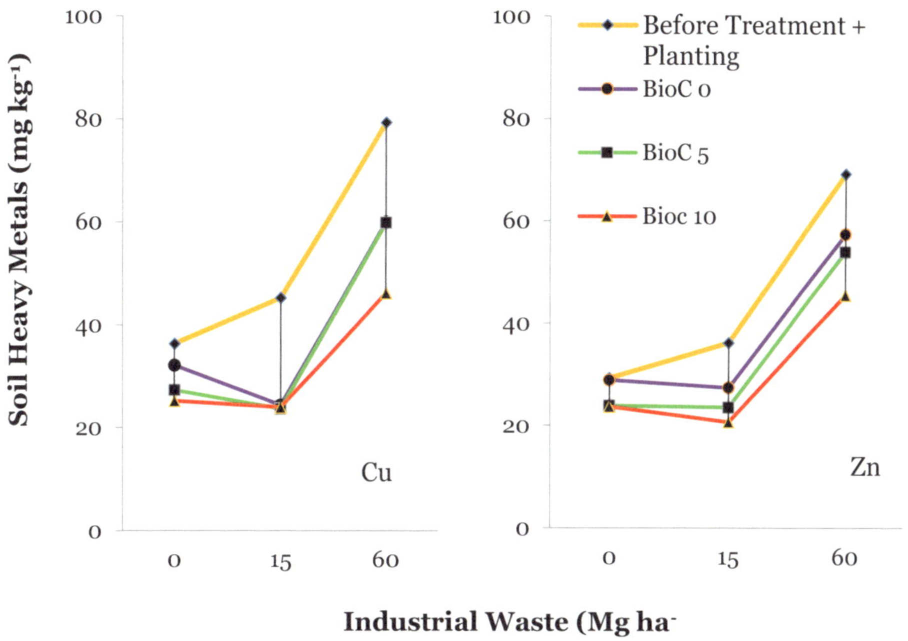

Figure 10.
The effect of thorny amaranth on the concentrations of Cu and Zn in a heavy-metal-polluted tropical soil treated with biochar (after [74] with permission).

of heavy metal removal may be higher at lower than that at higher levels of heavy metals. Therefore, the presence of biochar, which lowers the soil concentrations of heavy metals (**Figure 10**), may fasten the cleaning of heavy metals in soils by phytoremediation.

A similar trend with that in the growth of corn plants was observed in the plant root and shoot dry weights of thorny amaranth (**Figure 11**). The waste origin Cu and Zn may have disturbed the physiological functions in plant tissues and inhibited the growth of plant roots and shoots. It is clearly shown in **Figure 11** that, without biochar, waste treatments lowered the shoot dry weights by about 25.8% and 36.4% at waste treatment of 15 and 60 Mg ha^{-1}, respectively. These values were related to the increase of 8.90 (24.5%) and 43.0 mg kg^{-1} (116%) in Cu or 6.9 (23.5%) and

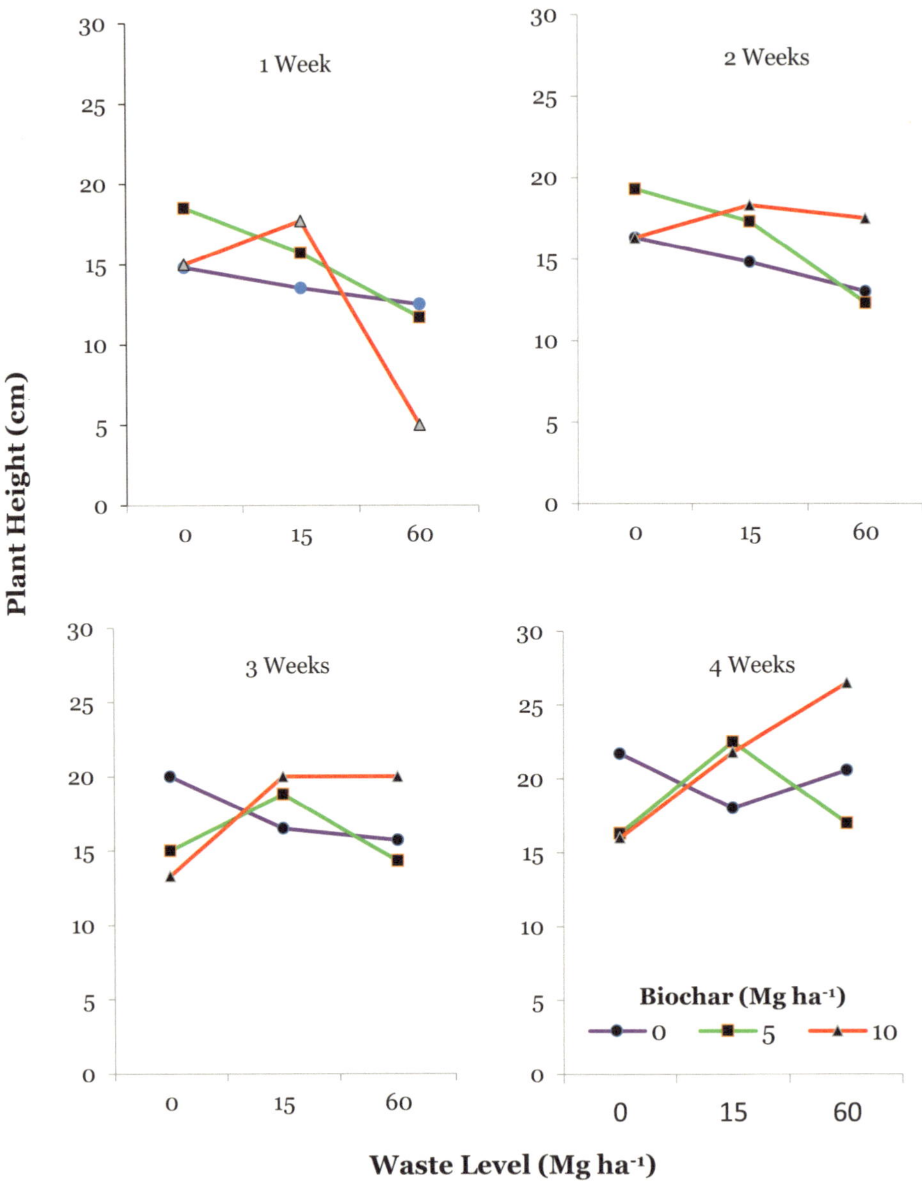

Figure 11.
The growth of thorny amaranth in heavy-metal polluted tropical soil treated with biochar (after [74] with permission).

32.9 mg kg^{-1} (112%) in Zn caused by the respective waste addition. The higher the soil Cu and Zn concentrations the more effective the heavy metal effect on plant shoot growth retardation. A similar trend was observed in the same soil samples for other plant species like caisim (*Brassica chinensis*), lettuce (*Lactuca sativa*), Napier grass (*Pennisetum purpureum*), and water spinach (*Ipomoea aquatica*) [63, 64, 75]. The growth of these plants was significantly retarded by the increase in the soil extracted Cu and/or Zn caused by waste treatment.

The root dry-weight increased by waste addition at 15 Mg ha^{-1} (**Figure 9**), suggesting that the growth of roots was more progressive under high concentrations of Cu, Zn and other heavy metals. This pattern was also reported by [74]. The study in [64] showed high correlation between the root/shoot of Napier grass with the soil concentration of Cu and/or Zn (**Figure 3**). However, high concentrations of heavy metals were found to decrease the root weight of thorny amaranth, suggesting that these plant roots were negatively affected by the higher concentration of Cu and Zn at a waste level of 60 Mg ha^{-1}.

Since it is reported to have high cation exchange capacity and high effect on soil pH [18, 35, 36], biochar was shown to improve the above agronomic responses of thorny amaranth (**Figures 11** and **12**). The presence of biochar may have increased the soil adsorption capacity and lowered the soil labile fractions of Cu and Zn, thereby alleviating their phytotoxicities and finally stimulating the plant growth. Numerous observations demonstrated that high soil Cu and Zn in general decreased with biochar treatment. Calculation shows that the extracted Cu at waste levels of 60 Mg ha^{-1} were 60.0, 59.8 and 46.1 mg kg^{-1} with biochar treatment of 0, 5 and 10 Mg ha^{-1}, respectively, and those for Zn were 57.4, 54.0 and 45.5 mg kg^{-1}, respectively. The increase in the soil adsorption capacity caused by the presence of biochar significantly decreased the soil labile Cu and Zn about 0.33 and 0.59%, respectively, at 5 Mg biochar ha^{-1} and 23.2 and 20.7% at 10 Mg biochar ha^{-1}, respectively. The increase in the soil adsorption capacity towards Cu and Zn was probably to be originated from the unique characteristic of biochar that possessed high amounts of organic functional groups that may provide abundant negative charges. Copper and Zn in biochar-treated soils were transformed into less soluble forms with higher bonding energy. The amount of stabilized heavy metals was determined by the biochar-treated soil-adsorptive surfaces. Therefore, biochar 10 Mg ha^{-1} was more effective than 5 Mg ha^{-1} in decreasing heavy metals at waste level of 60 Mg ha^{-1} (**Figure 10**). These changes may lower the negative effect of heavy metals on the growth of thorny amaranth. Therefore, the treatment of soil with biochar may improve the growth of thorny amaranth in heavy metal polluted soils.

The increase in soil pH induced by biochar treatment may have stimulated the enlargement of the soil adsorptive sites caused by the dissociation of biochar and soil colloid functional groups. However, as pointed out previously, a biochar level of 5 Mg ha^{-1} was probably not sufficient to handle heavy metals at a waste level of 60 Mg ha^{-1}, and the growth of plants at this treatment was in general not better than those without biochar (**Figure 12**). It is obvious that the effect of biochar was dependent on its level. The level of 5 Mg biochar ha^{-1} was effective at a waste level of 15 Mg ha^{-1} but not at a waste level of 60 Mg ha^{-1}. Biochar level of 10 Mg ha^{-1} was effective at waste levels of 15 and 60 Mg ha^{-1}. The improvement effect of biochar was also observed on plant shoot and root dry-weight (**Figure 12**). The improvement of shoot dry weight was clear; the effect of 5 Mg ha^{-1} was more effective than that of 10 Mg ha^{-1} as also that on root dry-weight (**Figure 12**).

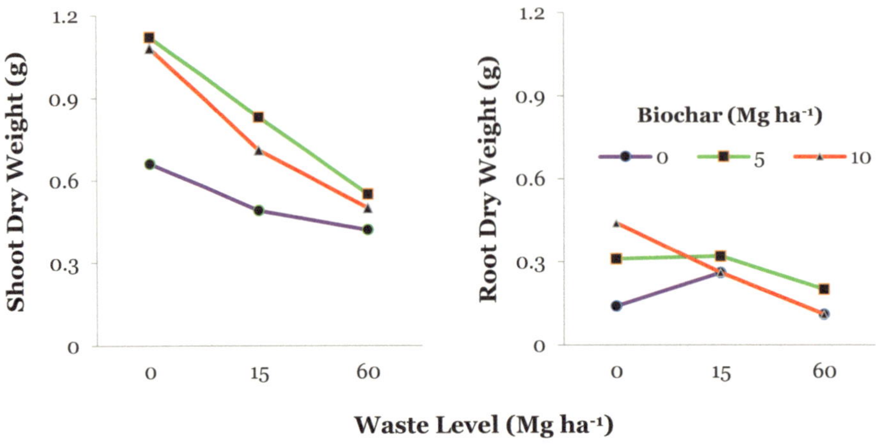

Figure 12.
The effect of biochar on the dry weights of thorny amaranth biomasses in tropical soil polluted with heavy metals (after [74] with permission).

6. Conclusions

The increase in the soil and environmental concentrations of heavy metals are reported from all over the world. The increase in heavy metal concentration may occur stimulated by industrialization. Since they are toxic and detrimental at high concentrations, the increase in the soil's heavy metal concentrations is reported to induce plant growth retardation. The presence of biochar that possesses high amounts of negative charges and may increase the soil pH may enlarge the soil's heavy metal cation retention. Therefore, the biochar application may increase the heavy metal immobilization in soil and cause a decrease in the soil available heavy metals. By these means, biochar application may also increase the growth of plants.

The biochar application may lower the soil concentration to the level at which plants may absorb heavy metals at suitable levels so that the absorption of heavy metals and the decrease of heavy metals in soil occur faster without physical and physiological disturbance. In phytoremediation, the use of biochar may accelerate the heavy metal absorption without physical and physiological disturbance on plant roots by the presence of high concentration of heavy metals.

However, in addition to its advantages to lower the concentrations of the polluting heavy metals in the environment, the use of biochar shows drawbacks, among which is the fact that biochar is bulky. The levels used in most experiments which were 5–10 Mg ha^{-1} are of great amount. It will cause difficulty in its field transportation and treatment. This needs further research to utilize biochar at lower levels without decreasing its effectiveness, for example by adjusting its particle size.

Author details

Abdul Kadir Salam
Faculty of Agriculture, Department of Soil Science, University of Lampung,
Bandar Lampung, Indonesia

*Address all correspondence to: abdul.kadir@fp.unila.ac.id

References

[1] Aksu A. Sources of metal pollution in the urban atmosphere (a case study: Tuzla, Istanbul). Journal of Environmental Health Science and Engineering. 2015;**13**(79):1-10. DOI: 10.1186/s40201-015-0224-9

[2] Salam AK. Management of Heavy Metals in Tropical Soil Enviroment. 1st ed. Bandar Lampung: Global Madani Press; 2017

[3] Adejoh IP. Assessment of heavy metal contamination of soil and cassava plants within the vicinity of a cement factory in north central, Nigeria. Advances in Applied Science Research. 2016;**7**(3):20-27

[4] Arshad N, Hamzah Z, Wood AK, Saat A. Determination of heavy metals concentrations in airborne particulates matter (APM) from Manjung district, Perak using energy dispersive X-ray fluorescence (EDXRF) spectrometer determination of heavy metals concentrations in airborne particulates Matt. AIP Conference Proceedings. 2015;**1659**:050008. DOI: 10.1063/1.4916878

[5] Asati A, Pichhode M, Nikhil K. Effect of heavy metals on plants: An overview. International Journal of Application or Innovation in Engineering & Management. 2016;**5**(3):56-66. DOI: 10.13140/RG.2.2.27583.87204

[6] Ayari F, Hamdi H, Jedidi N, Gharbi N, Kossai R. Heavy metal distribution in soil and plant in municipal solid waste compost amended plots. International Journal of Environmental Science and Technology. 2010;**7**(3):465-472

[7] Biasioli M, Kralj T, Grčman H, Díaz-Barrientos E, Madrid F, Ajmone-Marsan F. Potentially toxic elements contamination in urban soils. Journal of Environmental Quality. 2007;**36**(1):70. DOI: 10.2134/jeq2006.0254

[8] Cheng S. Effects of heavy metals on plants and resistance mechanisms. Environmental Science and Pollution Research. 2003;**10**(4):256-264. DOI: 10.1065/espr2002.11.141.2

[9] Chibuike GU, Obiora SC. Heavy metal polluted soils: Effect on plants and bioremediation methods. Applied and Environmental Soil Science. 2014;**ID 752708**:1-12. DOI: 10.1155/2014/752708

[10] Febriansyah MR, Septiana LM, Supriatin S, Salam AK. The patterns of lead and copper levels in the vicinity of heavy metal sources in Lampung, the southern part of Sumatra, Indonesia. IOP Conference Series: Earth and Environmental Science. 2021;**637**(012044):1-10. DOI: 10.1088/1755-1315/637/1/012044

[11] Gill M, Gill M. Heavy metal stress in plants : A review. International Journal of Advanced Research. 2014;**2**(6):1043-1055

[12] Hashim MA, Mukhopadhyay S, Narayan J, Sengupta B. Remediation technologies for heavy metal contaminated groundwater. Journal of Environmental Management. 2011;**92**:2355-2388. DOI: 10.1016/j.jenvman.2011.06.009

[13] Bendfeldt ES, Burger JA, Daniels WL. Quality of amended mine soils after sixteen years. Soil Science Society of America Journal. 2001;**65**:1736-1744. DOI: 10.2136/sssaj2001.1736

[14] Elbagermi MA, Edwards HGM, Alajtal AI. Monitoring of heavy

metals content in soil collected from city Centre and industrial areas of Misurata, Libya. International Journal of Analytical Chemistry. 2013;**2013**:18-23. DOI: 10.1155/2013/312581

[15] Fang H, Huang L, Wang J, He G, Reible D. Environmental assessment of heavy metal transport and transformation in the Hangzhou Bay, China. Journal of Hazardous Materials. 2016;**302**:447-457. DOI: 10.1016/j.jihazmat.2015.09.060

[16] Gaza T, Kugara J. Study of heavy metal air pollution, using a moss (Grimmia dissimulate) biomonitoring technique. Universal Journal of Chemistry. 2018;**6**(1):1-13. DOI: 10.13189/ujc.2018.060101

[17] Jankowski K, Malinowska E, Ciepiela GA, Jankowska J. Lead and cadmium content in grass growing near an expressway. Archives of Environmental Contamination and Toxicology. 2019;**76**:66-75. DOI: 10-1007/s00244-018-0565-3

[18] Jelecevic A, Sager M, Vollprecht D, Puschenreiter M, Liebhard P. Partitioning of heavy metals in different particle-size fractions of soils from former mining and smelting locations in Austria. Eurasian Journal of Soil Science. 2021;**10**(2):123-131. DOI: 10.18393/ejss.837139

[19] Li X, Zhao Z, Yuan Y, Wang X, Li Z. Heavy metal accumulation and its spatial distribution in agricultural soils : Evidence from Hunan province, China. RSC Advances. 2018;**8**:10665-10672. DOI: 10.1039/c7ra12435j

[20] Matthews-Amune OC, Kakulu S. Determination of heavy metals in forage grasses (carpet grass (*Axonopus Ompressus*), Guinea grass (*Panicum Maximum*) and elephant grass (*Pennisetum Purpureum*)) in the vicinity of Itakpe Iron ore mine, Nigeria. International Journal of Pure and Applied Science and Technology. 2012;**13**(2):16-25

[21] Miao X, Hao Y, Zhang F, Xie Z. Spatial distribution of heavy metals and their potential sources in the soil of Yellow River Delta: A traditional oil field in China. Environmental Geochemistry and Health. 2019;**5**. DOI: 10.1007/s10653-018-0234-5 [Published Online January 02, 2019]

[22] Mircea M et al. A study of heavy metals pollution in Italy with the atmospheric modelling system of the MINNI project. E3S Web Conference. 2013;**1**(03003):1-4

[23] Ng CC, Boyce AN. Heavy metals phyto-assessment in commonly grown vegetables : Water spinach (I. aquatica) and okra (*A. esculentus*). Springerplus. 2016;**5**:469. DOI: 10.1186/s40064-016-2125-5

[24] Ogunkunle CO, Fatoba PO, Oyedeji AO, Awotoye OO. Assessing the heavy metal transfer and translocation by *Sida acuta* and *Pennisetum purpureum* for phytoremediation purposes. Albanian Journal of Agricultural Sciences. 2014;**13**(1):71-80

[25] Page V, Feller U. Heavy metals in crop plants: Transport and redistribution processes on the whole plant level. Agronomy. 2015;**5**:447-463. DOI: 10.3390/agronomy5030447

[26] Payus C, Farhana A, Talip ABU, Hsiang TANWEI. Heavy metals accumulation in paddy cultivation area of Kompipinan, Papar District, Sabah. Journal of Sustainable Science and Management. 2015;**10**(1):76-86

[27] Richards BK, Steenhuis TS, Peverly JH, Mcbride MB. Metal mobility at an old, heavily loaded sludge

application site. Environmental Pollution. 1998;**99**:365-377

[28] Suguru PM. Simulated in-situ immobilization of heavy metals in selected soils of Karnataka. Fungal Genomics and Biology. 2015;**5**(1):1-6. DOI: 10.4172/2165-8056.1000122

[29] Sun L et al. Levels, sources, and spatial distribution of heavy metals in soils from a typical coal industrial city of Tangshan, China. Catena. 2018, 2019;**175**(July):101-109. DOI: 10.1016/j.catena.2018.12.014

[30] Tapilatu RF, Wona H, Siburian RHS, Saleda ST. Heavy metals contaminants in the eggs and temperatures of nesting beaches of sea turtles in Kaimana , West Papua, Indonesia. Biodiversitas. 2020;**21**(10):4582-4590. DOI: 10.13057/biodiv/d211016

[31] Timothy N, Williams ET. Environmental pollution by heavy metal: An overview. International Journal of Environmental Chemistry. 2019;**3**(2): 72-82. DOI: 10.11648/j.ijec.20190302.14

[32] Jamal Q, Khan K, Munir S, Anees M. Heavy metals accumulation and their toxic effects: Review heavy metals accumulation and their toxic effects : Review. Journal of Biomolecular Science. 2013;**1**(1-2):27-36

[33] Nascimento SS, Silva EB, Alleoni LRF, Grazziotti PH, Fonseca FG, Nardis BO. Availability and accumulation of lead for forage grasses in contaminated soil. Journal of Soil Science and Plant Nutrition. 2014;**14**(4):783-802. DOI: 10.4067/s0718-95162014005000063

[34] Popova E. Accumulation of heavy metals in the 'soil-plant' system. AIP Conference Proceedings. 1772;**050006**:2016. DOI: 10.1063/1.4964576

[35] Rahimi G, Kolahchi Z, Charkhabi A. Uptake and translocation of some heavy metals by rice crop (*Oryza sativa*) in paddy soils. Agriculture. 2017;**63**(4):163-175. DOI: 10.1515/agri-2017-0016

[36] Angelova V, Ivanova R, Pevicharova G, Ivanov K. Effect of organic amendments on heavy metals uptake by potato plants. In: 19th World Congress of Soil Science. 2010. pp. 84-87

[37] Chandra Shaha S, Kashem MA, Osman KT. Effect of lime and afrmyard manure on the concentration of cadmium in water spinach (*Ipomoea aquatica*). ISRN Agronomy. 2012;**2012**: 1-6. DOI: 10.5402/2012/719432

[38] Ghori SA et al. Wood-derived biochar influnces nutrient use efficiency of heavy metals in spinach (Spinacia oleracea) under groundwater and waste water irrigation. Journal of Environmental Engineering and Landscape Management. 2019;**27**(3):144-152. DOI: 10.3846/jeelm.2019.10792

[39] Nasser HM, Rahman MZ, Sultana S, Quddus MA, Haqque MA. Remediation of heavy metal polluted soil through organic amendments. Bangladesh Journal of Agricultural Research. 2017;**42**(4):589-598

[40] Quenea K, Lamy I, Winterton P, Bermond A, Dumat C. Interactions between metals and soil organic matter in various particle size fractions of soil contaminated with waste water. Geoderma. 2009;**149**(3-4):217-223. DOI: 10.1016/j.geoderma.2008.11.037

[41] Salam AK. A four year study on the effects of manipulated soil pH and organic matter contents on availabilities of industrial-waste-origin heavy-metals in tropical soils. Journal of Tropical Soils. 2000;**11**:31-46

[42] Salam AK et al. The residual copper and zinc in tropical soil over 21 years after amendment with heavy metal containing waste , lime, and compost. Applied and Environmental Soil Science. 2021;**2021**:7596840

[43] Stietiya MH, Wang JJ. Effect of organic matter oxidation on the fractionation of copper, zinc, lead, and arsenic in sewage sludge and amended soils. Journal of Environmental Quality. 2011;**40**:1162-1171. DOI: 10.2134/jeq2011.0008

[44] Lahori AH et al. Clays, limestone and biochar affect the bioavailability and geochemical fractions of cadmium and zinc from zn-smelter polluted soils. Sustainability. 2020;**12**(20):1-16. DOI: 10.3390/su12208606

[45] Wang Y, Gu K, Wang H, Shi B. Remediation of heavy-metal-contaminated soils by biochar : A review. Environmental Geotechnics. 2019;**9**(3):135-148. DOI: 10.1680/jenge.18.00091

[46] Salam AK. Enzymes in Tropical Soils. 1st ed. Bandar Lampung: Global Madani Press; 2014

[47] Atkinson CJ, Khan A, Zhang K. Biochar remediation of soil: Linking biochar production with function in heavy metal contaminated soils. Plant, Soil and Environment. 2021;**2021**(4):183-201

[48] Baltrenaite JKE. Biochar as adsorbent for removal of heavy metal ions [cadmium (II), copper (II), Lead (II), Zinc (II)] from aqueous phase. International Journal of Environmental Science and Technology. 2016;**13**:471-482. DOI: 10.1007/s13762-015-0873-3

[49] Buss W, Kammann C, Koyro H. Biochar reduces copper toxicity in *Chenopodium quinoa* Willd. in a sandy soil. Journal of Environmental Quality. 2011;**40**:X–X. DOI: 10.2134/jeq2011.0022

[50] Chen D et al. Effects of biochar on availability and plant uptake of heavy metals – A meta- analysis. Journal of Environmental Management. 2018;**222**:76-85. DOI: 10.1016/j.envman.2018.05.004

[51] Domingos D, Carlos J, Serra V, Zukowski JC. Biochar efficiency in the removal of heavy metals Eficiência de biocarvões na remoção de metais pesados. Acta Ortopedica Brasileira. 2019;**3**(3):131-138. DOI: 10.22571/2526-4338242

[52] Guo M, Song W, Tian J. Biochar-facilitated soil remediation: Mechanisms and efficacy variations. Frontiers in Environmental Science. 2020;**8**:1-23. Article: 521512. DOI: 10.3389/fenvs.2020.521512

[53] Hayyat A et al. Role of biochar in remediating heavy metals in soil. In: Ansari AA, editors. Phytoremediation. Switzerland: Springler Int Publ.; 2016. pp. 421-437

[54] Huang D et al. The effects of rice straw biochar on indigenous microbial community and enzymes activity in heavy metal-contaminated sediment. Chemosphere. 2017;**174**:545-553. DOI: 10.1016/j.chemosphere.2017.01.130

[55] Joner EJ. Biochar affects heavy metal uptake in plants through interactions in the rhizosphere. Applied Sciences. 2020;**10**:1-12

[56] Li Q et al. Removal of toxic metals from aqueous solution by biochars derived from long-root Eichhornia crassipes. Royal Society Open Science. 2018;**5**:180966. DOI: 10.1098/rso.180966

[57] Patra JM, Panda SS, Dhai NK. Biochar as a low-cost adsorbent for heavy metal

removal: A review. International Journal of Research in BioSciences. 2017;**6**(1):1-7

[58] Tan X, Liu Y, Zeng G, Wang X, Hu X, Gu Y. Chemosphere application of biochar for the removal of pollutants from aqueous solutions. Chemosphere. 2015;**125**:70-85. DOI: 10.1016/j.chemosphere.2014.12.058

[59] Wang S, Xu Y, Norbu N, Wang Z. Remediation of biochar on heavy metal polluted soils. IOP Conference Series: Earth and Environmental Science. 2018;**108**(042113):1-7. DOI: 10.1088/1755-1315/108/4/042113

[60] Liu J, Wang H, Ma N, Zhou B, Chen H, Yuan R. Optimization of the raw materials of biochars for the adsorption of heavy metal ions from aqueous solution. Water Science and Technology. 2022;**85**(10):2869-2881. DOI: 10.2166/wst.2022.158

[61] Palansooriya KN et al. Prediction of soil heavy metal immobilization by biochar using machine learning. Environmental Science & Technology. 2022;**56**:4287-4198. DOI: 10.1021/acs.est.1c08302

[62] Siddika A, Parveen Z. Heavy metal remediation from contaminated soil using biochars and modified biochars : A review. Indonesian Journal of Social and Environmental Issue. 2022;**3**(1):19-28. DOI: 10.47540/ijsei.v3i1.417

[63] Silva G, Aini SN, Bucharie H, Salam AK. Phytoextraction of Cu from tropical soil 21 years after treatment with heavy-metal containing waste. Journal of Tropical Soils. 2021;**26**(1):11-18. DOI: 10.5400/jts.2021.v26i1.11

[64] Salam AK, Hidayatullah MA, Supriatin S, Yusnaini S. The phytoextraction of Cu and Zn by elephant grass (*Pennisetum purpureum*) from tropical soil 21 years after amendment with industrial waste containing heavy metals. IOP Conference Series: Earth and Environmental Science. 2021;**637**(0124044):1-7. DOI: 10.1088/1755-1315/637/1/012044

[65] Salam AK, Helmke PA. The pH dependence of free ionic activities and total dissolved concentrations of copper and cadmium in soil solution. Geoderma. 1998;**83**(3-4):281-291. DOI: 10.1016/S0016-7061(98)00004-4

[66] Montiel-Rozas MM, Madejon E, Madejon P. Effect of heavy metals and organic matter on root exudates (low molecular weight organic acids) of herbaceous species : An assessment in sand and soil conditions under different levels of contamination *. Environmental Pollution. 2016;**216**:273-281

[67] Sun Y, Zhou Q, An J, Liu W, Liu R. Chelator-enhanced phytoextraction of heavy metals from contaminated soil irrigated by industrial wastewater with the hyperaccumulator plant (*Sedum alfredii* Hance). Geoderma. 2009;**150**:106-112. DOI: 10.1016/j.geoderma.2009.01.016

[68] Abdu N, Mohammed I. Adsorption-solubility equilibria and speciation of Pb, Cd , and Zn in a savanna soil. Spanish Journal of Soil Science. 2016;**6**(3):244-260. DOI: 10.3232/SJSS.2016.V6.N3.06

[69] El-maghrabi HH. Removal of heavy metals via adsorption using natural clay material removal of heavy metals via adsorption using natural clay material. Journal of Environment and Earth Science. 2016;**4**(2014):38-47

[70] He G, Zhang Z, Wu X, Cui M, Zhang J, Huang X. Adsorption of heavy metals on soil collected from Lixisol

of typical karst areas in the presence of CaCO3 and soil clay and their competition behavior. Sustainability. 2020;**12**(7315):1-19. DOI: 10.3390/su12187315

[71] Sdiri AT, Higashi T, Jamoussi F. Adsorption of copper and zinc onto natural clay in single and binary systems. International journal of Environmental Science and Technology. 2014;**11**(4):1081-1092. DOI: 10.1007/s13762-013-0305-1

[72] Ugwu EI et al. Adsorption mechanisms for heavy metal removal using low cost adsorbents: A review. IOP Conference Series: Earth and Environmental Science. 2020;**614**(1):1-12. DOI: 10.1088/1755-1315/614/1/012166

[73] Salam AK et al. The biochar-improved growth-characteristics of corn (*Zea mays* L.) in a 22-years old heavy-metal contaminated tropical soil. IOP Conference Series: Earth and Environmental Science (In production). 2022;**1034**:1-10. DOI: 10.1088/1755-1315/1034/012045

[74] Afrianti NA et al. The biochar-enhanced phytoextraction of heavy-metal-polluted tropical soils by thorny Amaranthus (*Amaranthus spinosus*). IOP Conference Series: Earth and Environmental Science, vol. (In production). 2022

[75] Rachman F, Supriatin S, Niswati A, Salam AK. Lime-enhanced phytoextraction of copper and zinc by land spinach (*Ipomoea reptans* Poir.) from tropical soils contaminated with heavy metals. AIP Conference Proceedings, vol. (In production). 2022

Chapter 6

Biochar Application in Soil Management Systems

Theophilus Olufemi Isimikalu

Abstract

Due to its potential for improving soil fertility and reducing greenhouse gas emissions, biochar is frequently used as a soil amendment. This chapter presents an overview of its application and soil conditioning mechanisms as a technique for long-term carbon sequestration and lower greenhouse gas emissions, as well as an option for improving soil fertility. It focuses on biochar amendment for improved soil properties that support plant nutrient uptake and crop yield improvement, soil properties and biochar carbon sequestration dynamics, biochar degradation processes, and soil interactions and conditioning mechanisms that influence biochar carbon stability in soils. Current biochar stability assessment techniques used in academic studies are also addressed, along with their suitability for use with various goals and situations.

Keywords: biochar, soil, management

1. Introduction

Sustainable soil management in agriculture aims at developing economically sound and environmentally safe crop management systems that build the quality of soils while being utilized for food production. Such systems are associated with efficient management of soil organic carbon (SOC) and soil fertility, the credible measurement of which Lal [1] regarded as an indicator of soil quality and health. Biochar, which the International Biochar Initiative defined as a solid material derived from the thermochemical conversion of biomass in an oxygen-limited environment, has received wide attention in the past two decades for its documented potential to improve soil fertility and mitigate greenhouse gas emissions.

Studies report that biochar application can enhance soil fertility, reduce greenhouse gas (GHG) emissions [2], increase stable carbon forms in soil [3], improve nutrient and water retention, reduce heavy metal toxicity [4], and increase soil ability to suppress soil-borne pathogens. Woolf et al. [5] stated that the use of biochar in soil could mitigate as much as 1.8–9.5 Pg (10^{15} g) carbon dioxide carbon emissions annually, globally.

Biochar's soil fertility improvement mechanism is through the manipulation of soil properties such as increased soil microbial activity, soil water holding capacity, soil porosity, soil reaction (pH), soil aggregation, soil organic carbon, among others. When these soil physical and chemical properties are improved, soil nutrient retention and uptake to support plant growth improve. Many studies have reported

 IntechOpen

increased agronomical crop performances following biochar amendment such as in Asai et al. [6]; likewise, others, including Butnan et al. [7], have reported none or unfavorable crop yield responses.

A suppression of greenhouse gases emission is another benefit of biochar addition to soil that has been widely proven in earlier research [8, 9]. Biochar's production in an oxygen-limited environment gives it a chemically recalcitrant carbon-rich solid property, being produced from biomass by heating in an oxygen-limited environment. Although biochar is expected to be largely resistant to biological degradation, research shows that some of its components are relatively easily biodegradable. Thus, several studies have examined its soil and crop yield improvement, and carbon sequestration potential, and widely varying responses have been reported [10, 11]. This has resulted in varying mean resident time (MRT) estimates of biochar-C, ranging from decadal to centennial scales.

The stability of biochar in soil is of high importance to its use as an organic amendment. Lehmann and Rondon [12] defined stability as the determining factor on how long C in biochar will be sequestered (remain in soil) to mitigate climate change and how long a biochar material will continue to benefit soil and plants. The wide variation in research observations has made it very hard to generalize findings on biochar-C stability in soils and thus makes it very important to study its stability in individual soils and under peculiar prevalent environmental conditions. Currently, variations in biochar effects in soils have been attributed mainly but not solely to soil properties such as soil texture and mineralogy, feedstock material, production conditions, environmental characteristics, and the interaction of these elements. Of these factors, biochar feedstock and production conditions are two factors more easily controllable in biochar use in soil.

1.1 Biochar production and basic properties

Biomass pyrolysis is generally classified according to the rate of reaction into slow, fast, and flash pyrolysis. Through the pyrolysis process, biomass can be transformed into bio-oil, syngas, and biochar (the percentage of each component depends on the pyrolysis condition). The two major thermal conversion processes widely used in biochar production, however, are slow and fast pyrolysis [13]. Slow pyrolysis is most widely used and carried out at lower temperatures (~350°C) and heating rates and longer residence times compared with fast pyrolysis (~1000°C), which optimizes biochar yields over energy production.

Lehmann [8] among other researchers found that the chemical and physical properties of biochar depend majorly on the properties of the original feedstock material and the production conditions (essentially temperature and charing time). Ogawa et al. [14] described biochar chemical structure as one containing different aromatic C structures and considered it a transitional form with intermediate properties between carbohydrate-based biomass and graphite carbon that can appear as a microcrystalline structure. The chemical structure also contains macro-, meso-, and micro-pores, which are derived from cellular fractures of plant cells. Downie et al. [15] similarly characterized biochar as having large surface area, which in addition to its chemical properties and structure gives it high sorption capacity as is the case with other organic compounds. Its composition is widely differentiated into a relatively recalcitrant C, labile (leachable C) and ash (**Figure 1**).

Schmidt and Noack [16] reported that the chemical difference between common OM sources and biochar is that it contains a higher proportion of aromatic carbon that has a fused structure, which differs from the aromatic structure seen in other

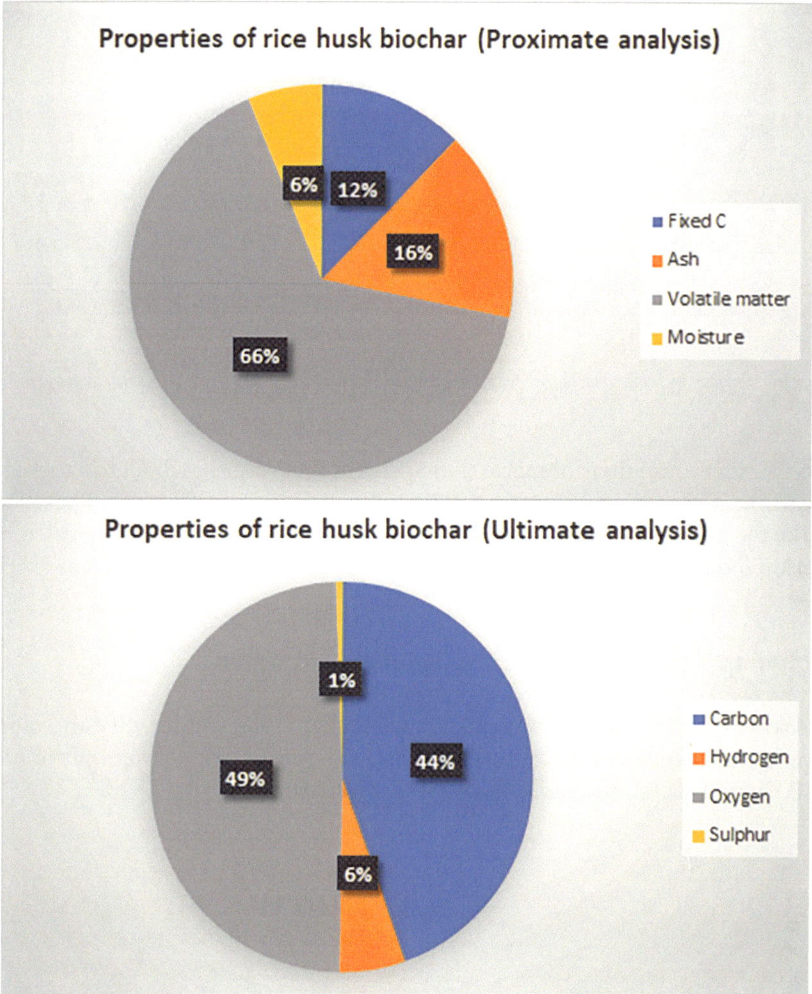

Figure 1.
Biochar properties analyzed using proximate and ultimate analytic procedures [9].

OM sources such as lignin. The fused aromatic structure can also vary, depending on the production temperature. Nguyen et al. [17] stated that these forms can include amorphous and turbostratic C, which occur at low and higher pyrolysis temperatures, respectively. It is this C structure that gives biochar the chemical stability that makes it hard for microorganisms to readily utilize its C, N, and possibly other nutrients it contains as energy source (**Figure 2**).

Lehmann and Joseph [19] reported that a fraction of biochar may be readily utilized or leached, and this fraction depends on the biochar type. Steiner [20] also noted that biochar may stimulate microbial activity and increase their abundance in soil due to its composition of essential macro- and micro-nutrients, which may serve as biological energy substrate. Some of the most important research applications of biochar-aiding soil functioning are as follows: (1) the improvement of soil fertility and adequate biomass production, (2) storage and cycling of carbon, and (3) alleviation of chemical toxicity and sustenance of soil biodiversity. Ever-increasing human populations and the attendant pressures on soil resources have resulted in extensive

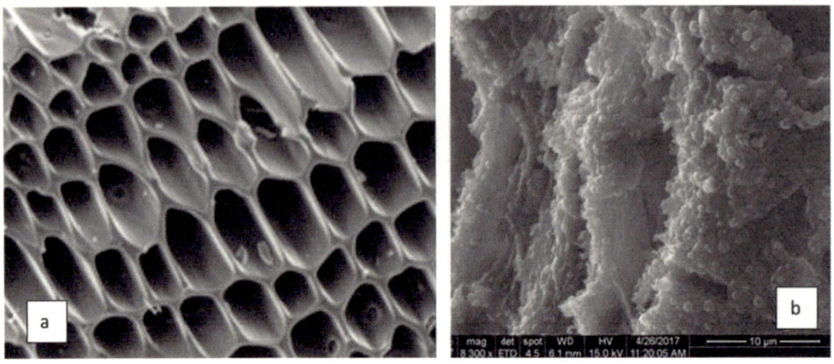

Figure 2.
a. Microscopic imagery of fresh wood biochar; b. imagery of the surface of aged wood biochar (image source: Joseph et al. [18]).

use of pesticides and other intensive management techniques, which has negative climate change impact, and threatens soil quality and human survival. These factors make the aforementioned potentials of biochar highly attractive in agricultural production today.

2. Biochar applications in soil management systems

This section discusses the applications of biochar along soil fertility and crop yield improvement, carbon storage and cycling, and soil remediation potentials of biochar. **Figure 3** shows biochar processes in the environment.

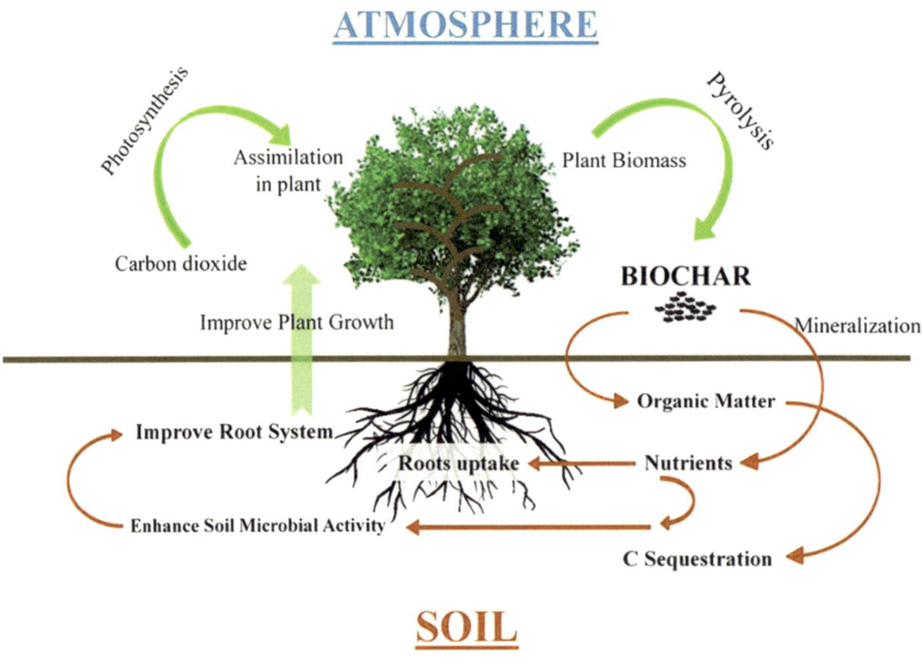

Figure 3.
Applications of biochar in soil (image source: [21]).

2.1 Biochar application for soil fertility and crop yield improvement

Improvement in crop yield following biochar amendment has been reported in many previous research studies such as that of Rondon et al. [22] in acidic and weathered tropical soils. Few numbers of research studies such as Husk and Major [23] have also reported positive effects in highly fertile temperate soils. In a meta-analysis, Biederman and Harpole [24] analyzed results of 371 individual studies and found that biochar amendment resulted in higher above-ground crop productivity, soil microbial biomass, K^+ concentration in plant tissue, rhizobial nodulation, soil N, P, K^+, and C in comparison with control conditions. There was, however, no obvious trend in soil productivity with biochar addition, and crop productivity varied with increase in application rates. **Figure 4** shows the properties of biochar from different feedstock materials.

In addition to the neutral or negative effect of biochar recorded in some previous studies, there also appear to be an upper limit beyond which biochar addition does not result in improved crop productivity. Lehmann et al. [25] reported that crop responses to biochar addition were positive at rates up to 55 t/ha, while a reduction in growth was recorded at higher application rates. Rondon et al. [22] on the other hand reported a much higher threshold of 165 t/ha. According to them, biochar application of >165 t/ha to a poor soil in a pot experiment resulted in yield decrease that equaled to that of unamended control.

Some other authors have reported yield decreases at lower levels of application. Asai et al. [6], for example, reported the highest rice yield at 4 t/ha biochar application rate in comparison with 8 and 16 t/ha. They reported that yields dropped to the level of the control treatment at 16 t/ha application rate. Jeffery et al. [26] reported that more positive responses from biochar addition to soil have been reported in pot than in field experiments, in acidic than in neutral soils, and in sandy than in loam and silt soils. Increases in yield in comparison with controls range from <10 to >200%.

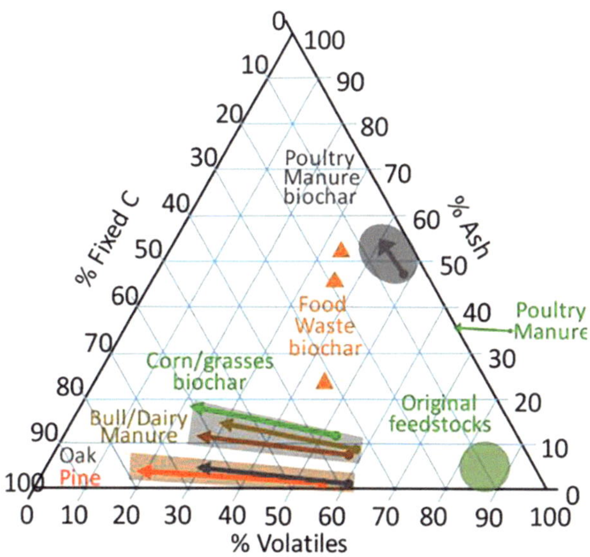

Figure 4.
Approximate properties of biochar derived from different feedstock materials (image source: Joseph et al. [18]).

2.2 Biochar carbon sequestration dynamics

Woolf et al. [5] and other authors have proposed biochar use in soil as a means of long-term C sequestration and reduced GHG emission. The main mechanism of biochar-C sequestration is through its incorporation into soil as a highly stabilized C produced through pyrolysis of biomass. Because pyrolysis progresses in the absence of oxygen, the C content of feedstock material is locked in the biochar, which is then applied to soil. Although Lehmann and Rondon [12] reported up to 50% loss of biomass C in biochar production, they reported that a considerably greater fraction of the locked stable C in biochar remained in soil for longer time periods in comparison with direct biomass input in agricultural fields.

Woolf et al. [5] also suggested another potential C negative benefit of biochar as the reduction in emission of CO_2 through reduced fertilizer demands to achieve crop yields. This idea is premised on the potential of biochar to improve soil water and nutrient retention capacity of soils. In addition to CO_2 emission reduction, Spokas et al. [27] reported reduced N_2O emission following biochar addition, and Leng et al. [28] reported that biochar addition resulted in reduced methane (CH_4) emission from agricultural soils through the improvement of soil aeration and reduction.

In a meta-analysis, Wang et al. [29] showed that biochar application could stimulate soil CO_2 emissions by as much as 28–32% and revealed that average biochar decomposition rate in studies lasting for <6 months was 0.023%/day. This suggests possible priming effects of biochar on SOC or other indirect interactions resulting in CO_2 emission from soil following biochar addition. CO_2 losses observed in previous research studies following biochar amendment vary widely, and attributed causes include variations in biochar feedstock, production conditions, duration of experiment, and environmental variables.

While Bruun et al. [30] reported cumulative C loss of 2.9 and 5.5% in a sandy loam amended with wheat straw biochar produced from slow and fast pyrolysis, respectively, some other studies such as Fang et al. [31] have reported lower biochar C mineralization rates of 0.1–3% of applied biochar-C mineralized per year. In summary, the carbon sequestration value of biochar is hung on its degradation in soil and the environmental factors that influence it.

2.3 Biochar-pesticide interactions in soil environments

Despite the fact that biochar was initially developed as a soil amendment because of its beneficial effects on carbon sequestration, greenhouse gas emissions reductions, and soil fertility improvement (Spokas et al., 2009) [32], it has recently drawn more attention for its potent ability to lower the bioavailability of pesticides [33, 34]. It has also been acknowledged that the presence of biochar in soil influences the nature of sorption mechanisms and the bioavailability of pesticide residues for living organisms in addition to improving the sorption of various pesticides [35].

By reducing the leaching of sprayed pesticides, the use of biochar in agricultural soils near bodies of water may also successfully lower the risk of pollution of subterranean water [33, 36]. Pesticide sorption ability of biochar have also been reported in previous research [37]. This is accomplished by using biochar's impacts on pesticide adsorption mechanisms and desorption behavior as a powerful tool to alter pesticide bio-accessibility and toxicological effects.

The repair of contaminated soils has been proposed using procedures such as soil washing, soil flushing, bioremediation, and soil vapor extraction. However, due to limited effectiveness, high maintenance costs, fertility loss, nutrient leaching, and soil

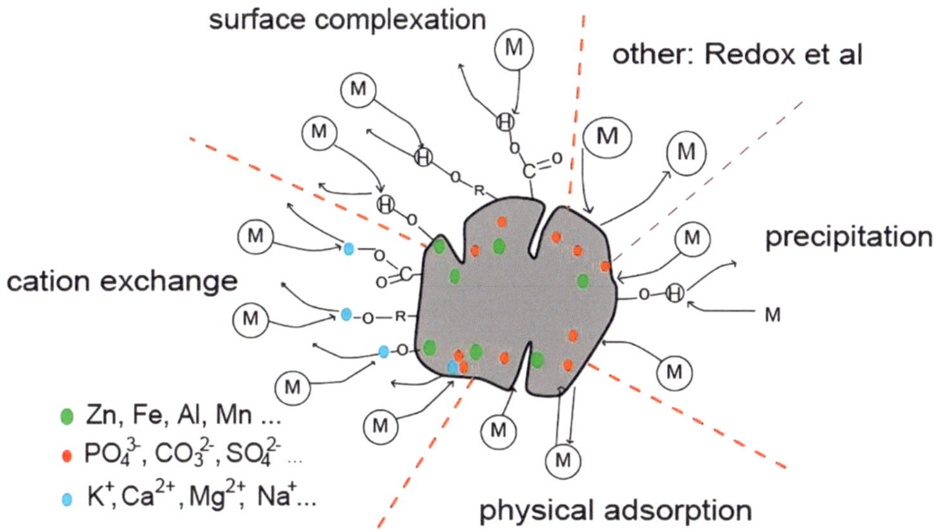

Figure 5.
The removal mechanism of heavy metals by biochar [41].

erosion, among other factors, these approaches are typically inapplicable in field settings [38]. Application of biochar as an *in situ* form of amendment for contaminated soil has thus shown promise as a method that represents a financially prudent alternative to address remedial demands [19].

By (1) binding pesticides to minimize their potential motility into water supplies and living beings, and (2) supplying nutrients to encourage plant growth and drive ecological restoration, biochar is a less disruptive approach of remediating pesticide-contaminated soil [39]. Additionally, applying biochar to soil requires only a small amount of pretreatment because it is an organic substance made from biological matter [34].

Khoram et al. [40] studied the functions of biochar in fundamental processes of pesticides in the environment and summarized those roles in remediating pesticide-contaminated soils as follows: (1) enhancing pesticide adsorption capacity; (2) reducing desorption and mobility of pesticides in soil layers; (3) reducing the amount of pesticides that are bioavailable in soil pore water, which is thought to be the portion that is bioavailable to soil organisms; (4) enhancing soil microbial activity by supplying necessary nutrients; and (5) enhancing soil physicochemical characteristics such as pH, CEC, and water holding capacity. Biochar amendment has also been shown to help in the remediation of heavy metal pollution in the environment (**Figure 5**).

The mode and other application variables of biochar to soil are a group of significant parameters that affect biochar reaction and stability in soil. It is thus important to be mindful that the complete lifecycle costs of handling and using biochar at scale must be kept as low as possible in order to maintain biochar management as a carbon-negative practice.

3. Mode, frequency, and rate of biochar application to soil

Wide-varying application rates have been used in previous research, ranging from <5 t/ha to >100 t/ha. IBI (2010) in its biochar fact sheet recommended rates

Figure 6.
Particles of biochar derived from different feedstocks.

of 2–22 t/ha in field trials and lower levels of 2–5 t/ha for large-scale agricultural use. Handling and application should generally determine particle size. The adsorption of ammonium and hexavalent chromium ions from aqueous solution was found in tests to be more effective with fine biochar particles [42, 43]. **Figure 6** shows biochar particles derived from different feedstock materials.

Comparative studies on soil fertility revealed that cowpea biomass production and nutrient uptake were unaffected by biochar particles with diameters of 1 or 20 mm [44] and 10 mm or less [45]. The specific surface area and the resulting accessibility of binding sites in biochar are, however, characteristics that are expected to vary depending on particle sizes and should be considered in biochar application to soil. A thorough understanding of the relationship between the properties of biochar and its applicability will allow for the establishment of appropriate process conditions to produce a biochar with the desired characteristics.

Currently, there are no standard application rates of biochar to soil for different agricultural aims due to varying responses from numerous tests. Variabilities result from biochar feedstock material and production conditions, among others factors as discussed earlier. These factors influence biochar characteristics including nutrient levels, ash content, carbon recalcitrance, etc., which all influence application rate. Due to the expected recalcitrance of biochar in soil, researchers such as Major et al. [46] suggest that a one-time application could provide positive benefits for more than one growing season.

Studies have, however, shown that unless a biochar material is derived from manure or is blended with nutrient-rich materials, it may not substitute for chemical fertilizers. Research has also shown that the level of biochar application to soil affects soil processes such as carbon dioxide (CO_2) emission rates [9], which is an important aspect of biochar use for carbon sequestration in soil.

4. Biochar-soil interaction and soil conditioning mechanism

Soil response to biochar has been shown to be a complex physical, chemical, and biological interaction. Kuzyakov et al. [47] among other authors report that the type and rate of interaction between biochar and soil depend on factors such as feedstock composition, conditions of the pyrolysis process, biochar particle size, soil properties, and local environmental conditions. Also, Mukherjee et al. [48] stated that biochar surface area has aromatic and aliphatic functional groups, which facilitate direct and indirect bonds with soil organic and mineral phases to form complexes in the inner core of biochar material. This complex formation may occur through specific bonding between biochar surface functional groups and soil mineral phase, sorption of soil OM on biochar-mineral phase, or through metal-organic cation bridging. Six et al. [49] in an earlier study showed that specific bonding of soil OM and minerals can inhibit the microbial decomposition of soil organic matter (SOM) and enhance aggregate formation.

To measure the influence of production conditions on biochar-C stability, Bamminger et al. [50] applied maize silage biochars produced through pyrolysis at 600°C and hydrothermal carbonization at 220°C to a forest and an arable soil. They reported that 13–16% of the hydrothermal-produced biochar was mineralized in 8 weeks, and the char exerted a positive priming effect on native SOM. On the other hand, 1.4–3% of the pyrolysis biochar was mineralized and a negative (−24 to −38%) priming effect on native SOM was recorded.

Due to the wide variations in mineralization rates of biochar in different research, biochar-C MRT varies widely in the literature. While Keith and Singh [3] in a 3-month soil-biochar incubation experiment reported MRT of 62–248 years, Murray et al. [51] reported half-life time of between 22 and 1506 years, and Wu et al. [52] reported MRT of 617–2829 years. The majority of differences in observations were attributed to influences of biochar, soil, and environmental properties.

4.1 Biochar-soil texture-soil mineralogy interactions

Kleber et al. [53] stated that clay type, functional groups and their distribution, the concentration and composition of cations and anions, and the polarity of soil compounds are some of the important factors that determine the interactions between OC and clay mineral surfaces in soil. They further highlighted the possible mechanisms of biochar/minerals interactions in soil such as cation bridging, ligand exchange, H bonding, and direct electrostatic interactions through hydrophobic and hydrophilic interactions. Lehmann and Sohi [54] also suggested that biochar-C may be concentrated within soil microaggregates, which supports the proposal of organo-mineral associations to enhance biochar-C stability in soil.

Brodowski et al. [55] reported higher stabilization of biochar-C in soils of higher clay content. Fang et al. [31] also observed the lowest biochar-C mineralization in a clayey Vertisol and higher mineralization in sandy clayey loam Entisol and sandy Inceptisol. They stated that oxides and oxyhydroxide minerals in an Oxisol contributed more to biochar-C stabilization than smectic minerals in the Vertisol. Research results in contrast to these findings have also been reported. Wattel-Koekkoek et al. [56], for example, in their study reported that there was no relationship between OM content in the clay-sized soil fractions and soil clay mineralogy in six kaolinite- and smectite-dominated soils obtained from different countries.

5. Biochar degradation processes

Many research studies have shown that the mechanisms of biochar degradation in soil are similar to those of other OC sources in soil. They categorized biochar degradation mechanisms into biotic and abiotic oxidative and nonoxidative degradation, and loss due to other phenomena. The biotic degradation path involves the breakdown of biochar materials by soil microorganisms, while the abiotic path involves the surface oxidation and bulk oxidation of biochar confirmed by the fact that CO_2 consumption correlates strongly with oxygen consumption during incubation experiments [57].

Another evidence of abiotic oxidative biochar degradation is the report of Bruun et al. [13], which showed that during incubation experiments, there is a lack of lag phase in the release of CO_2, which would be expected if soil microorganisms are inoculated in incubation samples. Investigations on the oxidative degradation of biochar by Nguyen et al. [17] also reported a permanent increase in C loss following temperature increase from 30 to 60°C. The increased biochar mineralization despite temperatures that are unfavorable for soils microorganisms suggests the presence of an abiotic degradation pathway.

Many research studies have shown that biochar addition to soil results in an immediate increased CO_2 emission that lasts for about 14 days after which it decreases exponentially. As is the case with mineral weathering, water availability, which plays a major role in soil processes such as hydration, hydrolysis, dissolution, carbonation, and decarbonation, is also expected to affect biochar weathering in soil and soil biota activities. Rates of these reactions are expected to depend on the nature of the reaction, biochar type and properties, and pedo-climatic conditions. This is demonstrated in the study by Isimikalu et al. [58], which evaluated the effect of soil moisture and temperature on biochar C degradation. In their study, they found that C mineralization declines under elevated moisture and concluded that C losses relating to soil water may be more connected to the leaching of dissolved organic carbon.

In a meta-analysis, Wang et al. [59] discovered that the amount of soil clay, the length of the experiment, the feedstock, and the temperature of the pyrolysis all significantly affect the pace of biochar breakdown. The MRTs of the labile and recalcitrant biochar C pools were calculated to be around 108 days and 556 years, respectively, with pool sizes of 3 and 97%. The findings demonstrated that only a tiny portion of biochar is accessible for degradation and that a significant portion (~97%) directly contributes to long-term carbon sequestration in soil. Additionally, they discovered that the mineralization of soil organic matter (SOM; overall mean: 3.8%, 95% CI = 8.1–0.8%) was modestly delayed by the addition of biochar in comparison with soil without the amendment.

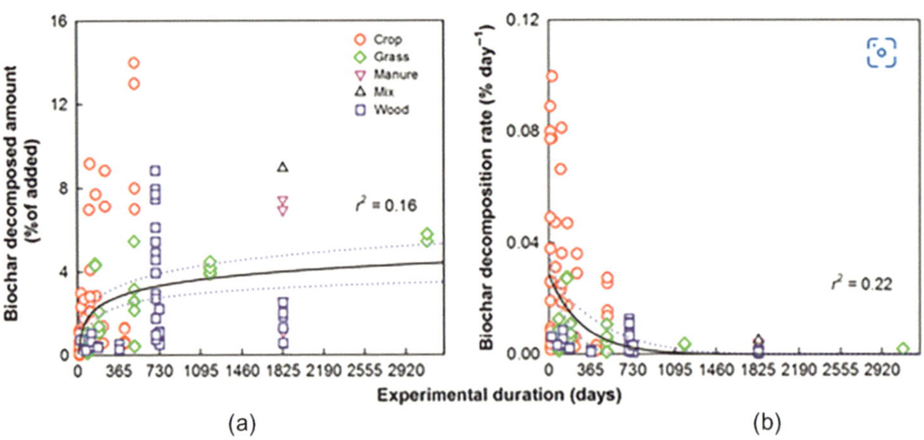

Figure 7.
Relationships between the decomposed amount (A) and rates of decomposition (B) using 128 observations of different feedstock biochar-derived CO_2 from 24 studies with stable (^{13}C) and radioactive (^{14}C) carbon isotopes. The dotted line indicates the 95% confidence band (source: Wang et al., [59]).

The C storage value of biochar materials is commonly assessed by the fraction of biochar-C that remains in soil for >100 years [60]. This proposition is based on the 100-year time horizon used in assessing the global warming potential of GHGs following IPCC [61], which is used in defining permanence in carbon offset. To determine longer-term stability from short-term measurements (research), data are extrapolated to 100 years' duration. **Figure 7** shows biochar decomposition trends in from several studies and feedstock types.

6. Biochar stability testing methods

IBI [60] categorized biochar stability testing methods into three: alpha, beta, and gamma methods based on the measuring techniques and working principles of the systems. Also, Leng et al. [62] categorized and ranked C stability testing methods into three as follows: analysis of biochar-C structure and composition, determination of biochar oxidation resistance, and evaluation of biochar persistence through incubation and modeling. This classification corresponds to alpha, gamma and beta methods, respectively, under IBI [60] categorization.

According to Leng et al. [62], only biochar persistence measurement through incubation and modeling gives the specific duration of biochar-C in soil and thus regarded it as the core method of biochar stability assessment, which serves as the basis of the other two methods. This is because analyzing biochar-C structure, composition, and oxidation resistance only shows a relative stability and not the actual persistence unless they are correlated with incubation and modeling data.

6.1 Alpha biochar stability measurement methods

Alpha methods are the cheapest and are used to execute routine estimations of biochar stability. The methods are time conserving—in the range of hours, and two of these alpha methods are mainly used in scientific research. These are the determination of volatile matter (VM) content and determination of hydrogen to OC (H/Corg) and oxygen to C (O/C ratio) molar ratio [60]. They are regarded as indirect methods of stability measurement and require calibration using a beta or gamma method.

6.2 Beta biochar stability measurement methods

Beta methods of biochar stabilization measurement currently most used in research studies are the laboratory and field-based incubations and to a lesser extent, the field-based chronosequence measurements [60]. Beta methods are applied in combination with modeling in order to estimate biochar loss and stability over a period much longer than the incubation duration. An attribute of these methods is that they directly quantify biochar loss over a certain time period. Using the knowledge gained by the beta techniques, an alpha method can be calibrated to provide a quick tool to estimate biochar stability. The time required to conduct a beta stability test is, however, much longer in comparison with an alpha test, and they consequently cost more.

6.2.1 Laboratory and field incubation studies

Incubation experiments could be executed in a laboratory environment or in natural field conditions. In laboratory incubation, soil samples are incubated in the absence of plant roots. In the field, however, CO_2 emissions may represent C decomposition and root respiration. In order to separate CO_2 sources in this type of studies, isotopic labeling of C is required, which both requires intensive instrumentation and costs that may not be readily available to researchers.

Zimmerman [57] showed that many studies use a simple evaluation that measures the total CO_2 efflux that does not require CO_2 source measurement. In such an assay, biochar-C mineralization is not separated from SOC mineralization, and the priming effect of each component on the other cannot be assessed. A common way of determining C loss from different sources in this type of trial is to deduct C loss under control treatment from losses under amended treatments.

6.2.1.1 Incubation duration

The trend of C mineralization from previous studies shows that C decomposition decreases until it reaches a constant rate 600–700 days after incubation. This, according to Chao et al. [63], possibly indicates that the biochar-C left may have a higher level of stability. Due to this phenomenon, incubation duration is seen as an important factor in biochar stability determination. The effect of this is that longer incubation time results in higher MRT owing to the lower rates of mineralization used for modeling. Generally, C mineralization experiments have lasted from 14 days to 8.5 years in the study by Kuzyakov et al. [64].

Kuzyakov et al. [64] in their 8.5 years' research reported that labile form of biochar-C was mineralized almost completely after about 3.5 years of incubation, and only about 6% of added biochar-C was mineralized in 8.5 years. Leng et al. [62] among other researchers therefore suggest that studies spanning less than 2 years may only reflect the mineralization of the labile component of biochar-C and recommend care in extrapolating C MRT with such data to avoid underestimation. Long duration of experimentation allows a long enough time to discriminate labile and recalcitrant C pools, which facilitates the use of a two-pool model for extrapolation, thereby taking care of the differences in mineralization rates of different OC pools.

6.2.2 Chronosequence measurements

Chronosequence measurements are taken from a sequence of soil samples at varying time intervals starting from the time biochar is applied [60]. Based on the obtained data, the long-term stability of biochar is estimated using a model. A disadvantage of this technique, however, is that results are affected by transport processes such as erosion and leaching. As such, the technique is less commonly used.

6.3 Gamma biochar stability measurement methods

As defined by IBI [60], gamma methods use measurements of molecular properties and chemical composition related to the long-term stability of an OC material. The equipment needed to perform these tests are very expensive, but require a short time to complete. Gamma methods are very reliable and are often used to calibrate alpha and to lesser extent beta methods, which can be used for routine analysis. Examples of gamma stability tests commonly used are different kinds of nuclear magnetic resonance (NMR) spectroscopy, analytical pyrolysis, and a method based on the amount of polycarboxylic acids.

Author details

Theophilus Olufemi Isimikalu
Faculty of Agriculture, Department of Agronomy, University of Ilorin,
Ilorin Kwara State, Nigeria

*Address all correspondence to: olufemith@gmail.com

References

[1] Lal R. Methods and Guidelines for Assessing Sustainable Use of Soil and Water Resources in the Tropics. Columbus, Ohio, Washington, D.C.: Department of Agronomy, Ohio Sate University; Soil Management Support Services, Soil Conservation Service, U.S. Department of Agriculture; 1994

[2] Waters D, Van Zwieten L, Singh BP, Downie A, Cowie L, Lehmann J. Biochar in soil for climate change mitigation and adaptation. In: Singh BP, Cowie AL, Chan KY, editors. Soil Health and Climate Change. Berlin Heidelberg: Springer-Verlag London Ltd.; 2011. pp. 345-368. DOI: 10.1007/978-3-642-20256-8_15

[3] Keith B, Singh BP. Interactive priming of biochar and labile organic matter mineralization in a smectite-rich soil. Environmental Science and Technology. 2011;**45**:9611-9618

[4] Park JH, Lamb D, Paneerselvam P, Choppala G, Bolan N, Chung JW. Role of organic amendments on enhanced bioremediation of heavy metal(loid) contaminated soils. Journal of Hazardous Materials. 2011;**185**:549-574

[5] Woolf D, Amonette JE, Street-Perrott FA, Lehmann J, Joseph S. Sustainable biochar to mitigate global climate change. Nature Communications. 2010;**1**(5):1-9. DOI: 10.1038/ncomms1053

[6] Asai H, Samson B, Stephen H, SongyikhangsuthorK,HommaK,KiyonoY, et al. Biochar Amendment techniques for upland rice production in northern Laos: 1. Soil physical properties, leaf SPAD and grain yield. Field Crops Research. 2009;**111**:81-84. DOI: 10.1016/j.fcr.2008.10.008

[7] Butnan S, Deenik JL, Toomsan B, Antal MJ, Vityakon P. Biochar characteristics and application rates affecting corn growth and properties of soils contrasting in texture and mineralogy. Geoderma. 2015;**237**(238):105-116

[8] Lehmann J. A handful of carbon. Nature. 2007;**447**:143-144

[9] Olaniyan JO, Isimikalu TO, Raji BA, Affinnih KO, Alasinrin SY, Ajala ON. Alasinrin and Olusegun Nathaniel Ajala. An investigation of the effect of biochar application rates on CO_2 emissions in soils under upland rice production in southern Guinea Savannah of Nigeria. Heliyon. 2020;**6**

[10] Junna S, Bingchen W, Gang X, Hongbo S. Effects of wheat straw biochar on carbon mineralization and guidance for large-scale soil quality improvement in the coastal wetland. Ecological Engineering. 2014;**62**:43-47

[11] Zhang A, Bian R, Pan G, Cui L, Hussain Q, Li L, et al. Effects of biochar amendment on soil quality, crop yield and greenhouse gas emission in a Chinese rice paddy: A field study of 2 consecutive rice growing cycles. Field Crops Research. 2012;**127**:153-160. DOI: 10.1016/j.fcr.2011.11.020

[12] Lehmann J, Rondon M. Biochar soil management on highly weathered soils in the humid tropics. In: Uphoff N, editor. Biological Approaches to Sustainable Soil Systems. Boca Raton, FL: CRC Press; 2006. pp. 517-530

[13] Bruun S, Stenberg B, Breland T, Gudmundsson J, Henriksen T, Jensen L, et al. Mpirical predictions of plant material C and N mineralization patterns from near infrared spectroscopy, stepwise chemical digestion and C/N

ratios. Soil Biology and Biochemistry. 2005;**37**(12):2283-2296. DOI: 10.1016/j.soilbio.2005.04.006

[14] Ogawa M, Okimori Y, Takahashi F. Carbon sequestration by carbonization of biomass and forestation: Three case studies. Mitigation and Adaptation Strategies for Global Change. 2006;**11**: 429-444

[15] Downie A, Crosky A, Munroe P. Physical properties of biochar. In: Lehmann J, Joseph S, editors. Biochar for Environmental Management, Science and Technology. London: Earthscan; 2009. pp. 13-32

[16] Schmidt MW, Noack AG. Black carbon in soils and sediments: Analysis, distribution, implications, and current challenges. Global Biogeochemical Cycles. 2000;**14**:777-794

[17] Nguyen BT, Lehmann J, Hockaday WC, Joseph S, Masiello CA. Temperature sensitivity of black carbon decomposition and oxidation. Environmental Science & Technology. 2010;**44**(9):3324-3331

[18] Joseph S, Taylor P, Rezende F, Draper K, Cowie A. The properties of fresh & aged biochar. The Properties of Fresh and Aged Biochar - Biochar for Sustainable Soils. 2022. Available from: https://biochar.international/guides/properties-fresh-aged-biochar [Accessed: May 23, 2022]

[19] Lehman J, Joseph S. Biochar for Environmental Management: Science and Technology. London & Sterling, VA: Earthscan; 2009

[20] Steiner C. Biochar carbon sequestration. Athens, GA 30602. In: Biorefining and Carbon Cycling Program. USA: University of Georgia; 2009

[21] Qadeer S, Anjum M, Khalid A, Waqas M, Batool A, Mahmood T. A dialogue on perspectives of biochar applications and its environmental risks. Water, Air & Soil Pollution. 2017;**228**(8):1-26. DOI: 10.1007/s11270-017-3428-z

[22] Rondon M, Lehmann J, Ramirez J, Hurtado M. Biological nitrogen fixation by common beans (*Phaseolus vulgaris* L.) increases with bio-char additions. Biology and Fertility of Soils. 2007;**43**:699-708

[23] Husk B, Major J. Commercial Scale Agricultural Biochar Field Trial in Québec, Canada, Over Two Years: Effects of Biochar on Soil Fertility, Biology, Crop Productivity and Quality. Quebec, Canada: BlueLeaf Inc.; 2010. Available from: http://www.blue-leaf.ca/main-en/report_a3.php

[24] Bierderman LA, Harpole WS. Biochar and its effects on plant productivity and nutrient cycling: A meta-analysis. GCB Bioenergy. 2013;**5**:202-214

[25] Lehmann J, Gaunt J, Rondon M. Bio-char sequestration in terrestrial ecosystems—A review. Mitigation and Adaptation Strategies for Global Change. 2006;**11**:395-419

[26] Jeffery S, Verheijen F, Van der Veldea M, Bastos AC. A quantitative review of the effects of biochar application to soils on crop productivity using meta-analysis. Agriculture, Ecosystems and Environment. 2011;**144**:175-187

[27] Spokas KA. Review of the stability of biochar in soils: Predictability of O:C molar ratios. Carbon Management. 2009;**1**(2):289-303

[28] Leng RA, Preston TR, Inthapanya S. Biochar reduces enteric methane and improves growth and feed conversion in local "Yellow" cattle fed cassava root chips and fresh cassava foliage. Livestock Research for Rural Development.

2012;**24**(11). Available from: http://www.lrrd.org/lrrd24/11/leng24199.htm

[29] Wang J, Dokohely ME, Xiong Z, Kuzyakov Y. Contrasting effects of aged and fresh biochars on glucose-induced priming and microbial activities in paddy soil. Journal of Soils and Sediments. 2016a;**16**:191-203

[30] Bruun EW, Petersen C, Strobel BW, Hauggaard-Nielsen H. Nitrogen and carbon leaching in repacked sandy soil with added fine particulate biochar. Soil Science Society of America Journal. 2012;**76**(4):11-42. DOI: 10.2136/sssaj2011.0101

[31] Fang Y, Singh B, Singh BP, Krull E. Biochar carbon stability in four contrasting soils. European Journal of Soil Science. 2015;**65**:60-71. DOI: 10.1111/ejss.12094

[32] Zhang H, Lin K, Wang H, Gan J. Effect of Pinus radiata derived biochars on soil sorption and desorption of phenanthrene. Environmental Pollution. 2010;**158**(9):2821-2825

[33] Ahmad M, Rajapaksha AU, Lim JE, Zhang M, Bolan N, Mohan D, et al. Biochar as a sorbent for contaminant management in soil and water: A review. Chemosphere. 2014;**99**:19-33

[34] Barrow CJ. Biochar: Potential for countering land degradation and for improving agriculture. Applied Geography. 2012;**34**:21-28

[35] Kookana RS, Sarmah AK, Van Zwieten L, Krull E, Singh B. Chapter three—Biochar application to soil: Agronomic and environmental benefits and unintended consequences. In: Donald SL, editor. Advances in Agronomy. Vol. Vol. 112. New York: Academic; 2011. pp. 103-143

[36] Laird D, Fleming P, Wang B, Horton R, Karlen D. Biochar impact on nutrient leaching from a Midwestern agricultural soil. Geoderma. 2010;**158**(3-4):436-442

[37] Sahoo S, Borpatragohain B, Rai A. Sorption, desorption, and degradation of pesticides in biochar amended agricultural soils. International Journal of Chemical Studies. 2019;7(3):5106-5111

[38] Kong LL, Liu WT, Zhou QX. Biochar: An effective amendment for remediating contaminated soil. Reviews of Environmental Contamination and Toxicology. 2014;**228**:83-99

[39] Vangronsveld J, Herzig R, Weyens N, Boulet J, Adriaensen K, Ruttens A, et al. Phytoremediation of contaminated soils and groundwater: Lessons from the field. Environmental Science and Pollution Research. 2009;**16**(7):765-794

[40] Khorram MS, Zhang Q, Lin D, Zheng Y, Fang H, Yu Y. Biochar: A review of its impact on pesticide behavior in soil environments and its potential applications. Journal of Environmental Science. 2016;**44**:269-279. DOI: 10.1016/j.jes.2015.12.027. PMID: 27266324.

[41] Li HB, Xiao L, Evando B, et al. Mechanisms of metal sorption by biochar: Biochar characteristics and modifications. Chemosphere. 2017;**178**:466-478

[42] Bai X, Li Z, Zhang Y, Ni J, Wang X, Zhou X. Recovery of ammonium in urine by biochar derived from faecal sludge and its application as soil conditioner. Waste and Biomass Valorization. 2017:1-10

[43] Han Y, Cao X, Ouyang X, Sohi SP, Chen J. Adsorption kinetics of magnetic biochar derived from peanut hull on removal of Cr (VI) from aqueous solution: Effects of production conditions and particle size. Chemosphere. 2016;**145**:336-341

[44] Lehmann J, da Silva JP, Steiner C, Nehls T, Zech W, Glaser B. Nutrient availability and leaching in an archaeological anthrosol and a ferralsol of the Central Amazon basin: Fertilizer, manure and charcoal amendments. Plant and Soil. 2003;**249**:343-357

[45] Steiner C, Teixeira WG, Lehmann J, Nehls T, de Macêdo JLV, Blum WEH, et al. Long term effects of manure, charcoal and mineral fertilization on crop production and fertility on a highly weathered Central Amazonian upland soil. Plant and Soil. 2007;**291**:275-290

[46] Major J, Lehmann J, Rondon M, Goodale C. Fate of soil-applied black carbon: Downward migration, leaching and soil respiration. Global Change Biology. 2010;**16**:1366-1379

[47] Kuzyakov Y, Subbotina I, Chen H, Bogomolova I, Xu X. Black carbon decomposition and incorporation into soil microbial biomass estimated by 14C labeling. Soil Biology and Biochemistry. 2009;**41**(2):210-219. DOI: 10.1016/j.soilbio.2008.10.016

[48] Mukherjee A, Zimmerman AR, Harris WG. Surface chemistry variations among a series of laboratory-produced biochars. Geoderma. 2011;**163**:247-255

[49] Six J, Conant RT, Paul EA, Paustian K. Stabilization mechanisms of soil organic matter: Implications for C-saturation of soils. Plant and Soil. 2002;**241**:155-176

[50] Bamminger C, Marschner B, J"uschke, E. An incubation study on the stability and biological effects of pyrogenic and hydrothermal biochar in two soils. European Journal of Soil Science. 2014;**65**(1):72-82

[51] Murray J, Keith A, Singh B. The stability of low- and high-ash biochars in acidic soils of contrasting mineralogy. Soil Biology and Biochemistry. 2015;**89**:217-225. DOI: 10.1016/j.soilbio.2015.07.014

[52] Wu H, Zeng G, Liang J, Chen J, Xu J, Dai J, et al. Responses of bacterial community and functional marker genes of nitrogen cycling to biochar, compost and combined amendments in soil. Applied Microbiology and Biotechnology. 2016;**100**(19):8583-8591. DOI: 10.1007/s00253-016-7614-5

[53] Kleber M, Sollins P, Sutton R. A conceptual model of organo-mineral interactions in soils: Self-assembly of organic molecular fragments into zonal structures on mineral surfaces. Biogeochemistry. 2007;**85**:9-24. DOI: 10.1007/s10533-007-9103-5

[54] Lehmann J, Sohi S. Comment on 'Fire-Derived Charcoal Causes Loss of Forest Humus. Science. 2008;**321**(5894):1295-1298

[55] Brodowski S, John B, Flessa H, Amelung W. Aggregate-occluded black carbon in soil. European Journal of Soil Science. 2006;**57**:539-546

[56] Wattel-Koekkoek EJ, van Genuchten PP, Buurman P, Van Lagen B. Amount and composition of clay-associated soil organic matter in a range of kaolinitic and smectitic soils. Geoderma. 2001;**99**:27-49

[57] Zimmerman AR. Abiotic and microbial oxidation of laboratory-produced black carbon (biochar). Environmental Science and Technology. 2010;**44**:1295-1301

[58] Isimikalu TO, Olaniyan JO, Alasinrin SY, Wahab AA, Salifu UM. Influence of environmental variables on carbon mineralization in diverse soil management systems under maize cropping in Kwara state Nigeria. Bulgarian Journal of Soil Science. 2022;7(1):34-48

[59] Wang J, Xiong X, Kuzyakov Y. Biochar stability in soil: Meta-analysis of decomposition and priming effects. GCB Bioenergy. 2016b;**8**(3):512-523. DOI: 10.1111/gcbb.12266

[60] International Biochar Initiative (IBI). Standardized Product Definition and Product Testing Guidelines for Biochar That is Used in Soil. Biochar-international; 2015. Available from: https://www.biochar-international.org/wp-content/uploads/2018/04/IBI_Biochar_Standards_V2.1_Final.pdf

[61] Intergovernmental Panel on Climate Change (IPCC). Summary for policymakers. In: Parry ML, Canziani O, Palutikof JP, van der Linden PJ, Hanson CE, editors. (IPCC), Climate Change 2007: Impacts, Adaptation and Vulnerability. Contribution of Working Group II to the Fourth Assessment Report of the Intergovernmental Panel on Climate Change. Cambridge, UK: Cambridge University Press; 2007. pp. 7-22

[62] Leng L, Xu X, Wei L, Fan L, Huang H, Li J, et al. Biochar stability assessment by incubation and modelling: Methods, drawbacks and recommendations. Science of the Total Environment. 2019;**664**:11-23

[63] Chao L, Zhang WD, Wang SL. Understanding the dominant controls on biochar decomposition using boosted regression trees. European Journal of Soil Science. 2018;**69**:512-520. DOI: 10.1111/ejss.12534

[64] Kuzyakov Y, Bogomolova I, Glaser B. Biochar stability in soil: Decomposition during eight years and transformation as assessed by compound-specific 14C analysis. Soil Biology and Biochemistry. 2014;**70**:229-236. DOI: 10.1016/j.soilbio.2013.12.021

Chapter 7

Aged Biochar for the Remediation of Heavy Metal Contaminated Soil: Analysis through an Experimental Case the Physicochemical Property Changes of Field Aging Biochar and Its Effects on the Immobilization Mechanism for Heavy Metal

Run-Hua Zhang, Lin-Fang Shi, Zhi-Guo Li, Guo-Lin Zhou, Yan-Lan Xie, Xing-Xue Huang, An-Hua Ye and Chu-Fa Lin

Abstract

Heavy metal inducing contamination soil has become a serious concern. Contaminated soil can cause physiochemical and biochemical changes into soil and the plants. Thus, the plant growth and the yield were affected. In additionally, that ultimately leads to the problem of food security and human health. In recent years, many kinds of ways were used for the remediation of heavy metal contaminated soil, such as isolation, phytoremediation, immobilization, extraction, and soil washing. As a new carbon-rich material, biochar has been applied to the remediation of heavy metal pollution in soil. As biochar is rich with porous structure, high cation exchange capacity, pH value, and surface function, it has become an adsorbent for soil heavy metal remediation. While, with time, the capacity of biochar to immobilize the heavy metals may be modified as the sorption sites may get occupied with native soil organic matter or competing contaminant, etc. And that the physicochemical properties of biochar changed significantly during field aging. Thus, to clarify the mechanism of field-aged biochar for the remediation of heavy metal contaminated soil, we analysis, through an experimental case, the physicochemical property changes of field-aged biochar and its effects on the immobilization mechanism for heavy metal.

Keywords: field-aged biochar, heavy metal, contaminated soil, remediation, adsorption mechanism

1. Introduction

Heavy metal pollution such as cadmium (Cd), lead (Pb), arsenic (As), and chromium (Cr) is characterized by high toxicity and biological enrichment in soil. Heavy metal contamination in soil poses a potential threat to human health and food security [1–3]. In addition, different types and concentrations of heavy metals often exist at the same time, making the polluted soil environment more complex. Therefore, it is urgent to seek remediation technology to remove heavy metals from contaminated soil. Fortunately, biochar, as an economical and efficient adsorption material, has opened up a new way for the immobilization of heavy metals [4, 5].

Biochar is an organic and pyrogenic material produced by pyrolysis of animal or plant-based feedstocks under oxygen limited conditions [6, 7]. In the pyrolysis process, the fatty carbon chain (c) in the raw material finally forms aromatic C, which is considered as fixed C and can exist in the soil for hundreds or thousands of years. The intermediate products between the fixed C surfaces are called active components. When biochar was applied to the soil, they are easily decomposed or weathered and oxidized by soil microorganisms, thus reducing the content of biochar in the soil.

At the present time, increasingly studies show that the properties of biochar will change significantly due to the influence of various environmental factors. This was identified as biochar field aging [8–10]. As far as we know, biochar mainly fixes heavy metals in soil or water through precipitation, surface complexation, cation exchange, electrostatic attraction, and cation-π interaction [11]. However, the field aging of biochar will cause the interaction between biochar and organic matter, minerals, and dissolved organic matter in soil [12]. These resulted in the change of specific surface area (SSA), cation exchange capacity (CEC), element composition, acidity, and O-containing functional group of biochar. These will further affect the ability to absorb heavy metals of biochar and its field application performance. The research showed that most biochar increased the content of O-containing functional groups after biochar artificial aging and enhanced the adsorption capacity of heavy metal [13]. However, Lin reported that water washing aging biochar and acidification treatment had a negative impact on the biochar's aluminum toxicity reduction and the improvement of acidic soil [14]. These contradictory results can be explained by many factors. To sum up, there is no consensus on whether the adsorption capacity of biochar for heavy metals changes with biochar field aging and how it changes with field aging.

Although the related research on the aging of biochar is booming, most of the research of the biochar aging is carried out by the simulated aging method under the controllable laboratory conditions [15]. It is not under the field conditions. Actually, there are great differences between artificial aging and field aging of biochar. That is, the changes of physical and chemical properties of biochar under field aging are different from under the artificial aging. So far, there are few studies on the characterization of physical and chemical properties of aged biochar (ABC) extracted from soil [16, 17]. Not to mention the influence of physical and chemical properties of field-aged biochar on heavy metal adsorption. Therefore, we studied the characteristics of field-aged biochar for the remediation of heavy metal contaminated soil by analysis through an experimental case the physicochemical property changes of field aging biochar and its effects on the immobilization mechanism for heavy metal.

2. Biochar

2.1 Preparation of biochar

Biochar is carbon-rich byproduct of pyrolyzing material (feedstocks) at high temperatures and low oxygen levels. It has porous structure, larger surface area, ample surface functional groups, and good CEC [17]. These unique physiochemical properties and distinct role of biochar give it to improve the soil biological and physiochemical properties, carbon sequestration, and remediation of heavy metals in soil. Biochar's physical and chemical residences rely on the categories of feedstocks and pyrolysis situations.

The feedstocks of biochar can classify into virgin sources, residues, and municipal solid wastes. The virgin sources include forest sources and oilseed/cereal crops. The residues are timber residues, agricultural residues, and wastes of livestock residues [18]. The wood, wood pellets, tea trash, coffee hulls, biodegradable sewage sludge, wheat straw, rice straw, macro and microalgae, maize fodder are unquestionable fantastic potentials as pyrolysis feedstocks so far.

Up to now, the pyrolysis situations of biochar are conventional pyrolysis, microwave-assisted pyrolysis, impregnation pyrolysis, co-precipitation, hydrothermal carbonization, etc. The conventional pyrolysis is the standard heating system in which heat is transferred from an external supply to the biomass through conductivity, radiation, and convection. It is inefficient and slow and dependent on the biomass thermal conductivity as well as the system's convection present day [17]. Microwave-assisted pyrolysis entails strength conversion rather than mere heating. At some point of this approach, electromagnetic energy is regenerated into thermal energy by using dielectric heating. And the temperature of the feedstock at its center is higher than the temperature of the components. This system accelerates chemical reactions and shortens their duration, saving energy surface and time [18]. This approach is among the foremost promising strategies of fast and improving chemical reactions. The magnetic biochar was obtained from the impregnation-pyrolysis system. It has been studied and used extensively at present. Firstly, the biomass was impregnated in a solution that has transition metal salt and without the solvent. Then the dried residue is pyrolyzed in inert or anoxic atmosphere to get the magnetic biochar [19]. The synthesis of magnetic biochar by co-precipitation was more sophisticated than the impregnation-pyrolysis methodology. However, it is more manageable, allowing the magnetic medium to be stably adhered to the biochar matrix. In co-precipitation system, the biochar was firstly dispersed into a solution containing transition metals with the pH range of 9–11 for a while at a given temperature. And then, the supernatant of the solution was removed. Next, the residue was washed and dried at room temperature, and the magnetic biochar was obtained [20]. Different from the pyrolysis, the hydrothermal carbonization system is reacted at lower temperature with many kinds of biomass in a metal particle solution. These reactions are relatively milder reaction situations than the abovementioned methods [21].

2.2 Biochar properties for remediation of contamination soil

Biochar displays a tough morphological surface with honeycomb under the microscope. The inherent micropores engender biochar a comparatively excessive intrapore volume and low envelope density. Biochar consists frequently of amorphous, aromatic carbon and possesses over abundant oxygen containing surface functional groups (C=O, –COOH, and –OH) and a disorderly stacked graphene sheet shape. Biochar largely has negatively charged surfaces that will increase the surface assimilation capability of cation species [22]. Thus, it plays a crucial position in improving nutrient retention in soil. The biochar capabilities and applications largely depend on their structural and physicochemical properties. So, it is vital to represent the structural and physiochemical properties of biochar before its use. Different feedstocks and pyrolysis method situations contribute to different structural and physical traits of biochar, including structural complexity, extent, porosity, particle size distribution, density, and mechanical strength [23]. During pyrolysis, biomass feedstock undergoes a variety of physical, chemical, and molecular changes. Pyrolysis circumstance and feedstock type considerably affect the structural and physicochemical traits of the ensuing biochar product. Such as the aromaticity of biochar usually increases, whereas the surface practicality decreases due to the fact that pyrolysis temperature is elevated. It is often an outcome of the innovative losses of aliphatic C–H, olefinic C=C, carbonyl, carboxyl, and hydroxyl groups at a higher temperature [24].

Biochar has been explored for mitigating soil heavy metal contamination. Many reported analysis indicates that biochar is capable of efficaciously immobilizing heavy metallic elements in soil and sorbing heavy metallic cations from water. Thus, biochar serves as a promising amendment for decreasing the eco-toxicity of heavy metal contaminated soils [24]. Biochar's high sorption capability together with high surface area applicable to immobilize contaminants. It is assumed that the contaminants will not be freed into the matrix until the biochar is degraded [25]. Biochar can also immobilize heavy metals through reduction. The oxygen functional group on the surface of biochar reduces the hexavalent chromium (Cr (VI)) to the trivalent (Cr (III)) via influencing its redox response. Cr (III) is usually nontoxic and tightly attached to soil particles, whereas Cr (VI) is very poisonous and mobiles [26].

3. Heavy metals contaminated soil

3.1 Sources of heavy metal contamination to soil and heavy metal toxic effects

The metals with relatively higher specific density compared to water (>5 g/cm^3) are considered as heavy metals [27]. Naturally, heavy metals are found over the earth's crust in trace amount, so also considered as trace metals [28]. In past decades, the amounts of heavy metals are found to be increasing dramatically besides their natural occurrence. In the meantime, public health concerns due to the toxicity of these metals increased worldwide. Heavy metals such as lead (Pb), mercury (Hg), arsenic (As), lithium (Li), titanium (Ti), antimony (Sb), cadmium (Cd), chromium (Cr) are the most toxic metals with highly detrimental effects on human and animal organs and plant system [29]. Exposure to these metals beyond their permissible limit has life-threatening effects on biological world. Hg, Pb, As, Cd, and Cr are classified as top-priority toxic metal pollutants of significant concern.

Normally, the soil parent material itself contains most of the heavy metals in trace amount, which is not bioavailable. Rather anthropogenically added heavy metals have high bioavailability. There are different identified sources adding heavy metal contamination to soil, including agricultural practices, industrial and domestic effluents, natural and atmospheric sources. Sufficient Cd, Zn, Cr, and Ni will be generated due to wastewater, industrial wastes, and deposited sludge released from industrial applications such as milling, electroplating, etching, tanning, textile and dye, metal casting and smelting, wood preservation and processing, photography, pharmaceutical printed circuit board (PCB), glass processing, manufacturing, etc. [30].

Most of the mismanaged anthropogenic activities are responsible for rapid contamination of soil with various toxic metals. Phosphate and nitrate fertilizers also contain variable amount of Cr, Cd, Ni, Pb, Hg, and Zn in which Cd is of main concern due to its accumulation in plant leaves. Pesticides, fungicides extensively used in agriculture, horticulture, and animal husbandries are the mixture of different compounds containing metals such as Cu, Hg, Fe, Pb, Zn. All these practices contribute to elevate the background concentration of heavy metals in soil. The agriculture practices, such as cattle manure, pig manure, and livestock manure, can add a large amount of Cu, Mn, Cd, Cr, Pb, and Zn to the soil. These ascribed to the compounds of livestock were containing various metals as the animal feed in the pig and poultry industry. And the feces containing metals of these animals were reused for land application. In the long run, these heavy metals will cause a large accumulation in the soil [31].

3.2 Heavy metal removal mechanism of biochar in soil

The heavy metal contamination soils and its management are a challenging issue. Because it is hard to mineralize them into other forms and their persistence. The remediation mechanism of biochar is different for different heavy metal pollutants. And for the same heavy metal ion, the adsorption mechanism is different when the biochar is different [32]. The removal mechanism of biochar acting on the bioavailable fraction of soil heavy metals is as follows: complexation, physical sorption, electrostatic attraction, ion exchange precipitation, etc. That can be able to reduce also their leachability [33]. Biochar rich in oxygen containing acidic functional groups (phenolic, carbonyl, lactonic, carboxylic, phenolic, and hydroxyl) plays significantly important role in binding (complexation) of heavy metals and metalloids onto the biochar surface as well as inner pores. Physical adsorption involves the removal of heavy metals by diffusion of metal ions into the pores of sorbent. Since biochar is the carbon material with a well-distributed pore networks including micropores (<2 nm), mesopores (2–50 nm), and macropores (>50 nm) [34]. Biochar surface is negatively charged. Heavy metal immobilization via electrostatic attractions takes place between metal ions and biochar's surface charge. This is the electrostatic attraction. Ion exchange from soil matrix to the surface of biochar is another method of metal fixation by biochar. The size of functional groups and metal species on the surface of biochar are the most important factors affecting the residual efficiency of heavy metals in the ion exchange process [35]. Precipitation is considered as the most common accountable mechanism for heavy metal immobilization by biochar. During the sorption process formation of solid(s), either in solution or on a surface is known as precipitation.

4. Experimental case study: Materials and methods

4.1 Field aging of biochar and separation of aged biochar particles

The work of the field aging of biochar was carried out in Wuhan City, Hubei Province (30°28′N, 114°25′E). The biochar used in this work was got from Zhengzhou

Lishe Environmental Protection Co., Ltd. in China. The biochar was made from the 3–5 mm corn straw pyrolysis at 500°C. Using the abovementioned biochar as raw material, the biochar field aging test was started in 2015. The biochar was added to the soil at a ratio of 1% (w/w) as biochar treatment, and the soil without biochar addition was as control (CK). Seven years later, we got the field-aged biochar by manually separating the biochar particles with a diameter greater than 3 mm from the soil by tweezers. We planted vegetables in the soil with or without biochar as usual during the past 7 years. The field-aged biochar is carefully washed with deionized water to remove soil particles attached to its surface and then dried at 35° C for 8 hours to eliminate moisture. The field-aged biochar (ABC) was obtained. Comparatively, the biochar purchased from Lishe Environmental Protection Company is relatively named fresh biochar (FBC). The appearance of soil, FBC, and ABC is shown in **Figure 1**.

4.2 Biochar characterization

4.2.1 pH and EC of biochar

We examined the EC and pH values of the biochar samples by deionized water at the ratio of 1: 5 (W/V). Conductivity meter (DDS-307A, Rex, China) and PH meter (phs-3c, Rex, China) were equipped respectively.

4.2.2 Pore diameter and specific surface area (SSA) of biochar

A scanning electron microscopy (SEM, Tianmei, SU8010, China) was used for observing the surface morphology of biochar. Before imaging, biochar samples were sprayed with gold for 2 minutes to improve conductivity and imaging quality of the biochar. The pore diameter and the SSA were analyzed by a pore size analyzer and an automated surface area (Mike ASAP2020, USA). About 0.5 g of each biochar sample was degassed at 125° C for 3 hours, and then SSA was determined using Brunauer–Emmett–Teller (BET) equation according to N_2 adsorption/desorption data.

Figure 1.
*Appearances of FBC (a), ABC (d), original soil (b), the aged biochar-soil mixture with soil (c), and the particle size of FBC (e) and ABC (f). **Note:** FBC: Fresh biochar; ABC: Field-aged biochar.*

4.2.3 The crystal structures and functional groups of biochar

X-ray powder diffractometer (XRD, Brooke, D8 ADVANCE) equipped with Cu Kα radiation (λ = 1.5406 Å) was used for determining the crystal structures of biochar samples. The XRD pattern was acquired at 0.02° step size, 5°/min scanning speed, and in the 2θ range of 5 ~ 90°. Fourier transform infrared spectroscopy (FTIR) (Thermo Fisher, Nicolet in10) and KBr were used for characterizing the functional groups on the surface of biochar. The spectral regions from 4000 to 400 cm^{-1} were recorded at a resolution of 2 cm^{-1}.

4.2.4 The elements contents of C, N, O, H, and S of biochar

The Elemental Analyzer (EA, Vario el cube) was used for analyzing the contents of C, N, O, H, and S elements in FBC and ABC samples with argon as a carrier gas. X-ray photoelectron spectroscopy (XPS, thermo escalab250xl, USA) was used to study the combined states of major elements in biochar. The elemental binding energy was corrected to C1s (284.8 eV) obtained in the experiment.

4.3 Batch adsorption experiment

4.3.1 Adsorption kinetics

In this experiment, 20 mg/L Cd^{2+} and Pb^{2+} solutions were used for subsequent adsorption experiments. Using 0.01 mol/l $NaNO_3$ solution as background electrolyte, dissolve $CdCl_2$ and $PbCl_2$ respectively to prepare 1 g/L Cd^{2+} and Pb^{2+} stock solutions, respectively. Then we adjust the initial pH of the stock solution to 5.0 ± 0.2 with 0.1 mol/L NaOH or HCl solution. The adsorption kinetics experiment was carried out at room temperature. About 0.15 g of ABC or FBC respectively was added to 150 ml of 20 mg/L Cd^{2+} and Pb^{2+} solution, shaken well at 180 RPM/min. The collection time intervals of all samples were 0.083, 0.25, 0.5, 1, 2, 4, 6, 8, 12, and 24 h, respectively. There were three replicates per treatment. All samples were filtered through a 0.45 μm membrane in the adsorption experiments. Inductively coupled plasma atomic emission spectrometry (ICP-AES) (Icpe-9000, Shimadzu Corporation, Japan) was used to determine the concentrations of Pb^{2+} and Cd^{2+} and the changes of substituted Na^+, Ca^{2+}, K^+, and Mg^{2+}.

The adsorption capacities of biochar for Cd^{2+} and Pb^{2+} were calculated according to the formula (1):

$$Q_e = \frac{(C_0 - C_e)V}{m} \tag{1}$$

C_e(mg/L): the remaining concentration of $Pb^{2+}$and Cd^{2+} in the solution; C_0 (mg/L): the initial concentration of $Pb^{2+}$and Cd^{2+} in the solution; V (mL): the volume of heavy metal solution; m (g): the dosage of biochar.

Two different models were used to fit the adsorption kinetic data. The formula is as follows:

$$Q_t = Q_e\left(1 - e^{-k_1 t}\right) \tag{2}$$

$$Q_t = \frac{Q_e^2 k_2 t}{1 + Q_e k_2 t} \tag{3}$$

Q_t: the adsorption capacity of Cd^{2+} or Pb^{2+} on biochar at time t; Q_e (mg/g): the adsorption capacity of Cd^{2+} or Pb^{2+} on biochar at time equilibrium; t (h) is the adsorption time; k_1 (h^{-1}): the rate constant of the pseudo-first-order kinetic equation; k_2 (g/mg/h) the rate constant of pseudo-second-order kinetic equation.

4.3.2 Biochar isotherms adsorption

The adsorption isotherms were conducted as follows: add 0.02 g of biochar to 20 ml of Cd^{2+} and Pb^{2+} Solution for adsorption isotherm, and the initial concentration is 5–120 mg/L. Equilibrium adsorption was performed at room temperature for 24 h. The rest of the operation is the same as the adsorption kinetics.

The adsorption isotherms was disclosed using Langmuir and Freundlich models. The equations are listed as follows:

$$Q_e = \frac{Q_m b C_e}{1 + bC_e} \tag{4}$$

$$R_L = 1/(1 + bC_0) \tag{5}$$

$$Q_e = K_f C_e^{1/n} \tag{6}$$

Q_e (mg/g): the amounts of Cd^{2+} or Pb^{2+} adsorbed on biochar; Q_m (mg/g): the maximum saturated adsorption capacity of biochar; C_0 (mg/L): the initial concentration of Cd^{2+} or Pb^{2+}; C_e (mg/L): the equilibrium concentration of Cd^{2+} or Pb^{2+}; b (L/mg) and K_f ((mg/g) $(mg/L)^{-n}$): the corresponding constants of Langmuir and Freundlich; n: the Freundlich constant related to the surface site heterogeneity; R_L: the dimensionless constant separation factor.

4.3.3 Biochar saturated adsorption

Biochar saturation adsorption experiments were executed on the basis of adsorption kinetics. About 0.5 g biochar and 100 mL 100 mg/L Cd^{2+} or Pb^{2+} solution were mixed and shaken for 24 h at180 rpm/min at 25°C. The mixture liquid was filtered by 0.45 μm acetate membrane. ICP-AES was used for determining the concentrations of Cd^{2+} and Pb^{2+} in the mixed solution before and after adsorption by biochar.

4.3.4 Toxicity characteristic leaching procedure (TCLP)

The TCLP test was with the solution. The extraction solution was obtained by dissolving glacial acetic acid with 5.7 mL into the deionized water to a total volume of 100 mL. And the solution's initial pH was adjusted at 2.9 ± 0.05. Next, the 0.25 g of biochar was added into the extraction solution with 10 ml and shaken the solution at 180 rpm/min for 18 h at25 °C. Then, the Cd^{2+} and Pb^{2+} contents in the extraction solution were analyzed using ICP-AES after being filtered through a 0.45 μm acetate membrane.

$$TCLP\text{–}Cd/_{Pb} = \frac{q_1}{q_2} \times 100\% \tag{7}$$

q_1 (mg/g): the contents of Cd^{2+} and Pb^{2+} in the extraction solution of TCLP; q_2 (mg/g): the biochar's total saturated sorption of Cd^{2+} and Pb^{2+}.

4.3.5 Statistical analysis

The statistical analysis in the paper is using the SPSS23.0 software (SPSS Inc., USA). Data are presented as mean ± standard deviation. One-way analysis of variance (ANOVA) with a least significant difference (LSD) at a 0.05 significance level was used for the standard analysis. XPS and XRD data were analyzed with Advantage and Jade6.5 software, respectively. Origin Pro 2018 (OriginLab, USA) and Microsoft Excel 2016 (Microsoft Corporation, USA) were used for drawing the figures and tables.

5. Experimental case study: results and discussion

5.1 The capabilities of field-aged biochar

5.1.1 The microstructure of field-aged biochar

Compared with FBC, the surface of ABC became smoother after 7 years of application in the soil as show in **Figure 2(c)** and **(d)**. And there were no small particles attached on its surface. However, the surface of FBC was rough and has some attached aggregates shown in **Figure 2(a)** and **(b)**. These SEM results illustrated that the small particles attached on FBC surface may be labile fractions or soluble substances (e.g., $CaCO_3$ as shown in **Figure 3**) from the pyrolysis processes of corn straw [36].

5.1.2 The surface morphology of field-aged biochar

The SSA data demonstrated that, compared with the FBC, the ABC's pore volume, the SSA, and size were 0.02 ml/g, 8.32 m^2/g, and 9.62 nm. While the FBCs were 0.0017 ml/g, 2.98 m^2/g, and 2.32 nm, respectively (**Table 1**). These demonstrated that the SSA, pore volume, and pore size of ABC increased by 179%, 1076%, and 314% compared with FBC.

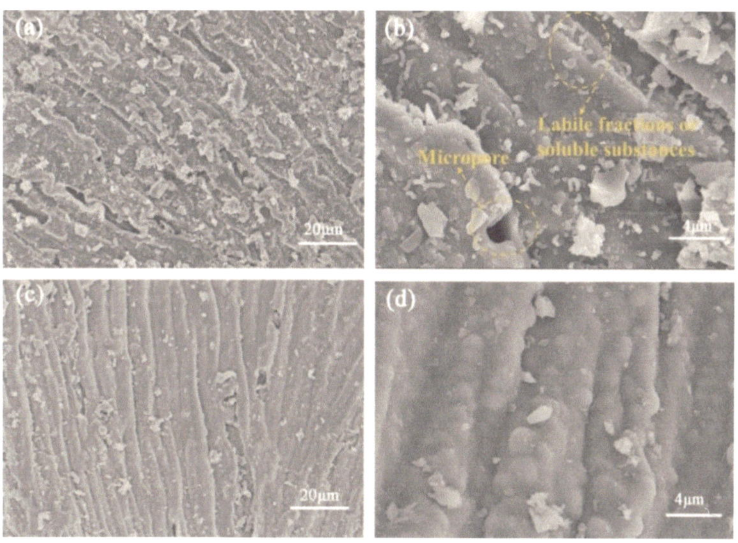

Figure 2.
Scanning electron micrographs of FBC (a, b) and ABC (c, d).

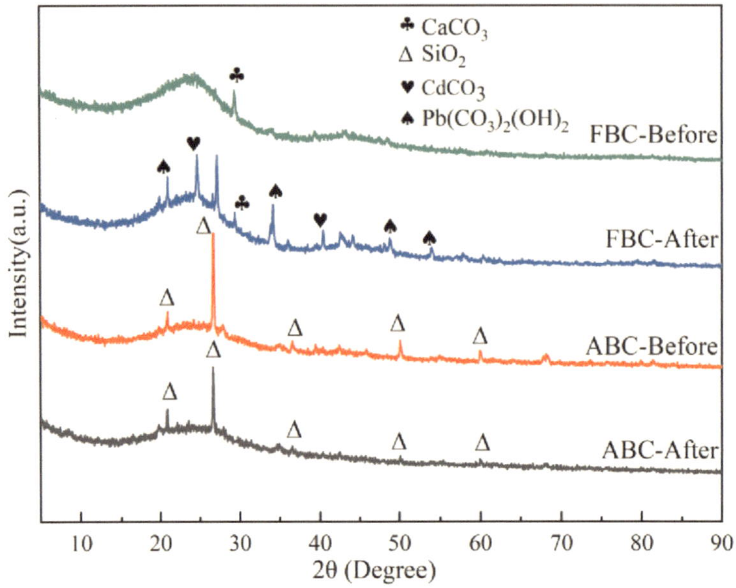

Figure 3.
XRD patterns of FBC and ABC before and after adsorption of heavy metals.

5.2 The chemical properties of field-aged biochar

5.2.1 Potential of hydrogen (pH) of field-aged biochar

The properties of biochar are presented in **Table 1**. The pH of ABC was 5.83 as acidic while FBC was 8.87 displayed alkaline. This demonstrated that biochar in field aging process remarkably decreased the pH.

5.2.2 The elemental analysis of field-aged biochar

Table 2 shows the elemental contents of ABC and FBC. The C content in ABC was 46.89%, deceased considerably compared with the content in FBC, which is 81.01. The C content in ABC was 21.26% while was 12.47% in FBC. The atomic ratios, including H:C O:C and (N+O+S):C, increased notably with aging, suggesting that ABC was highly oxidized and exhibited lower aromaticity than FBC.

5.2.3 The crystal structures of field-aged biochar

Figure 3 shows the crystal structures of FBC and ABC before and after the reaction with heavy metal, which were investigated by XRD and analyzed using Jade6.5 PDF cards. In ABC, there is a strong peak appearing at 26.6° where was assigned to the characteristic diffraction peak of SiO_2 either before or after the absorption of heavy metal [37]. However, there is a peak located at 29.5° where is the crystalline structures of $CaCO_3$ in FBC in the XRD pattern before the absorption of heavy metal. And interestingly, the diffraction peaks appearing at 2θ = 30.33°, 20.97°, and 24.79° after the heavy metal ions are adsorbed by FBC can be ascribed to $CdCO_3$, Pb $(CO_3)_2(OH)_2$, and $PbCO_3$ [20]. Totally, the intensity of the characteristic peaks of $CaCO_3$ and SiO_2 in FBC and ABC was obviously weakened or disappeared after the absorption with heavy

Sample	FBC	ABC
C (%)	81.01	46.89
H (%)	2.77	3.02
O (%)	12.47	21.26
N (%)	0.40	6.61
S (%)	0.00	1.85
H:C	0.03	0.06
O:C	0.15	0.45
(N + O + S):C	0.16	0.63
SSA (m^2/g)	2.98	8.32
Total pore volume (cm^3/g)	0.0017	0.02
Average pore diameter (nm)	2.32	9.62
pH	8.87 ± 0.01	5.83 ± 0.02
EC (μS/cm)	142.20 ± 0.00	252.00 ± 0.01

Values are mean ± SD (n = 3).

Table 1.
The physicochemical properties of FBC and ABC.

Binding energy (eV)	Area of samples (%)				Attribution
	FBC	FBC-Cd,Pb	ABC	ABC-Cd,Pb	
284.8 ± 0.1	57.33	58.2	42.92	51.6	C-C/ C=C
285.4 ± 0.1	19.38	16.94	27.34	24.69	C-O-C/C-OH
286.4 ± 0.3	18.24	18.17	22.61	18.43	C=O
289.1 ± 0.2	5.05	6.7	7.13	5.28	O-C=O/Carbonate

Table 2.
The data of XPS analysis of C1s for samples.

metal. We can conclude that precipitation may be one of the main mechanisms for Cd^{2+} and Pb^{2+} removal by FBC.

5.2.4 The functional groups of field-aged biochar

Figure 4 shows the FTIR spectra of ABC and FBC before and after the adsorption. There were some differences in the types and intensity of surface functional groups on ABC and FBC. The main functional groups on ABC were including C=O [38], COO^- [39], Si–O–Al, C-H, Al–O–Si [40], Si-O-Si [37]. While the functional groups on FBC were C=C, COOH/CHO, phenolic –OH bending, CO2–3, C-O, and C-C, the aromatic ring C-H, the aromatic C-H [40]. The -OH and C-H were in all samples. There were also changes of FBC/ABC between before and after the adsorption of heavy metals. After FBC adsorbs metal ions, the peak waves of C = C, C = O and aromatic hydrocarbons were shifted. This may be the result of chelation of heavy metal ions with

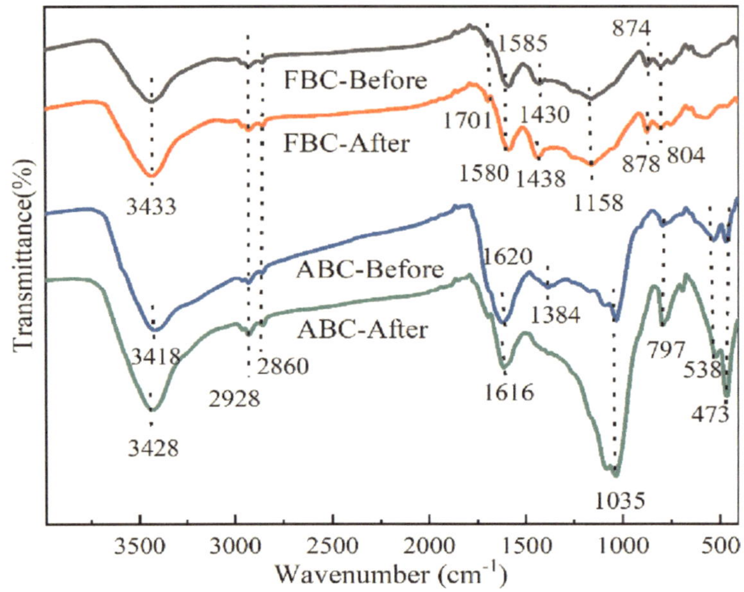

Figure 4.
FTIR spectroscopy of FBC and ABC before and after adsorption. Note: Peak at 3418–3433 cm^{-1}*: -OH; peak at 2950 cm*$^{-1}$ *and 2860 cm*$^{-1}$*: Saturated C-H; peak at 1620 cm*$^{-1}$*: C=O; peak at 1585 cm*$^{-1}$*: Aromatic C=C; peak at 1430 cm*$^{-1}$*: -COOH/CHO; phenolic –OH bending; CO2–3; peak at 1384 cm*$^{-1}$*: COO*$^{-}$*; peak at 1158 cm*$^{-1}$*: C-O and C-C; peak at 1158 cm*$^{-1}$*: Si–O–Si or Si–O–Al asymmetric; peak at 874 cm*$^{-1}$*:Aromatic ring C-H bonds; peak at 804 cm*$^{-1}$*: Aromatic C-H; peak at 797 cm*$^{-1}$*: Aromatic C-H stretching; peak at 538 cm*$^{-1}$*: The symmetric of Al–O–Si; peak at 538 cm*$^{-1}$*: The antisymmetric of Si-O-Si.*

aromatic structure of biochar (cation-π interaction). In addition, we also found that the atomic ratio of (N+O+S): C and H: C in ABC increased significantly, which can measure the polarity and aromaticity of biochar. These indicated that the aromaticity of ABC decreased while the polarity increased. This would reduce the stability of heavy metal ions, which passivated by biochar and the heavy metal would be released to the soil inducing secondary pollution. These results were consistent with the increasments of TCLP leaching rate of Cd^{2+} and Pb^{2+} in ABC.

5.2.5 The combination state of main elements in field aging biochar

Figure 5 shows that the C, O, N, S, Si, and Al were observed in ABC (**Figure 5(c)**), while only C and O elements were in FBC (**Figure 5(a)**) before the heavy metal absorption. And O element is more in ABC than in FBC interestingly. That the O:C in ABC was 0.45 while was 0.15 in FBC. This ascribed to the O-containing functional groups of biochar markedly increased after aging in field. Data as showed in **Table 2**, the C1s spectrum were divided into four peaks: 285.4 ± 0.1 EV (C—O), 284.8 ± 0.1 EV (C—C/C═C), 289.1 ± 0.2 EV (O—C═O or carbonate), and 286.4 ± 0.3 EV (C═O). The relative proportion of C—C/C═C, C-O, C═O, and O—C═O/carbonate in ABC were 42.92%, 27.34%, 22.61%, and 7.13% respectively, while they were 57.33%, 19.38%, 18.24%, and 5.05% in FBC. That is, the ABC's relative contents of C—O, C—O, and O—C═O functional groups were dramatically higher than that of FBC. During the field aging process, due to the oxidation or weathering of persistent free radicals in the soil, oxygen-containing functional groups, including phenolic groups and carboxyl, were

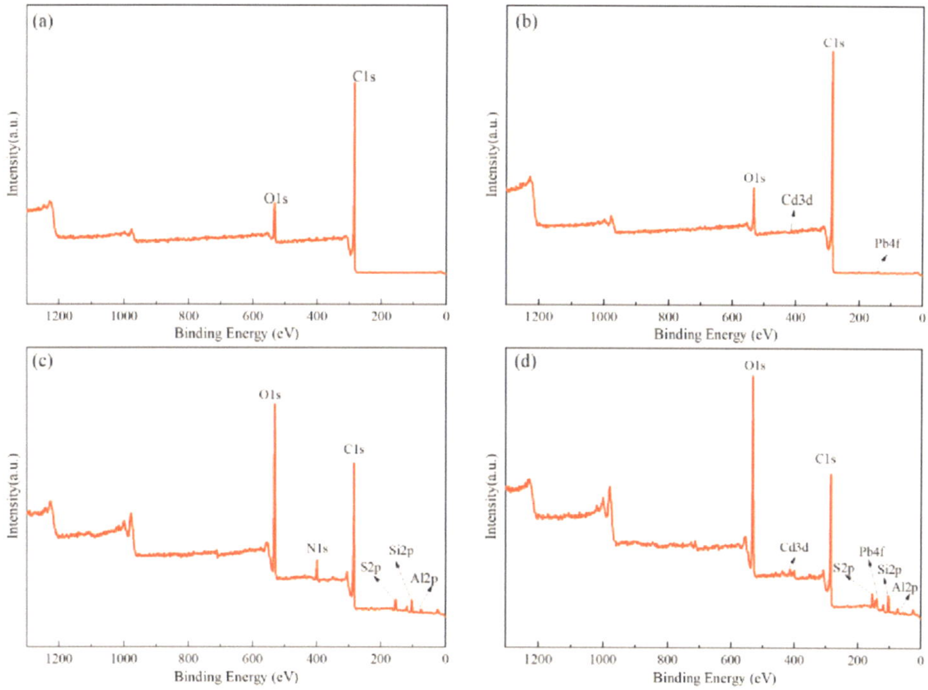

Figure 5.
XPS spectra for typical survey scan of FBC (a, b) and ABC (c, d) before and after adsorption.

formed on the surface of biochar [41]. This will contribute to an increase in the relative contents of C=O, O—C=O, and C—O. **Figure 6** shows that the peaks of Cd3d and Pb4f were observed in FBC and ABC before and after the adsorption of heavy metal. It was concluded that Cd^{2+} and Pb^{2+} were successfully adsorbed on the surface of biochar. It has been proved that the peaks of 412.5 EV and 405.6 EV belong to Cd $3d^{3/2}$ and Cd $3d^{5/2}$, respectively. This indicated that Cd exists in the form of Cd—O through complexation with hydroxyl (—OH) or diprotic oxygen (—O—) on the surface of biochar [42, 43]. The peaks of Pb4f appeared at around 144.4 eV and 139.4 eV, which can be attributed to Pb—O—C and Pb—O [44].

5.2.6 The discussion of chemical properties of biochar

The pH of biochar was reduced by three units after being aged in the agricultural field for 7 years. This change can be attributed to the alkaline substances in FBC leached out [5]. During aging, the alkaline substances were dissolving. In addition, the decrease in basicity of biochar may be due to O-containing functional groups' formation on the biochar's surface. These can be well explained by elemental analysis and XPS. The analysis of elements showed that the O:C of FBC is 0.15 while the O:C of ABC was increased to 0.45. That is a representative of the oxidation level of biochar. XPS results indicated that the relative content of C=O, C—O, and O—C=O functional groups of FBC were dramatically lower than that of ABC. During the field aging process, due to the oxidation by persistent free radicals in soil or weathering effects, including phenolic groups and carboxyl, oxygen-containing functional groups were

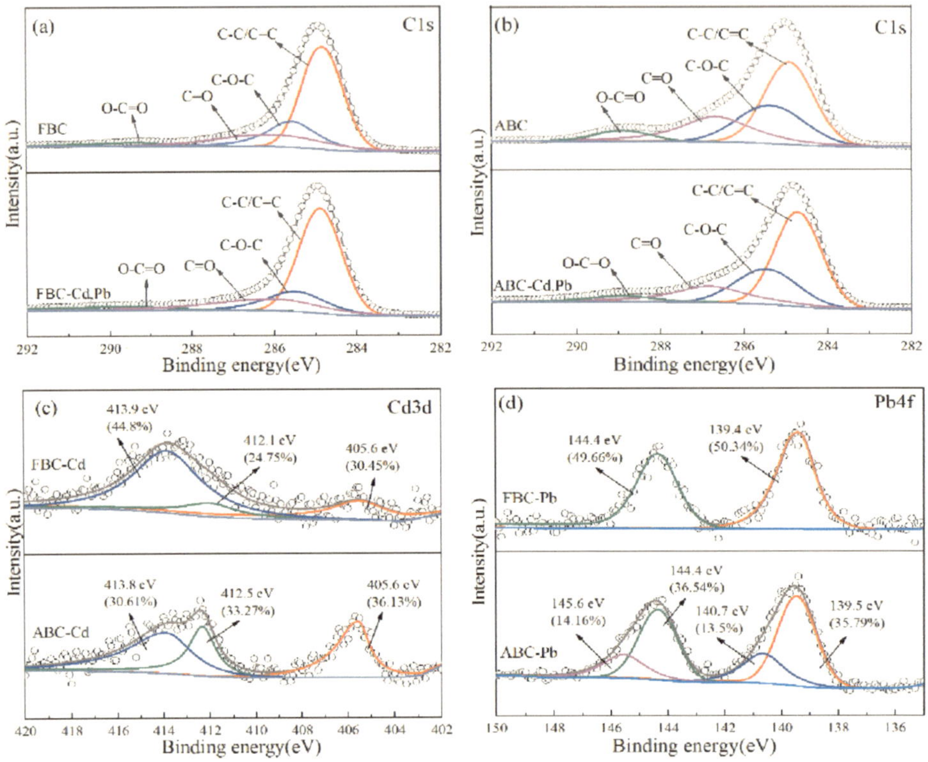

Figure 6.
XPS spectra of C1s, Cd3d, and Pb4f for FBC and ABC before and after adsorption.

formed on the surface of biochar. These lead to the increase of the relative content of O—C=O, C=O, and C—O. However, the formation of acidic functional groups on the surface of biochar, such as carboxyl and phenol groups, can cause the biochar's lower pH value [12].

It was further confirmed that more O-containing functional groups were formed on aged biochar surface during aging in field [5]. Biochar has large SSA and abundant pore structure. In our study, the average pore size and SSA of ABC increased by 4331.39% and 279.19% compared with FBC, respectively. Even more interesting is that the total pore volume in ABC increased more than 9.9-fold compared with FBC. The freeze–thaw cycles and rainfall events of biochar in soil maybe related with these changes. The expansion and elimination of water molecules also occur in biochar during freeze–thaw cycles [12]. As a result, the SSA of the biochar increases. As well, aging in field may also influence the functional groups and elemental composition of biochar surface. In this study, the relative contents of S, O, and N in ABC increased significantly compared with FBC. While the relative content of C decreased in ABC compared with FBC. These results were in accordance with the results of the prevenients [8]. It was found that the O content of biochar's surface increased while the C content decreased after 5 years in the field soil, and the o content increased. These indicated the dissolution of unstable C during the aging process of biochar [12, 45]. Furthermore, the results of XPS indicate that the amounts of Al and Si in ABC increased, implying that soil minerals could have been attached on biochar surface during field aging. This result was also reported in previous studies.

5.3 Biochar sorption experiments

5.3.1 The kinetic adsorption of field-aged biochar

Figure 7 shows the adsorption kinetics of Cd^{2+} and Pb^{2+} on FBC and ABC in the single-metal (**Figure** 7(**a** and **b**)) and binary-metal (**Figure** 7(**c, d**)) systems. Correspondingly, **Table 3** gives the parameters fitted by pseudo first-order model and

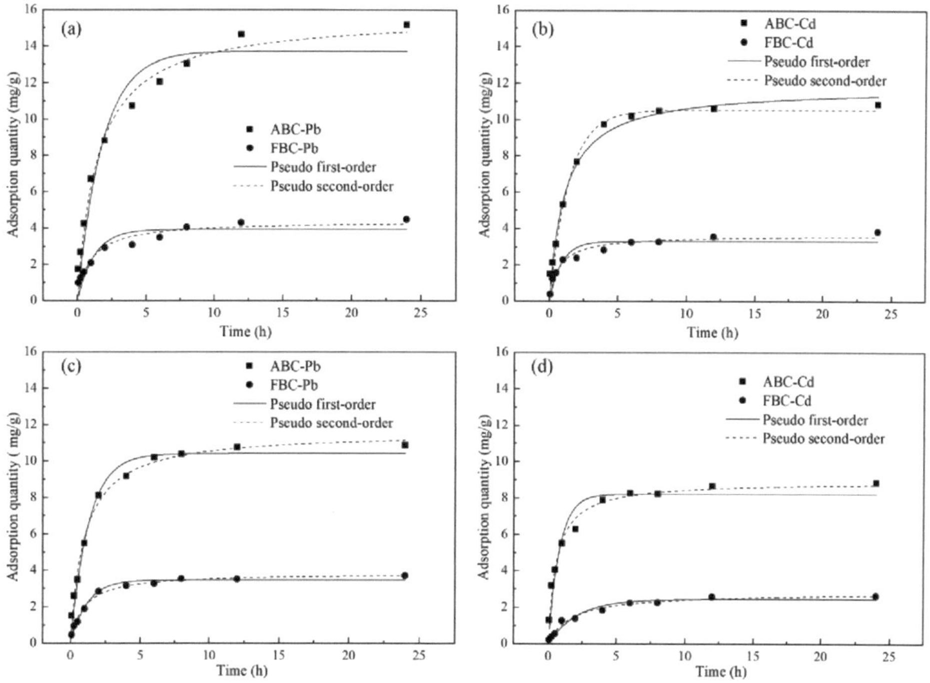

Figure 7.
Kinetic adsorption of Cd^{2+} and Pb^{2+} by PBC and ABC in single- (a, b) and binary-metal (c, d) systems.

Adsorption	Biochars	Pseudo first-order			Pseudo second-order		
		R^2	k_1 (h)	Q_e (mg/g)	R^2	k_2 (g/h/mg)	Q_e (mg/g)
Single	FBC-Pb	0.8623	0.8183	3.92319	0.9754	0.2432	4.3774
	ABC-Pb	0.95141	0.5487	13.70752	0.98771	0.0435	15.6564
	FBC-Cd	0.9160	1.1416	3.2934	0.96737	0.40082	3.6364
	ABC-Cd	0.98975	0.07578	11.78350	0.99512	0.69917	10.52060
Binary	FBC-Pb	0.9794	0.8554	3.45187	0.9901	0.2871	3.8336
	ABC-Pb	0.9809	0.7792	10.3918	0.9917	0.0870	11.5773
	FBC-Cd	0.96701	0.47213	2.45183	0.98320	0.20870	2.81340
	ABC-Cd	0.94195	1.22718	8.20247	0.98758	0.19306	8.90370

Table 3.
Fitting parameters of pseudo-first-order and pseudo-second-order kinetics Cd^{2+} and Pb^{2+} by FBC and ABC single- and binary-metal systems.

pseudo first-order model. From the perspective of R^2 in **Table 3**, the pseudo second-order kinetic model (R^2 = 0.9674–0.9917) was better fit for the Cd^{2+} and Pb^{2+} adsorption kinetics data for FBC and ABC in the single-metal and binary-metal systems, compared with the pseudo first-order model (R^2 = 0.8623–0.9898). In addition, the calculated Q_e values based on the pseudo second-order model were approximately the experimental Q_e values. Overall, the adsorption of two metal ions on ABC or FBC increased dramatically within 2.5–3.0 hours and then approached to a flat with the augment of reaction time shown in **Figure** 7(**a–d**). And the adsorption quantity on Cd^{2+} and Pb^{2+} of ABC was stronger than FBC either in the single-metal or binary-metal system. It was interesting that the capacity of quilibrium adsorption in single-metal system was higher than that of the binary-metal system. While the capacity of its total adsorption was weaker. That is, ABC and FBC reach the adsorption equilibrium at 8 h in the binary-metal system, while at 12 h in the bimetallic system. These were indicating that there was a competitive relationship between Cd^{2+} and Pb^{2+}.

5.3.2 The isothermal adsorption of field-aged biochar

Figure 8 shows the isothermal adsorption of Cd^{2+} and Pb^{2+} on FBC and ABC in the single-(**Figure 8(a, b)**) and binary-metal (**Figure 8(c, d)**) systems. Correspondingly, **Table 4** gives the fitting parameters of the Langmuir and Freunlich isothermals for Cd^{2+} and Pb^{2+} by FBC and ABC in single- and binary-metal systems. As shown in **Figure 8**, Langmiur model due to its higher correlation coefficient (R^2 = 0.96013–0.9910) was more reasonable than Freunlich model (R^2 = 0.7924–0.9679) in this isothermal adsorption experimental data analyses. **Figure 8** shows that at low initial concentrations, the adsorption capacity of eight ABCs or FBCs increased with the

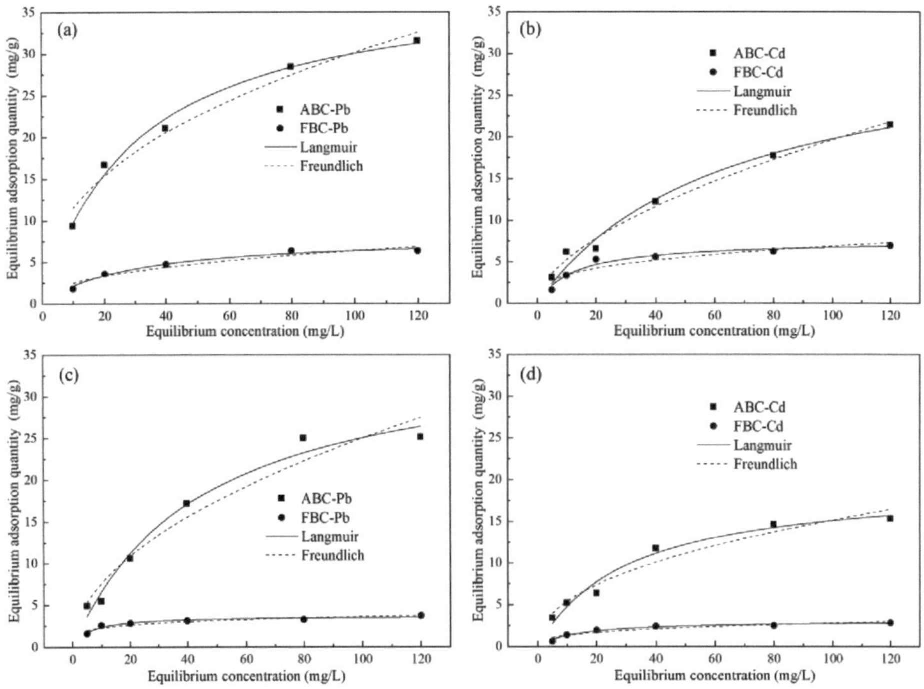

Figure 8.
Isothermal adsorption of Cd^{2+} and Pb^{2+} by FBC and ABC in single- (a, b) and binary-metal (c, d) systems.

Adsorption	Biochars	Langmuir parameters			Freundlich parameters		
		R^2	b	Q_m	R^2	K_f	1/n
Single	FBC-Pb	0.9796	0.03572	8.1343	0.91864	1.0021	0.40162
	ABC-Pb	0.9914	0.03267	39.2158	0.97437	4.3564	0.42020
	FBC-Cd	0.96033	0.07847	7.57289	0.86456	1.6577	0.30922
	ABC-Cd	0.98236	0.01584	32.1566	0.98122	1.4110	0.57164
Binary	FBC-Pb	0.94327	0.17694	3.68653	0.87602	1.47112	0.19475
	ABC-Pb	0.98103	0.00298	36.2279	0.95295	2.30530	0.51751
	FBC-Cd	0.97325	0.07456	3.09823	0.87805	0.65300	0.31578
	ABC-Cd	0.97655	0.03210	19.74660	0.95364	1.933256	0.44694

Table 4.
Fitting parameters of Langmuir and Freundlich isotherms for Cd^{2+} and Pb^{2+} by FBC and ABC in single- and binary-metal systems.

increasing of initial concentration of Cd^{2+} and Pb^{2+} and then gradually slowed down when the biochar reached saturation. It has been known that in Langmuir isotherm the separation factor R_L commonly used to evaluate the affinity between the adsorbent and the adsorbed material [46]. They are as follows: when $R_L > 1$, adsorption is unfavorable; when $R_L = 1$, adsorption is linear; when $0 < R_L < 1$, adsorption is favorable; when $R_L = 0$, adsorption is nonlinear; and when $R_L < 0$, adsorption is irreversible [14, 47]. In this experiment, the initial concentrations of Cd^{2+} and Pb^{2+} were ranged from 5 to 120 mg/L (**Table 4**). The R_L values of FBC-Pb, FBC-Cd, ABC-Pb, and ABC-Cd were between 0 and 1. These results indicated that both ABC and ABC were favorable for Cd^{2+} and Pb^{2+} adsorption. In addition, **Table 4** shows that all 1/n values were in the range of 0–1 in this study, indicating that adsorption is also favorable [48]. The above analysis illustrated that the adsorption of Cd^{2+} and Pb^{2+} on FBC and ABC was monolayer adsorption.

5.3.3 The metal leachability and bioavailability of field-aged biochar

In order to understand the metal leachability and bioavailability of ABC, we take the leaching characteristics of Cd^{2+} and Pb^{2+} in FBC and ABC by TCLP method. The results showed that the concentration of Cd^{2+} and Pb^{2+} in TCLP leachate was 39.21% and 28.62% in ABC while was 24.08% and 21.24%, respectively in FBC. This implied that the adsorption mechanisms of ABC for Cd^{2+} and Pb^{2+} were different from FBC. At same time, the increase of TCLP leachability of Cd^{2+} and Pb^{2+} suggesting the stability of ABC to immobilized heavy metals was significantly reduced.

5.3.4 The discussion of biochar sorption

The results of isothermal and kinetic adsorption experiments showed that the pseudo-secondary kinetic and Langmuir model were more fitted with the adsorption of metal ions by FBC and ABC. FBC and ABC immobilize Cd^{2+} and Pb^{2+} in binary metal system as a chemical reaction as confirmed by the above results [20, 49, 50]. ABC adsorbs more heavy metal ions than FBC. These ascribed to the aged biochar having more oxygen-containing functional groups and a larger SSA. These indicated that ABC surface has more chemisorption active sites. Additionally, the biochar can well serve as a habitat for microorganisms, such as bacteria and fungi, due to its abundant porous structure.

In addition, the active component carbon and mineral nutrients in biochar can be used as its energy source by microorganisms in the soil environment [8, 36]. Therefore, it is likely that microorganisms will attach to the ABC surface after it is applied to the soil. Furthermore, the adsorption of heavy metals may be promoted by the beneficial microorganisms immobilized on biochar. It has been reported that the combined application of biochar and bacteria can improve the adsorption of heavy metals [33]. There are some possible mechanisms for the interaction between bacteria, biochar, and heavy metal ions. First, the respiration of bacterial cells attached to the surface of biochar form metal carbonate precipitates. Second, new adsorption sites were formed by bacteria colonized of biochar [13]. Third, bacterial cells are as transport carriers between heavy metals and biochar. That is, the heavy metal ions in the soil solution are first transferred to the cells, and then the cells adsorbed on the biochar and actively pumped out of the bacterial cells [51].

6. Conclusions

In conclusion, under field conditions, the physicochemical properties of biochar have changed in soil after 7 years of field aging. The pore volume and SSA of biochar increased with field aging, owing to the dissolution of unstable carbon or carbides in biochar. FTIR and XPS results proved that there were abundant O-containing functional groups on the surface of aged biochar. The results of adsorption kinetics and adsorption isotherm showed that the adsorption of heavy metal ions on ABC and FBC surface was controlled by chemical adsorption. FBC immobilizes Cd^{2+} and Pb^{2+} mainly through cation exchange, co-precipitation, and cation-π interaction. Whereas the main mechanism of ABC removing Cd^{2+} and Pb^{2+} may be the cation exchange and surface complexation. Compared with that of FBC, the adsorption performance of ABC for Cd^{2+} and Pb^{2+} is improved due to the increases of O-containing functional groups and SSA in ABC. Nevertheless, the stability of ABC to immobilized heavy metals was significantly reduced.

Acknowledgements

This study was financially supported by the Projects of National Natural Science Foundation of China (31201618), China Agriculture Research System (CARS-23-G27).

Author details

Run-Hua Zhang[1*], Lin-Fang Shi[1], Zhi-Guo Li[2], Guo-Lin Zhou[1], Yan-Lan Xie[1], Xing-Xue Huang[1], An-Hua Ye[1] and Chu-Fa Lin[1]

1 Wuhan Academy of Agricultural Sciences, Wuhan, China

2 Key Laboratory of Aquatic Botany and Watershed Ecology, Wuhan Botanical Garden of Sciences, Wuhan, China

*Address all correspondence to: rhzhanag0508@126.com

References

[1] Gong X, Huang D, Liu Y, Zou D, Hu X, Zhou L, et al. Nanoscale zerovalent iron, carbon nanotubes and biochar facilitated the phytoremediation of cadmium contaminated sediments by changing cadmium fractions, sediments properties and bacterial community structure. Ecotoxicology and Environmental Safety. 2021;**208**:111510. DOI: 10.1016/j.ecoenv.2020.111510

[2] Li X, Xiao J, Salam MMA, Ma C, Chen G. Impacts of bamboo biochar on the phytoremediation potential of Salix psammophila grown in multi-metals contaminated soil. International Journal of Phytoremediation. 2021;**23**:387-399. DOI: 10.1080/15226514.2020.1816893

[3] Raeisi S, Motaghian H, Hosseinpur AR. Effect of the soil biochar aging on the sorption and desorption of Pb2+ under competition of Zn2+ in a sandy calcareous soil. Environmental Earth Sciences. 2020;**79**(6):148. DOI: 10.1007/s12665-020-8891-y

[4] Rathnayake D, Rego F, Van Poucke R, Bridgwater AV, Masek O, Meers E, et al. Chemical stabilization of Cd-contaminated soil using fresh and aged wheat straw biochar. Environmental Science and Pollution Research International. 2021;**28**:10155-10166. DOI: 10.1007/s11356-020-11574-6

[5] Xu Z, Xu X, Tsang DCW, Cao X. Contrasting impacts of pre- and post-application aging of biochar on the immobilization of Cd in contaminated soils. Environmental Pollution. 2018; **242**:1362-1370. DOI: 10.1016/j.envpol.2018.08.012

[6] Lam YY, Lau SSS, Wong JWC. Removal of Cd(II) from aqueous solutions using plant-derived biochar: Kinetics, isotherm and characterization. Bioresource Technology Reports. 2019;**8**:100323. DOI: 10.1016/j.biteb.2019.100323

[7] Su Y, Wen Y, Yang W, Zhang X, Xia M, Zhou N, et al. The mechanism transformation of ramie biochar's cadmium adsorption by aging. Bioresource Technology. 2021; **2021**(330):124947. DOI: 10.1016/j.biortech.2021.124947

[8] Dong X, Li G, Lin Q, Zhao X. Quantity and quality changes of biochar aged for 5 years in soil under field conditions. Catena. 2017;**159**:136-143. DOI: 10.1016/j.catena.2017.08.008

[9] Nguyen, Wallace HM, Xu C-Y, Zwieten L(V), Hanweng Z, Xu Z, et al. The effects of short term, long term and reapplication of biochar on soil bacteria. Science of the Total Environment. 2018; **636**:142-151. DOI: 10.1016/j.scitotenv.2018.04.278

[10] Wu Z, Zhang X, Dong Y, Xu X, Xiong Z. Microbial explanations for field-aged biochar mitigating greenhouse gas emissions during a rice-growing season. Environmental Science and Pollution Research International. 2018; **25**:31307-31317. DOI: 10.1007/s11356-018-3112-x

[11] Zhang, Huang Q. Effect mechanism of biochar's zeta potential on farmland soil's cadmium immobilization. Environmental Science and Pollution Research International. 2019;**26**: 19738-19748. DOI: 10.1007/s11356-019-05298-5

[12] Wang L, O'connor D, Rinklebe J, Ok YS, Tsang DCW, Shen Z, et al. Biochar aging: Mechanisms, physicochemical changes, assessment, and implications for field applications. Environmental Science & Technology.

2020;**54**:14797-14814. DOI: 10.1021/acs.est.0c04033

[13] Tan L, Ma Z, Yang K, Cui Q, Wang K, Wang T, et al. Effect of three artificial aging techniques on physicochemical properties and Pb adsorption capacities of different biochars. Science of the Total Environment. 2020;**2020**(699):134223. DOI: 10.1016/j.scitotenv.2019.134223

[14] Lin Q, Zhang L, Riaz M, Zhang M, Xia H, Lv B, et al. Assessing the potential of biochar and aged biochar to alleviate aluminum toxicity in an acid soil for achieving cabbage productivity. Ecotoxicology and Environmental Safety. 2018;**161**:290-295. DOI: 10.1016/j.ecoenv.2018.06.010

[15] Mia S, Dijkstra FA, Singh B. Aging induced changes in Biochar's functionality and adsorption behavior for phosphate and ammonium. Environmental Science & Technology. 2017;**26**:1-26. DOI: 10.1021/acs.est.7b00647

[16] He E, Yang Y, Xu Z, Qiu H, Yang F, Peijnenburg W, et al. Two years of aging influences the distribution and lability of metal(loid)s in a contaminated soil amended with different biochars. Science of the Total Environment. 2019; **673**:245-253. DOI: 10.1016/j.scitotenv.2019.04.037

[17] Godwin PM, Pan Y, Xiao H, Afzal MT. Progress in preparation and application of modified biochar for improving heavy metal ion removal from wastewater. Journal of Bioresources and Bioproducts. 2019;**4**: 31-42. DOI: 10.21967/jbb.v4i1.180

[18] Murtaza G, Ahmed Z, Usman M. Feedstock type, pyrolysis temperature and acid modification effects on physiochemical attributes of biochar and soil quality. Arabian Journal of Geosciences. 2022;**15**:305. DOI: 10.1007/s12517-022-09539-9

[19] Yang F, Zhang S, Sun Y, Du Q, Song J, Tsang D. A novel electrochemical modification combined with one-step pyrolysis for preparation of sustainable thorn-like iron-based biochar composites. Bioresource Technology. 2019;**274**:379-385. DOI: 10.1016/j.biortech.2018.10.042

[20] Zhou N, Zu J, Feng Q, Chen H, Li J, Zhong ME, et al. Effect of pyrolysis condition on the adsorption mechanism of heavy metals on tobacco stem biochar in competitive mode. Environmental Science and Pollution Research International. 2019;**26**:26947-26962. DOI: 10.1007/s11356-019-05917-1

[21] Yi Y, Huang Z, Lu B, Xian J, Fang Z. Magnetic biochar for environmental remediation: A review. Bioresource Technology. 2019;**298**:122468. DOI: 10.1016/j.biortech.2019.122468

[22] Lou K, Rajapaksha AU, Ok YS, Chang SX. Pyrolysis temperature and steam activation effects on sorption of phosphate on pine sawdust biochars in aqueous solutions. Chemical Speciation & Bioavailability. 2016;**28**:42-50. DOI: 10.1080/09542299.2016.1165080

[23] Plaimart J, Acharya K, Mrozik W, Davenport RJ, Vinitnantharat S, Werner D. Coconut husk biochar amendment enhances nutrient retention by suppressing nitrification in agricultural soil following anaerobic digestate application. Environmental Pollution. 2021;**268**:115684. DOI: 10.1016/j.envpol.2020.115684

[24] Guo D, Li Y, Cui B, Hu M, Luo S, Ji B, et al. Natural adsorption of methylene blue by waste fallen leaves of Magnoliaceae and its repeated thermal regeneration for reuse. Journal of

Cleaner Production. 2020;**267**:121903. DOI: 10.1016/j.jclepro.2020.121903

[25] Hilber I, Bastos AC, Loureiro S, Soja G, Marsz A, Cornelissen G, et al. The different faces of biochar: Contamination risk versus remediation tool. Journal of Environmental Engineering and Landscape Management. 2017;**25**:86-104. DOI: 10.3846/16486897.2016.1254089

[26] Choppala G, Bolan N, Megharaj M, Chen Z, Naidu R. The influence of biochar and black carbon on reduction and bioavailability of chromate in soils. Journal of Environmental Quality. 2012; **41**:1175-1184. DOI: 10.2134/jeq2011.0145

[27] Järup L. Hazards of heavy metal contamination. British Medical Bulletin. 2003;**68**:167-182

[28] Tchounwou PB, Yedjou CG, Patlolla AK, Sutton DJ. Heavy metal toxicity and the environment. Molecular, Clinical and Environmental Toxicology. 2012;**101**:133-164. DOI: 10.1093/bmb/ldg032

[29] Nagajyoti PC, Lee KD, Sreekanth T. Heavy metals, occurrence and toxicity for plants: A review. Environmental Chemistry Letters. 2010;**8**:199-216. DOI: 10.1007/s10311-010-0297-8

[30] Barakat M. New trends in removing heavy metals from industrial wastewater. Arabian Journal of Chemistry. 2011;**4**:361-377. DOI: 10.1016/j.arabjc.2010.07.019

[31] Wuana RA, Okieimen FE. Heavy metals in contaminated soils: A review of sources, chemistry, risks and best available strategies for remediation. International Scholarly Research Notices. 2011;**2011**: 1-20. DOI: 10.5402/2011/402647

[32] Wang Q, Wang B, Lee X, Lehmann J, Gao B. Sorption and desorption of Pb (II) to biochar as affected by oxidation and pH. Science of the Total Environment. 2018;**634**: 188-194. DOI: 10.1016/j.scitotenv.2018.03.189

[33] Paz-Ferreiro J, Lu H, Fu S, Mendez A, Gasco G. Use of phytoremediation and biochar to remediate heavy metal polluted soils: A review. Solid Earth. 2014;**5**:65-75. DOI: 10.5194/se-5-65-2014

[34] Lehmann J, Joseph S. Biochar for environmental management: An introduction. In: Lehmann J, Joseph S, editors. Biochar for environment management. 1st ed. Earthscan; 2009. p.1-9

[35] Inyang M, Gao B, Yao Y, Xue Y, Zimmerman AR, Pullammanappallil P, et al. Removal of heavy metals from aqueous solution by biochars derived from anaerobically digested biomass. Bioresource Technology. 2021;**110**:50-56. DOI: 10.1016/j.biortech.2012.01.072

[36] Wang L, Gao C, Yang K, Sheng Y, Xu J, Zhao Y, et al. Effects of biochar aging in the soil on its mechanical property and performance for soil CO2 and N2O emissions. Science of the Total Environment. 2021;**782**:146824. DOI: 10.1016/j.scitotenv.2021.146824

[37] Tan L, Ma Z, Yang K, Cui Q, Wang K, Wang T, et al. Effect of three artificial aging techniques on physicochemical properties and Pb adsorption capacities of different biochars. Science of the Total Environment. 2020;**699**:134223. DOI: 10.1016/j.scitotenv.2019.134223

[38] Xue C, Zhu L, Lei S, Liu M, Hong C, Che L, et al. Lead competition alters the zinc adsorption mechanism on animal-derived biochar. Science of the Total Environment. 2020;**713**:136395. DOI: 10.1016/j.scitotenv.2019.136395

[39] Bai S, Wang L, Ma F, Zhu S, Xiao T, Yu T, et al. Self-assembly biochar colloids mycelial pellet for heavy metal removal from aqueous solution. Chemosphere. 2020;**242**:125182. DOI: 10.1016/j.chemosphere.2019.125182

[40] Zhang W, Tan X, Gu Y, Liu S, Liu Y, Hu X, et al. Rice waste biochars produced at different pyrolysis temperatures for arsenic and cadmium abatement and detoxification in sediment. Chemosphere. 2020;**250**: 126268. DOI: 10.1016/j.chemosphere.2020.126268

[41] Yuan C, Gao B, Peng Y, Gao X, Fan B, Chen Q. A meta-analysis of heavy metal bioavailability response to biochar aging: Importance of soil and biochar properties. Science of the Total Environment. 2021;**756**:144058. DOI: 10.1016/j.scitotenv.2020.144058

[42] Khan ZH, Gao M, Qiu W, Song Z. Properties and adsorption mechanism of magnetic biochar modified with molybdenum disulfide for cadmium in aqueous solution. Chemosphere. 2020; **255**:126995. DOI: 10.1016/j.chemosphere.2020.126995

[43] Teng D, Zhang B, Xu G, Wang B, Mao K, Wang J, et al. Efficient removal of Cd(II) from aqueous solution by pinecone biochar: Sorption performance and governing mechanisms. Environmental Pollution. 2020;**265**:115001. DOI: 10.1016/j.envpol. 2020. 115001

[44] Zhou Z, Xu Z, Fe NQ, Yao D, Yu J, Wang D, et al. Effect of pyrolysis condition on the adsorption mechanism of lead, cadmium and copper on tobacco stem biochar. Journal of Cleaner Production. 2018:996-1005. DOI: 10.1016/j.jclepro.2018.03.268

[45] Ren X, Sun H, Wang F, Zhang P, Zhu H. Effect of aging in field soil on biochar's properties and its sorption capacity. Environmental Pollution. 2018; **2018**(242):1880-1886. DOI: 10.1016/j.envpol.2018.07.078

[46] Yang YQ, Cui MH, Ren YG, Guo JC, Zheng ZY, Liu H. Towards understanding the mechanism of heavy metals immobilization in biochar derived from Co-pyrolysis of sawdust and sewage sludge. Bulletin of Environmental Contamination and Toxicology. 2020;**104**:489-496. DOI: 10.1007/s00128-020-02801-4

[47] Reguyal F, Sarmah AK, Gao WJJOHM. Synthesis of magnetic biochar from pine sawdust via oxidative hydrolysis of FeCl2 for the removal sulfamethoxazole from aqueous solution. Journal of Hazardous Materials. 2017;**321**:868-878. DOI: 10.1016/j.jhazmat.2016.10.006

[48] Quan G, Fan Q, Cui L, Zimmerman AR, Wang H, Zhu Z, et al. Simulated photocatalytic aging of biochar in soil ecosystem: Insight into organic carbon release, surface physicochemical properties and cadmium sorption. Environmental Research. 2020;**183**: 109241. DOI: 10.1016/j.envres.2020.109241

[49] Vithanage M, Rajapaksha AU, Ahmad M, Uchimiya M, Dou X, Alessi DS, et al. Mechanisms of antimony adsorption onto soybean Stover-derived biochar in aqueous solutions. Journal of Environmental Management. 2015;**151**: 443-449. DOI: 10.1016/j.jenvman.2014.11.005

[50] Vithanage M, Rajapaksha AU, Ahmad M, Uchimiya M, Dou X, Alessi DS, et al. Mechanisms of antimony adsorption onto soybean Stover-derived biochar in aqueous solutions. Journal of Environmental Management. 2015;**2015** (151):443-449. DOI: 10.1016/j.jenvman.2014.11.005

[51] Park JH, Yun JJ, Kang SW, Kim SH, Cho JS, Wang JJ, et al. Removal of potentially toxic metal by biochar derived from rendered solid residue with high content of protein and bone tissue. Ecotoxicology and Environmental Safety. 2021;**208**:111690. DOI: 10.1016/j.ecoenv.2020.111690

Chapter 8

Sustainable and Eco-Friendly Biomass Derived Biochars for the Removal of Contaminants from Wastewater: Current Status and Perspectives

Uplabdhi Tyagi and Neeru Anand

Abstract

Human activities and rapid modernization have affected the ecological and economical aspects worldwide resulting in alarming situations such as global warming and the accumulation of waste disposal and toxic contaminants in water. Contaminants in water are toxic and carcinogenic, posing a serious threat to the environment. Water is a precious and limited resource and hence it is highly imperative to utilize effective remediation strategies for the removal of pollutants. Several competitive remediation techniques have been proposed due to their distinctive features including ease of operation, inexpensiveness and universal nature. The present chapter highlights the potential of ecofriendly biomass-derived biochars as adsorbents for the effective removal of toxic contaminants. This includes biochars derived from industrial solid wastes, agricultural wastes, clays minerals and municipal wastes. Biomass-derived biochars are found to be highly efficient, alternative and carbon-neutral precursors and provide a new approach to the modular adsorption process. The present chapter also includes conversion of waste materials into efficient bio-adsorbents followed by their applications for the purification of wastewater. Besides, attempts are made to discuss the techno-economic and future perspectives of eco-friendly and low-cost biochars for the treatment of wastewater.

Keywords: adsorption, waste management, green synthesis, biomass engineering

1. Introduction

Freshwater is a basic demand for human activities including industrial, agricultural and domestic activities. These activities produce a huge amount of contaminated water resulting from the discharge of undesirable toxic and carcinogenic contaminants (inorganic/organic/biological agents/radioactive wastes) into water bodies that impose a serious concern on the environment and living species. According to World Health Organization (WHO) and the literature available [1, 2], the majority of water

 IntechOpen

on earth is salty, requiring treatment before it can be used. The rest of the freshwater is in glaciers and underground reservoirs. Industrial activities (automobile manufacturing, textile, dyeing, paint, paper and pulp, tannery and leather industry) and Agricultural activities (excess use of fertilizers and pesticides, antibiotics, processed wastes of crop plantation) and unwanted environmental changes (damages to sewer system due to high rainfall, soil runoff, use of pesticides and fertilizers) are the major cause of water pollution [3–5]. Hence, the preservation of freshwater, as well as the quality improvement of contaminated water (decontamination of pollutants from water), is a growing challenge.

Literature reports several feasible and popular conventional separation techniques for the treatment of polluted water such as chemical precipitation, adsorption, ion-exchange, flotation, coagulation and flocculation, ultrafiltration, nanofiltration, reverse osmosis, electrochemical process, evaporation and photo-catalysis [6, 7]. Each technique is effective in its own way and offers several advantages for one process but at the same time imposes several restrictions on other processes. However, amongst these popular conventional separation techniques, the chemical and electrochemical treatment processes are ineffective even at very low pollutant concentrations, due to excessive amount of chemical usage, sensitivity towards variable wastewater input and producing a large amount of sludge that needs further treatment before releasing it to the environment [8]. Other processes such as ultrafiltration, nanofiltration, reverse osmosis and ion exchange are the most expensive to treat a large amount of wastewater adding demerits to explore at the industrial scale.

Safe drinking water demand at a reasonably low price with an effective and sustainable treatment approach is a prime focus of industrialists and academicians. Currently, adsorption is used for wastewater treatment and is gaining wide attention due to its effectiveness and feasibility. In this regard, biomass-derived biochars are gaining attention due to their high potential, sustainability, carbon neutrality, low cost, mobile capability and wide availability in nature. Synthesis of biochars from inexpensive matters (living and non-living biomass) leads to significant cost reduction in waste disposal [9, 10]. These biochars can be obtained from various sources including industries and agricultural activities, plant wastes, fruit wastes, naturally occurring inorganic materials and living and dead biomass [11]. Literature reports a wide variety of biomass-derived biochars for effective wastewater treatment such as date pits, *vermiculite* plants, coconut shell and husk, bamboo waste, rice husk, ground nutshell, shells of almond, wheat bran and *Heveabrasiliensis* seed coat [12, 13]. These waste materials not only balance the environmental problems but an unutilized and a potential resource is also managed during the process. Also, utilization of these biochars resolves several major challenges associated with up-scaling technology including pollutant selectivity, regeneration, sludge formation and pollutant recovery and also exhibits excellent adsorption ability. Many factors affect the adsorption capacity of these biochars including physical and chemical properties of pollutants (i.e. molecular weight, oxidation state and ionic radius), characteristics of biochar and the process parameters (i.e., quantity of bio-sorbent, pH, temperature and sorbate concentration). Besides influencing the dissociation of pollutant sites and solution chemistry, pH plays a crucial role in the speciation and biosorption affinity of pollutants. Other factors include the composition of biomass (cellulose, hemicellulose, lignin and extractives), pore structure, surface charge and heteroatom content in the biochar. The adsorption capacity of biochar is highly dependent on the chemical compositions and carbohydrate contents of biomass which may vary from source to source and species to species. The biomass exists in diverse forms and has distinctive physical and chemical compositions, carbohydrate and lignin fractions as summarized in **Table 1**. In comparison to other wastes, agricultural and forestry wastes have high percentage composition of carbohydrates and lignin [14, 15]. Utilization

Type of waste	Cellulose	Hemicellulose	Lignin
Forestry waste			
Softwood			
Douglas fir	39	23	28
Scoots pine	40	25	28
Scandinavian pine	44	26	29
Scandinavian spruce	43	27	29
Hardwood			
Babool	48.5	18.3	20.9
Catalpa	54	16.6	15.94
Indian rosewood	39	25	25
Chinaberry	40	21	30
Oil palm frond	49.8	17	20.5
Hardwood barks	22-40	20-38	30-55
Softwood barks	18-28	15-33	30-60
Canola plant species			
C. caperata	46	16.9	15
B. adusta	44.28	15.61	13.83
F. gilva	46.04	16.99	15.75
P. tuberregium	44.99	16.59	11.09
Agricultural waste			
Olive husk	24	23.6	48.4
Cotton seed hairs	80-95	5-20	0
Corn cob	50.5	31	15
Stored refuse	60	20	20
Corn stover	37-42	20-28	18-22
Coastal Bermuda grass	25	35	6.4
Swine waste	6	28	0
Solid cattle manure	1.6-4.7	1.4-3.3	2.7-5.7
Municipal waste			
Pineapple leaf	73.4	25	10.5
Banana steam	60.6	12.4	18.6
Coconut shell	44.2	56.3	32.8
Hazelnut shell	28.8	30.4	42.9

Type of waste	Cellulose	Hemicellulose	Lignin
Newspaper	40-55	25-40	28-30
Primary wastewater solids	8-15	NA	25-29
Industrial waste			
Sugar beet waste	26.3	18.5	2.5
Sugarcane baggase	26-50	24-34	10-26
Rice straw	32-41	15-24	10-18
Sawdust	45.1	28.1	24
Cotton flax	80-95	5-20	NA
Sorghum straw	32	24	13

Table 1.
Percentage Composition in Biomass (Cellulose, Hemicellulose, Lignin).

of such biochars not only enhances the removal efficiency of various pollutants but simultaneously helps in the reduction of atmospheric carbon dioxide via the processing of waste biomass for a wide range of applications such as the synthesis of biofuels (i.e. biobutanol, bioethanol, and biomethanol), energy storage and soil refinement [16]. Apart from biochars, hydrochars have gained significant importance. Hydrochar is a char which is made by hydrothermal carbonization (a process where biomass is heated to a temperature range of 200–300°C in the presence of water), and is comprised of two phases: liquid and solid. Hydrochars offer advantages like low oxygen and ash content, zero hazardous chemical waste generation, high production yield (approximately 30-60 wt%), mild temperature processing (180-250°C), large surface areas and porosity. These materials offer several applications in many areas including soil amelioration, energy storage and water purification.

2. Remediation techniques employed for the removal of contaminants from wastewater

Several techniques are commercially available, to remove various contaminants including (inorganic and organic chemicals in dissolved and non-dissolved forms, biologically active agents, radioactive substances, polychlorinated biphenyls and pesticides) from wastewaters and some are summarized in **Table 2**. The following **Table 2** lists the most common commercially available techniques used in different sectors for pollutant removal due to their distinctive characteristics such as low-cost operation, flexibility and design simplicity.

2.1 Ion exchange

The ion exchange reaction is a reversible chemical reaction that involves the removal of dissolved ions from a solution and their replacement with other ions of the same or similar electrical charge. This process uses an insoluble matrix (or support structure) which is in the form of small microbeads (0.25–1.43 mm radius), usually white or yellowish, and are fabricated from an organic polymer substrate. This process has been widely employed for the separation of ionic dyes and heavy metal ions from aqueous streams. The widely used materials for this process are ion-exchange

Type of wastewater	Removed pollutant	Type of adsorption	Adsorbents		Performance		References
			Novel	Commercial	Novel (adsorption capacity)	Commercial (adsorption capacity)	
Industrial	Fluoride ions (7.5 PPM)	Adsorption	Carbon slurry	Powder Activated carbon	4.86 mg/g	1.10 mg/g	[4]
Industrial (Textile Wastewater)	Direct Blue 85 (450 PPM)	Oxidation	Metal oxide/ hydroxide sludge	Powder Activated carbon	339 mg/g	7.69–18.7 mg/g	[13]
Industrial (Medical discharge, Surface treating wastewater and automotive discharge)	Chromium (VI) (55 PPM)	Adsorption	Calcinated cereal and other crops by-product	Powder Activated carbon	90.37%	89.85%	[16]
Industrial (Textile Wastewater)	Safranin-T (30 PPM)	Coagulation, adsorption, flocculation and reverse osmosis	Chemically activated rice and wheat husks	Powder Activated carbon	0.014 mol/g	0.526 mol/g	[8]

Table 2.
Performance summary of different adsorption techniques using different wastewater.

resins that can be natural or synthetic having the ability to exchange their cations with the solutes present in the aqueous streams. Several parameters affecting the ion-exchange process are temperature, solution pH, initial metal concentration, contact time and ionic charges. Zeolites (silicate minerals) are most abundant in nature and have been extensively used to separate heavy metal ions from aqueous streams under different conditions [16, 18]. Although natural Zeolites show good performance in a few cases scale-up of the process at an industrial level is still restricted. In contrast, synthetic resins show high efficiency in comparison with natural resins. Literature reports that macroporous anion exchangers (MP62, weak basic and S6328a, strong basic) are more effective with higher affinity and adsorptive capacity to separate pollutants from wastewaters originating from textile industries [19, 20].

2.2 Advanced oxidation processes (AOP)

Advanced Oxidation Process (AOP) is a treatment technology designed to remove organic matter from wastewater by oxidation through a reaction with hydroxyl radicals. As opposed to direct oxidation, AOPs usually consume less energy. In AOPs, a sufficient amount of hydroxyl radicals are produced that impact water purification. Hybrid advanced oxidation processes such as photocatalytic fenton, photo-fenton, H_2O_2/O_3/photocatalysis and photo-electrocatalysis have drawn the attention of industrialists and academicians due to their efficiency and cost-effectiveness [18, 19]. Currently, several nano-particle supported AOPs have been discovered for the remediation of several contaminants from wastewater such as methyl orange, methylene blue, 2,4-dichlorophenol and pentachlorophenol. Recently, research has also been carried out to explore the activity of photo-Fenton and/or heterogeneous Fenton catalysts for the simultaneous removal of multiple contaminants from waste streams. TiO_2 photocatalyst mixed with fly ash has also been employed for simultaneous separation of Cd^{+2} ion and methyl orange dye from an aqueous stream (removal efficiencies of Cd^{+2}: 88% and methyl orange: 70%) [21]. Similarly, heterogeneous catalyst $Fe^{II}Fe_2^{III}O_4$ nanoparticles supported on activated carbon have been utilized in a photo-Fenton process for remediation of pollutants (aniline and benzotriazole) and maximum removal efficiency for aniline was found to be 70.4% and benzotriazole to be 99.5% [22]. Although, the studies based on nano-particle supported AOPs proved to be promising at the pilot-scale, this process has no valid evidence to prove its cost-effectiveness and its eco-friendly operation due to the toxicity of nanoparticles. Also, there is no reliable information available about the commercialization of AOPs for the simultaneous treatment of multi-component pollutant systems.

2.3 Flotation

The flotation technique has been extensively used to remove inorganic heavy metal ions from aqueous streams. Flotation is a separation process that works on the introduction of gas bubbles as the transport medium. Suspended particulate matter, being hydrophobic or adhering to gas bubbles and move towards the water solution surface—i.e., contrary to the direction of gravity. In this technique, heavy metal ions are made hydrophobic by the use of some hydrophobic agents such as surfactants (surface-active chemicals) and separated with the assistance of air bubbles. The surface-active agents consist of a hydrophilic head (water-loving part, polar) and hydrophobic carbon chains (non-polar, water-hating part). The air bubbles loaded by solutes float over the water surface and are separated as a metal-rich froth [17].

This process is highly effective for the removal of sulfide minerals. Despite several advantages (i.e. almost all minerals can be removed, surface properties are highly governed and controlled by flotation agents used), this process has some disadvantages such as high cost and complex. According to recent research, open tank settling clarifiers are currently used as primary, secondary and tertiary clarifiers. This is primarily due to their reluctance to embrace new technologies in the development of dissolved air flotation (DAF), especially in paper mills. The specific clarification is limited to 0.5 GPM per square foot. Chemical treatment improves specific load and transparency while the residence time of settling is still 60-200 minutes.

2.4 Adsorption

Adsorption is a well-established separation process used widely for the removal of inorganic and organic compounds from wastewater. This process is proved to be superior as compared to other remediation techniques due to its ease of operation. It is simple and flexible in design, capable to treat dye wastewater effectively even at higher concentrations and also insensitive to the toxicity of contaminants [3, 5, 23]. The adsorption technique is dependent upon the affinity of contaminants towards the adsorbing materials. It is influenced by many other factors such as specific surface area of adsorbent, interactions between pollutant and sorbent, particle size distributions, solution pH, system temperature and contact time. The proper selection criteria of any adsorbent for separation are based on several characteristics such as adsorption capacity of adsorbent, selectivity, regeneration power, mechanical strength and low cost. Several adsorbents have been extensively utilized and show high sorption capacity for simultaneous removal of organic and inorganic solutes from wastewater as shown in **Table 3**. For instance, Fly ash has been successfully utilized for the separation of heavy metals and dyes from a multi-component aqueous solution; $Ca(PO_3)_2$-modified carbon can be used for the separation of heavy metal ions and dye (acid blue 25); Nano-particles (TiO_2) for removal of organic dye, copper and silver heavy metals; Graphene oxide nano-composite can be used for separation of cadmium and ionic dyes; Magnetic metal–organic frameworks composite i.e. (Cu-MOFs/Fe_3O_4) have been used for separation of malachite green dye and lead ions; Zr-based magnetic Composites i.e. Zr-MFCs and Amino-decorated for separation of lead and methylene blue [17, 24]. The use of this technology for the treatment of textile wastewaters is still limited due to excessive maintenance cost, high regeneration cost, issues regarding proper disposal of used adsorbents and the requirement of pretreatment to

Preparation methods	Reaction time	Heating rate	Temperature (°C)	Yield (%)			References
				Solid	Liquid	Gas	
Fast pyrolysis	Seconds	Fast	<1000	10	70	20	[3, 5]
Hydro-carbonization	Minutes to hours	Slow	< 350	50–80	—	—	[20]
Flash pyrolysis	Seconds	Faster	775–1025	10–15	70–80	5–20	[19]
Slow pyrolysis	Hours	Slow	< 700	35	30	35	[12]
Gasification	Seconds to Minutes	Faster	700–1500	10	5	85	[3, 5]

Table 3.
Summary of adsorption capacity of various biomass-derived biochar with different operating conditions.

reduce suspended solids into feed for acceptable operational range. Thus, the adsorption technique shows promising outcomes at a commercial scale and resolves several challenges associated with waste disposal and regeneration.

2.4.1 Utilization of biochar as an adsorbent

All the above processes show their advantage and disadvantage concerning process efficiency, high costs (capital or operational), adsorbents, process conditions and removal percentage of pollutants. In this regard, biochars are receiving increasing attention and are highly recommended as a bio-adsorbent since they can both mitigate climate change by capturing carbon dioxide from the atmosphere into soil and increase the removal of organic pollutants. Biochar is defined as a carbon-rich material produced during the pyrolysis process that is a thermochemical decomposition of biomass with a temperature of about ≤700°C in the absence or limited supply of oxygen. As it is having a high-carbon content (approximately 60–90%), the application of biochar for the removal of a wide variety of contaminants from wastewater is considered a significant and long-term approach to sink atmospheric CO_2 in terrestrial ecosystems. Several kinds of biomass can be used as sources of biochar, such as wood chips, animal manure, and crop residues. Biochars have the ability to enhance the recycling of agricultural and forestry wastes. Biochar adsorbents are relatively cost-effective, environment-friendly and will be a beneficial tool for environmental remediation. Thus biochar research is gaining attention.

2.4.1.1 Characteristics of Biochar

The properties of biochar are determined by the pyrolysis temperature, the residence time, the feedstock considered, and the technology used for conversion. These factors influence the effectiveness of contamination removal. It was found that the amount of carbonized matter, the surface area, the pores, and the hydrophobicity of biochar increased with increasing temperature, consequently increasing the affinity of organic pollutants for adsorption. The presence of a high amount of carbonized matter in biochar favours the adsorption of contaminants, especially for the compounds having oxygen and hydrogen functional groups. According to research, activated carbon derived from wheat residue at 500-700°C was well carbonized and had a high surface area (>300 m^2/g), whereas charcoal made at 300-400°C was partially carbonized and had a lower surface area (<200 m^2/g) [25]. Hence, the former material exhibits high sorption capability for the removal of organic pollutants. Biochar can be made of diverse materials exhibiting different properties. The change of properties of biochar can be correlated to their function. Additionally, improving the adsorption capability of biochar through different treatments, such as chemical activation and surface modifications are found to be effective in improving its properties. This may be due to the enhanced porous structure and sorption properties that occurs after activation process [25]. Apart from activation of biochar, magnetization is also a useful method to improve biochar property. **Tables 4** and **5** summarizes the different preparation methods of biochar under different operating conditions.

2.4.1.2 Biochar adsorption mechanism

Adsorption is a surface phenomenon with a common mechanism for the removal of organic and inorganic pollutants. When a solution containing an adsorbent solute comes into contact with a solid with a very porous surface structure, the intermolecular

Material	Process type	Concentration range	Adsorbate	Contact time	Adsorption capacity	Percentage adsorption	Reference
Juniper wood	Fast Pyrolysis	—	Cd (II)	30 min	24.8–28.3 μ molg^{-1}	—	[26]
Charfines, bituminous coal and lignite coal	Slow Pyrolysis	50 mg L^{-1}	Direct brown	60 min	6.4, 2.04 and 4.1mg g^{-1}	—	[27]
Lignite-lignin	Slow Pyrolysis	—	Cu (II), Ni (II) and Pb (II)	Cu (II) and Ni (II): 40–70 min Pb (II): 10–30 min	17.8, 13.0 and 56.7 mg g^{-1}	67%	[28]
Peat	Slow Pyrolysis	100–500 mgdm^{-3}	Pb	4 h	27–106 mg g^{-1}	—	[29]
Pink bark	Fast Pyrolysis	≤ 400 mg L^{-1}	Cu (II), Ni (II) and Cd (II)	24 h	0.149, 0.107 and 0.126, mmolg^{-1}	—	[25]
Sphagnum peat moss	Fast Pyrolysis	35–210, 10–100 and 25–200 mg L^{-1}	Pb, Ni and Cu	—	24.6, 7.5 and 14.3 mg g^{-1}	—	[24]
Starch graft copolymer	Slow Pyrolysis	—	Cu (II) and Pb (II)	2 h	2.12 and 2.09 mmol g^{-1}	—	[17]
Bentonite	Gasification	100 μg ml^{-1}	Cu (II)	180 min	4.75 mg g^{-1}	85%	[25]
Chitosan bead (Chemically crosslinked)	Fast Pyrolysis	—	Reactive blue 2, reactive yellow reactive yellow 2 and Reactive red 2	5 days	86 1911 2498, 2436 and 2422 mg g^{-1}	—	[29]
R. arrhizus and *C.vulgaris*	Fast Pyrolysis	1996.2 and 387 mg L^{-1}	Iron (III)-cyanide complex	—	612.2 and 387 mg g^{-1}	—	[24]

Material	Process type	Concentration range	Adsorbate	Contact time	Adsorption capacity	Percentage adsorption	Reference
Anodonta shell	Gasification	—	Reactive green 12 and direct green	15 days	260.436 and11.3 mg g^{-1}	—	[29]
Pinus sylvestris bark	Slow Pyrolysis	5–20 mg L^{-1}	Cr (III)	24 h	9.77 mg g^{-1}	≥9	[30]
Natural clay	Fast Pyrolysis	10–50 ppm	Ni (II)	45 min	12.5 mg g^{-1}	—	[30]
Saw dust: walnut	Gasification	50–1000 and 50–500 mg L^{-1}	Methylene blue and Acid blue 25	60–180 min	59.17, 36.98 mg g^{-1}	—	[25]
Peanut hull	Fast Pyrolysis	≤ 1000 mg L^{-1}	Pb (II), Zn (II), Cu (II) and Cd(II)	4 h	30, 9, 8 and 6 mg g^{-1}	—	[31]
Sawdust	Slow Pyrolysis	1–50 mg L^{-1}	Cu (II)	60 min	4.40–0.16 mg g^{-1}	—	[25]
Peanut hull carbon	Gasification	10–20 mg L^{-1}	Hg (II)	5–180 min	109.89 mg g^{-1}	—	[31]
Kraft lignin	Fast Pyrolysis	5–200 mg L^{-1}	Cu (II)	3 h	3.38 mg g^{-1}	—	[24]
Alkali-treated straw	Gasification	—	Cr (III)	60 min	3.91 mg g^{-1}	—	[17]
Orange peel	Fast Pyrolysis	—	Direct red 23 and Direct red 80	15 min	10.72 and 21.05 mg g^{-1}	—	[31]
Hazelnut shell	Slow Pyrolysis	0.1–2.0 mmoL^{-1}	Cd^{2+}, Cr (IV), Zn^{2+} and Cr (III)	5 h	5.42, 3.99, 1.78 and 3.08 gKg^{-1}	92.4, 97.8, 87.9 and 94.6	[31]

Table 4.
Several preparation methods of biochar under different operating conditions.

Synthesis method	Process summary	Merits	Demerits	Reference
Slow pyrolysis	Prolonged residence time ranging from 1 to 6 hr. of lignocellulosic biomass with low process temperature (<700°C) at atmospheric pressure	Equal fractions of products (liquid, solid & gas). Large pore size, high ash and lignin content. Cost effective, robust and modular	Highly endothermic	[26]
Fast pyrolysis	Fast heating rate with temperature (<1000°C) with shorter reaction time ranging from few seconds to minutes.	Major products are non-condensable gases Produced bio-oil can be utilized as a feedstock for the production of energy.	Low biochar yield	[30]
Sol–gel method	Heating the solution of 0.5 mol of Citric acid and 0.5 M Aluminium nitrate at 65°C for 120 min at 180 rpm to form gel	High adsorption efficacy High surface area and available active site High thermal stability	High cost of the feedstock Large shrinkage of volume and heavy cracking during drying Produces non-uniform crystal defects	[29]
Ball milling method	Mixing of 3.30 g raw material with 330 g of agate spheres with 60 g of distilled water at 300 rpm for 12 h. Direction of mixing may be changed every 3 h. Finally, the solution can be centrifuged for 5 min at 9000 rpm. The resultant solid biochar is dried at 80°C for 12 h.	Larger pore volume Efficient adsorbent	Loud noise & strong vibration during its working process Time consuming	[22]
Co-precipitation method	Immersed the raw material with a solution of Magnesium Chloride and Aluminium Chloride (3:1) with uniform stirring for 12 h. Resultant suspension can be added to a beaker and stabilized at 60°C for 12 hr. with uniform stirring at 400 rpm. Finally, filter the resultant and washed thoroughly with distilled water and dried overnight at 80°C.	Promotes surface adsorption Generates high-capacity adsorbent	Some impurities get precipitated with the product Expensive	[28]

Synthesis method	Process summary	Merits	Demerits	Reference
Hydrothermal synthesis method	Solution A → Dissolve 0.01 mole Iron nitrate and 0.02 mole Magnesium nitrate in 50 mL ultra-pure water. Solution B → Dissolve 0.01 mol Sodium carbonate and 0.03 mol Sodium hydroxide in 30 mL aqueous solution. Add 2 g raw material in Solution B followed by Solution A with uniform stirring for 30 min. Transfer the solution to Teflon-lined high-pressure reactor and aged for 6 hr. at 120°C. Finally, filter the resultant product and washed with distilled water and dried for 8 h at 70°C.	Good magnetic properties	Unable to see the growing crystals Expensive	[24]
Solvothermal method	Mix Iron chloride, PEG 4000 and Sodium acetate in 80 mL of Ethylene glycol with constant stirring for 30 min followed by the addition of raw material. Autoclave the resultant solution for 8 hr. at 200°C and quench the mixture to room temperature. Collect the obtained black precipitate using magnet followed by washing with Ethanol and distilled water and kept the sample in oven for 8 h at 70°C.	Uniformly dispersed magnetic nanoparticles with controllable particle size synthesized High product purity	Unable to see the growing crystals	[17]
Succinylation	Add 5 g raw material to 500 mL of xylene, 14 mL of trimethylamine & 10 g of succinic anhydride and heat the solution for 8 h at 120°C for Succinylation. Filter the resultant solution using micro syringe and washed with acetone several times to remove the residues of xylene.	Production of efficient biochar	Low stability at high temperatures	[20]

Table 5.
Overview of synthesis of biochar-based sorbents.

attractive force between the liquid and the solid causes some of the solute molecules to concentrate from the solution or deposit on the solid surface. The mechanisms for the removal of organic pollutants with biochar involves surface sorption, cation/ion exchange, electrostatic interactions, precipitation and complexation [28]. All these mechanism as an individual or together plays important role and show great effect on adsorption capacity.

Surface sorption: In this process, metal ions diffuse into the pores of the sorbent to form chemical bonds. The pore volume and the surface area of the sorbent (biochar) depend upon the carbonization temperature.

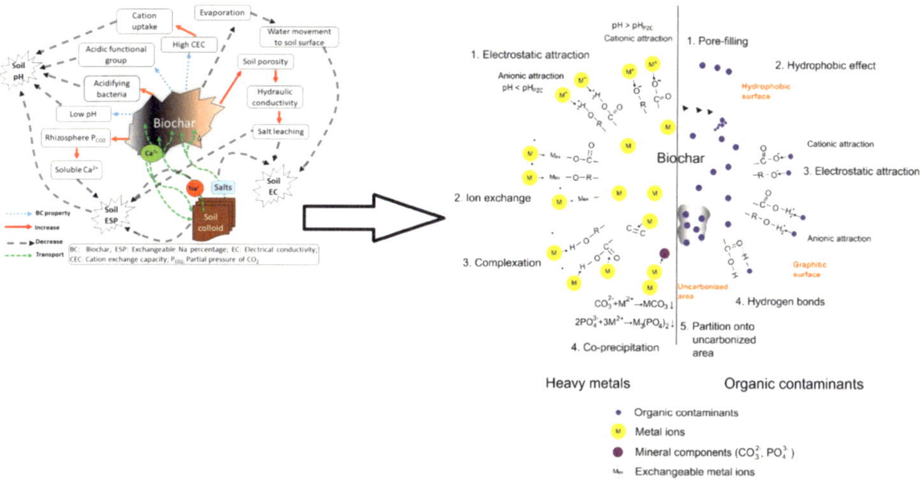

Figure 1.
Properties and mechanism of biochar functioning.

Electrostatic interaction: It is a mechanism that uses electrostatic interaction between the charged biochar particles and the metal ions to prevent metal ion mobilization.

Cation/ion exchange: The major principle of this mechanism is the exchange between protons and ionized cations on the surface of the biochar. As a result, its ability to remove heavy metals depends on the size of the contaminated surface and the surface functional groups of the biochar.

Precipitation: It is one of the main mechanisms that can be used to remove inorganic pollutants from biochar. As a result, mineral precipitates are formed either within the solution or on the surface of the sorbing material. In particular, this occurs for biochar produced from pyrolysis of cellulose and hemicelluloses with a temperature exceeding 300°C and with an alkaline property.

Complexation: Metal complexation involves the formation of multi-atom structures through the interaction of specific metal ligands. Due to the oxygen-containing functional groups present in low-temperature biochar such as phenolic, lactonic, and carboxyl, it can bind with heavy metals. The oxygen content of the biochar can lead to an increase in surface oxidation and metal complexation.

Biochar's remediation effect is achieved by these mechanisms as shown in **Figure 1** and the nature of bonding working together, rather than acting separately. The nature of the bonding depends on the type of species interaction while the adsorption process is usually classified as physisorption (characteristic of weak Van Der Waals forces) or chemisorption (characteristic of covalent bonding) [29].

3. Development of economic and sustainable biomass derived biochar

3.1 Preparation methods of biochar

Several techniques including pyrolysis, gasification, hydro carbonization have been used for the synthesis of biochar affecting the adsorption capacity and are discussed in **Table 4**. Pyrolysis of biomass is found to be the most widely used technique and can be carried out in the absence of oxygen at high temperature. Pyrolysis process can be classified as slow, fast and flash depending on the temperature and residence time. A slower heating rate and a lower pyrolysis temperature can result in the high yield of solid products [27]. It was found that slow pyrolysis results in the formation of ~35% solid yield indicating the effectiveness of the process among other three-pyrolysis techniques. Hydrothermal carbonization (HTC) is another important technique used for the synthesis of biochar. Biochar obtained from HTC exhibit superior adsorption properties with zero production of toxic substances. The main limitations of this method are requirement of high pressure, reactor cost and high temperature that limits the practical applications. Recent literature shows that the treatment of sewage sludge is found to be more effective and feasible using HTC as compared to other thermochemical processes due to low energy consumption and high thermal and mechanical stability of biochar. In addition to slow pyrolysis and HTC, other methods such as rapid pyrolysis, flash pyrolysis and gasification are also efficient and cost effective. However, such methods have low product yield and are typically used to produce bio-oil or gaseous materials.

There is a strong relationship between the preparation method and the physicochemical properties of biochar as shown in **Figure 2**. Biochar can be produced from wide range of biomass such as municipal, agricultural, aquatic or forestry having different physical, chemical and structural properties. There are several factors affecting the physicochemical properties of biochar including type of the raw material, source of biomass, pyrolysis type (slow, rapid or flash), duration of pyrolysis, size of the substrate, temperature and heating rate [26]. These operating parameters results in the number of surface functional groups including hydroxyl, carbonyl, methyl and carboxyl. In addition, several factors affect the structure of biochar including oxygen-containing aromatic functional groups, high carbon content, surface area and high porosity. These factors significantly favour the adsorption of pollutants onto the surface of biochar.

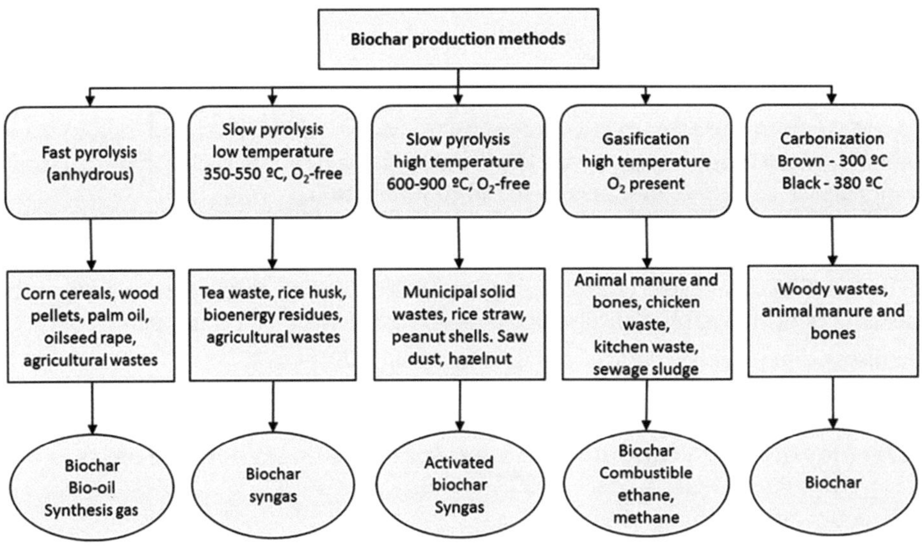

Figure 2.
Biochar preparation methods and its applications.

3.2 Biochar's properties influencing its activity

As discussed above, properties of biochar are influenced by pyrolysis temperature, residence time, feedstock, and the thermal conversion technology. The variations in these parameters results in the variation in the removal efficiency of pollutants as shown in **Figure 3**. The selection of biochar for a specific purpose depends on several factors such as mechanical strength, adsorption efficiency, cost, regeneration, ease of synthesis, selectivity for different pollutants, reusability and rate of adsorption and desorption. Due to high porosity, sorption ability of biochar is highly dependent on the surface area. The high surface area enhances the ability to adsorb the pollutants on the surface of biochar. This can be done either by physical modification (such as purging of steam and gas) or by chemical activation using various chemical reagents (concentrated or diluted). In addition to porosity, several other factors including pH, temperature, adsorbent dose, and agitation speed affects the adsorption process [20]. pH is the most crucial parameter that affects the dissociation of functional groups and the charge on the active sites, thereby affecting the adsorption capacity. Another significant parameter is the biochar dosage. Significant increase in the adsorption of pollutants has been found with increase in the biochar dosage due to the availability of sufficient active sites on the surface of biochar. While further increase in the biochar dose than the optimal dosage declines the adsorption of pollutants due to the saturation or blockage of active sites. Generally, adsorption processes are endothermic in nature thus on increasing the temperature, increase in the adsorption of pollutants was observed. It has been observed that high temperature leads to the degradation of molecules that results in the decline of adsorption capacity [18]. Thus, maintaining an optimum temperature is highly essential. Agitation speed is another critical parameter that influences the adsorption capacity and reaction mechanism. With increase in the agitation speed, gradual increase in the adsorption capacity has been observed.

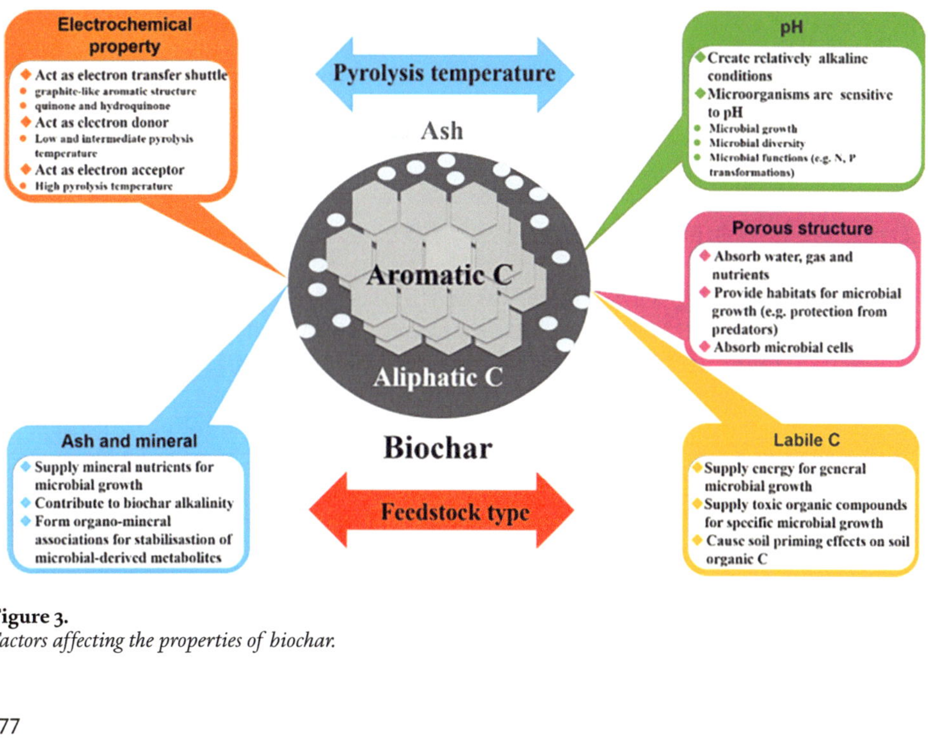

Figure 3.
Factors affecting the properties of biochar.

This may be due to the increase in the turbulence and reduction in the thickness of the boundary layer around the biochar that improves the interaction between adsorbate and adsorbent. According to the literature, the boundary layer and intraparticle diffusion are the controlling steps for the adsorption mechanism, and the optimum speed for adsorption process is usually in the range of 120 rpm to 200 rpm.

4. Future perspectives

It is evident from the above studies that the biochars are potential and economical candidates for water purification. This study covers the advancement in the field of biochars followed by their utilization in various fields. However, detailed research is still required in terms of physical and chemical modifications to enhance porosity and surface area of biochar. Further, polymorphs of biomass for the production of biochar and their effect on multicomponent systems still needs exploration. A more underlying mechanistic approach is required to understand the role and performance of individual components i.e. cellulose, hemicellulose and lignin as these polymers provide heterogeneity to the biomass matrix. Differences in the magnitude of adsorption capacities using different biochars having the same origin and composition is an indicative of unexplained correlation between morphological patterns and molecular structure of biochars. Besides this, a critical investigation is required to determine the effectiveness of surface area, porosity and functional groups of biochars. Many studies cover the technical performance of biochars while the economic feasibility and environmental impact is neglected. Studies need to be carried out in detail to suggest an effective binding mechanism of several pollutants with biochars. Also, no study has been reported on the removal of anions, radionuclides and pesticides using biochars. Further, limited data is available for the competitive adsorption of contaminants especially on phenols and dyes. In addition, some biochars are incapable to perform under neutral conditions (pH 7.0) and at low concentrations (μg/mL), therefore it is essential to develop biochars which are effective at normal temperature and short residence time. Despite the limited price information and widespread utilization, scale-up technology of biochars is strongly recommended due to their engineering applicability, easy availability and techno-economic feasibility.

5. Drawbacks of biochar

Although there is a growing consensus on the benefits of biochar in various areas and at the same time different point of view exists. Several concerns have been raised on the sustainability and carbon neutrality in the utilization of biochar. Some of the challenges which limit the usage of biochar for scale-up production include (i) incompetence while supressing the emission of greenhouse gases (ii) effectiveness of biochar for all type of organic pollutants (iii) toxicity of biochar. For instance, production of biochar from different raw material may contain chlorinated organic compounds such as polyvinyl chloride or pentachlorophenol and may result in the formation of polychlorinated biphenyl-p-dioxins, PAH and furans. However, if there is a sudden increment in the level of such compounds in the biochar then it imposes threat to the environment and human health. Therefore, it is essential to suitably select the feedstock and synthesis conditions including temperature, residence time and technology that could control the concentrations of potentially toxic compounds in the desired biochar. Safe usage of biochar materials ensuring human health and environment benefit along with comprehensive life cycle analysis and environmental risk assessment is recommended.

6. Conclusion

This chapter attempts to cover wide range of low-cost biochars for the effective removal of toxic contaminants from wastewater. These materials offer several advantages including technical feasibility and engineering applicability and serves as a boon for the environmental scientists and government authorities. The suitable selection of biochars not only minimizes the cost inefficiency but also improves profitability and adds promising benefits for the scale-up technologies in future. In addition, some biomass derived materials with and without prior pretreatment can be used as biochars in non-industrialized sectors. The purpose is to implement sustainable development policies at local and national levels. With few exceptions, it appears from the literature that biochars having good carbon content are usually versatile adsorbents that can be successfully used to remove contaminants from wastewater. Besides the technological progression, some limitations that still need to be overcome are (i) low surface area of biochar (ii) critical balance between pH and operating temperature during adsorption (iii) relationship between composition and constituents of the biochars is essential. Last but not the least, exploration of the possibility of recovering or reusing adsorbed substances needs attention.

Acknowledgements

The authors acknowledge Guru Gobind Singh Indraprastha University, New Delhi, India.

Declaration of competing interests

The authors declare that they have no known competing financial interests or personal relationships that could have appeared to influence the work reported in this chapter.

Author details

Uplabdhi Tyagi and Neeru Anand*
University School of Chemical Technology, Guru Gobind Singh Indraprastha University, India

*Address all correspondence to: neeruanand@ipu.ac.in

References

[1] Ghosh D, Maiti SK. Biochar assisted phytoremediation and biomass disposal in heavy metal contaminated mine soils: A review. International Journal of Phytoremediation. 2021;**23**(6):559-576

[2] Shukla P, Giri BS, Mishra RK, Pandey A, Chaturvedi P. Lignocellulosic biomass-based engineered biochar composites: A facile strategy for abatement of emerging pollutants and utilization in industrial applications. Renewable and Sustainable Energy Reviews. 2021;**152**:111643

[3] Singh E, Kumar A, Mishra R, You S, Singh L, Kumar S, et al. Pyrolysis of waste biomass and plastics for production of biochar and its use for removal of heavy metals from aqueous solution. Bioresource Technology. 2021a;**320**:124278

[4] Godwin PM, Pan Y, Xiao H, Afzal MT. Progress in preparation and application of modified biochar for improving heavy metal ion removal from wastewater. Journal of Bioresources and Bioproducts. 2019;**4**(1):31-42

[5] Singh A, Sharma R, Pant D, Malaviya P. Engineered algal biochar for contaminant remediation and electrochemical applications. Science of the Total Environment. 2021b;**774**:145676

[6] Tomczyk A, Sokołowska Z, Boguta P. Biomass type effect on biochar surface characteristic and adsorption capacity relative to silver and copper. Fuel. 2020;**278**:118168

[7] Yaashikaa PR, Kumar PS, Varjani S, Saravanan A. A critical review on the biochar production techniques, characterization, stability and applications for circular bioeconomy. Biotechnology Reports. 2020;**28**:e00570

[8] Liu W, Dong Y, Lin H, Shi Y. Synthesis strategies, mechanisms, and potential risks of biomass-based adsorbents (BAs) for heavy metal removal from aqueous environment: A review. Water, Air, & Soil Pollution. 2021;**232**(10):1-22

[9] Yin W, Dai D, Hou J, Wang S, Wu X, Wang X. Hierarchical porous biochar-based functional materials derived from biowaste for Pb (II) removal. Applied Surface Science. 2019;**465**:297-302

[10] Zhao J, Shen XJ, Domene X, Alcañiz JM, Liao X, Palet C. Comparison of biochars derived from different types of feedstock and their potential for heavy metal removal in multiple-metal solutions. Scientific Reports. 2019;**9**(1):1-12

[11] Qiu B, Tao X, Wang H, Li W, Ding X, Chu H. Biochar as a low-cost adsorbent for aqueous heavy metal removal: A review. Journal of Analytical and Applied Pyrolysis. 2021;**155**:105081

[12] Kwon G, Bhatnagar A, Wang H, Kwon EE, Song H. A review of recent advancements in utilization of biomass and industrial wastes into engineered biochar. Journal of Hazardous Materials. 2020;**400**:123242

[13] Kaya N, Arslan F, Yildiz Uzun Z. Production and characterization of carbon-based adsorbents from waste lignocellulosic biomass: Their effectiveness in heavy metal removal. Fullerenes, Nanotubes, and Carbon Nanostructures. 2020;**28**(10):769-780

[14] Gao L, Goldfarb JL. Heterogeneous biochars from agriculture residues and coal fly ash for the removal of heavy metals from coking wastewater. RSC Advances. 2019;**9**(28):16018-16027

[15] Zhang Z, Zhu Z, Shen B, Liu L. Insights into biochar and hydrochar production and applications: A review. Energy. 2019;**171**:581-598

[16] Kim JY, Oh S, Park YK. Overview of biochar production from preservative-treated wood with detailed analysis of biochar characteristics, heavy metals behaviors, and their ecotoxicity. Journal of Hazardous Materials. 2020;**384**:121356

[17] Mahdi Z, El Hanandeh A, Yu QJ. Preparation, characterization and application of surface modified biochar from date seed for improved lead, copper, and nickel removal from aqueous solutions. Journal of Environmental Chemical Engineering. 2019;**7**(5):103379

[18] Yuan P, Wang J, Pan Y, Shen B, Wu C. Review of biochar for the management of contaminated soil: Preparation, application and prospect. Science of the Total Environment. 2019;**659**:473-490

[19] Hopkins D, Hawboldt K. Biochar for the removal of metals from solution: A review of lignocellulosic and novel marine feedstocks. Journal of Environmental Chemical Engineering. 2020;**8**(4):103975

[20] He J, Strezov V, Kumar R, Weldekidan H, Jahan S, Dastjerdi BH, et al. Pyrolysis of heavy metal contaminated Avicennia marina biomass from phytoremediation: Characterisation of biomass and pyrolysis products. Journal of Cleaner Production. 2019;**234**:1235-1245

[21] Shi J, Yang Z, Dai H, Lu X, Peng L, Tan X, et al. Preparation and application of modified zeolites as adsorbents in wastewater treatment. Water Science and Technology. 2018;**2017**(3):621-635

[22] Huang J, Kankanamge NR, Chow C, Welsh DT, Li T, Teasdale PR. Removing ammonium from water and wastewater using cost-effective adsorbents: A review. Journal of Environmental Sciences. 2018;**63**:174-197

[23] Wang M, Zhu Y, Cheng L, Andserson B, Zhao X, Wang D, et al. Review on utilization of biochar for metal-contaminated soil and sediment remediation. Journal of Environmental Sciences. 2018;**63**:156-173

[24] Bhattacharjee C, Dutta S, Saxena VK. A review on biosorptive removal of dyes and heavy metals from wastewater using watermelon rind as biosorbent. Environmental Advances. 2020;**2**:100007

[25] Jain AK, Gupta VK, Bhatnagar A, Suhas. A comparative study of adsorbents prepared from industrial wastes for removal of dyes. Separation Science and Technology. 2003;**38**(2): 463-481

[26] Ayala FE, Reyes-Vidal Y, Bacame-Valenzuela J, Pérez-García J, Palomares AH. Natural and synthetic zeolites for the removal of heavy metals and metalloids generated in the mining industry. In: New Trends in Removal of Heavy Metals from Industrial Wastewater. Elsevier; 2021. pp. 631-648

[27] Abdulsalam J, Mulopo J, Bada S, Oboirien B. Natural gas storage properties of adsorbents synthesised from three different coal waste in South Africa. Fuel. 2020;**267**:117157

[28] Atun G, Hisarli G. Adsorption of carminic acid, a dye onto glass powder. Chemical Engineering Journal. 2003;**95**(1-3):241-249

[29] Tingting LI, Ruixue LI, Zheng MA, Anting YANG, Chenlu JIAO, Jian WANG. Preparation of cellulose-sodium alginate-sepiolite porous bead and its application in adsorption of methylene

blue. Acta Materiae Compositae Sinica. 2021;**38**(12):4273-4281

[30] Hubadillah SK, Othman MHD, Harun Z, Ismail AF, Rahman MA, Jaafar J. A novel green ceramic hollow fiber membrane (CHFM) derived from rice husk ash as combined adsorbent-separator for efficient heavy metals removal. Ceramics International. 2017;**43**(5):4716-4720

[31] Li J, Xing X, Li J, Shi M, Lin A, Xu C, et al. Preparation of thiol-functionalized activated carbon from sewage sludge with coal blending for heavy metal removal from contaminated water. Environmental Pollution. 2018;**234**:677-683

Section 3

Biochar Uses in Energy Sector and Chemical Productions

Chapter 9

Prospects of Biochar as a Renewable Resource for Electricity

Ariharaputhiran Anitha and Nagarajan Ramila Devi

Abstract

To face the change in energy paradigm, we need to devise technology that utilizes renewable resources and eventually realizes sustainability. Fuel cells generate electricity in a greener way, the efficiency and its cost-effectiveness depend mainly on the electrode material. Biochar serves as the promising electrode material, fuel, and separator membrane for fuel cells by being cheap, renewable, and possessing excellent electrochemical performance. The chapter is expected to provide a database of knowledge on how biochar with diversified physical and chemical features and functionalities can be effectively utilized for the possible application as electrode material for energy systems. The chapter appreciates the immense wealth of choice of biochar available with us for an important application in the area of energy as electrode material, fuel, and separator membrane for fuel cells.

Keywords: biochar, biomass carbon, fuel cells, electrode, separator

1. Introduction

The enormous usage of fossil fuels leads to harmful environmental damage, viz., global warming, depletion of energy resource, and lack of sustainable growth. To overcome this situation, now the world is in need of pollution-free green energy source. Moreover, the energy source should be available anytime anywhere in order to be a perennial source and to have a sustainable development. In regard of this, waste has to be used in a vast amount as an energy source instead of traditional fossil fuels which liberates huge amount of carbon dioxide, i.e., waste to wealth conversion. The energy derived from biomass termed biomass energy is the fourth largest energy source next to three fossil fuels, viz., coal, petroleum and natural gas. About 5% of the United States' primary energy need is fulfilled by biomass in 2021. As biomass is a carbon-neutral resource, the energy produced out of it is considered clean green energy.

It is noteworthy to know about the biomass and biochar. Biomass is the matter from biological organisms and biochar is the product obtained by the thermal/chemical processing of biomass. The sources of biomass include forest residue, agricultural crops and residues, domestic waste, municipal waste, marine waste, and industrial waste [1]. Biomass though worthless, but it is a great source for valuable biochar. The biochar finds its multifarious applications such as:

- Adsorbent for toxic pollutants [2]
- Soil amendment to improve soil health
- Catalyst support for electrolysis [3, 4]
- Electrode material for electrochemical energy devices, viz., lithium-ion batteries [5], supercapacitors [6, 7], and fuel cells [8, 9].

The widespread utility of these electrochemical energy devices is hampered due to the high cost of the electrode materials. There arises revolution in the field of energy due to the utility of biochar as electrode material in electrochemical energy devices as it replaces the costlier electrode materials, thereby paves the way to the production of electrical energy at low cost. In addition to that, it gives value-added utilization of biomass in the field of energy.

A brief description of fuel cells is worthwhile here. Fuel cell converts the chemical energy of fuel and oxidizing agent into electricity by electrochemical redox reactions. It serves as an endless power source for space vehicles and submarines. The main components of fuel cells are the cathode, anode, and electrolyte which facilitate the passage of ions. Fuel cells are of many types, such as proton exchange membrane fuel cell (PEMFC), solid oxide fuel cell (SOFC), microbial fuel cell (MFC), and direct carbon fuel cell (DCFC).

In this chapter, detailed survey of utility of biochar as electrode material and separator in MFC and as fuel in DCFC is given in systematic way. This enables us to understand the value-added utilization of biomass in the field of energy and to explore many other biomass for its utility in the near future.

2. Microbial fuel cell (MFC)

Microbial fuel cell (MFC), a bioelectrochemical device, employs organic waste to produce electrical energy [10, 11]. In MFC, proton exchange membrane separates the anode and cathode. Electrodes used in MFC are crucial in determining its efficiency [12]. Electrogenic bacteria on the surface of the anode oxidizes the organic matter to generate electrons and protons [13, 14]. The electrons generated flow reaches the cathode to combine with an electron acceptor via an external circuit [15, 16]. Due to superior electrochemical oxidation capacity, great abundance, and clean reaction product, oxygen is the most widely used electron acceptor [17, 18]. However, the poor cathode oxygen reduction reaction and high oxygen mass transfer resistance significantly lower the performance of MFC [19]. The working mechanism of MFC is given as follows:

Acetate and glucose in organic compounds are oxidized in MFCs, which generate electrons, flow down to an external circuit, and produce electricity, whereas organic compounds are anaerobically oxidized and result in the evolution of protons, electrons, and CO_2. In MFC, water is generated in cathode by the reduction of protons and electrons with the usage of oxygen supplied from outside. The protons and electrons thus liberated reach the cathode through an electrical circuit in presence of electrolyte medium. The formation of water in MFC is represented by the equation given as follows [20]:

$$CH_3COO^- + 4\,H_2O \rightarrow 2HCO_3^- + 9\,H^+ + 8\,e^-. \quad (1)$$

$$2O_2 + 8H + +8e^- \rightarrow 4H. \quad (2)$$

In MFC, oxidation occurs at the anode and reduction at the cathode, thereby creates the potential difference between the electrodes, leading to the generation of bioelectricity as shown in **Figure 1**.

Biochar can act as electrodes (anode/cathode), electrocatalysts, and proton exchange membranes b [21]. Biochar acting as electrodes should possess high porosity, rich carbon content, excellent electrical conductivity, large surface area, and cost-effective. Moreover, it should be nonbiodegradable, biocompatible, and pave way to waste to wealth conversion. **Table 1** lists the biochar derived from various biomass, its method of preparation, processing temperature, its utility as electrode material for MFC, and the power density derived from it.

2.1 Biochar as separator

Proton exchange membrane (PEM) in MFC is superior in its performance when it possesses large proton conductivity, minimal oxygen, and substrate crossover, decreased biofouling rate, and low cost. As the source of biochar is abundant and easily available, the biochar possesses strong cation exchange properties, a high concentration of surface-active sites, and excellent porous nature supports its use in PEMs. In PEM fuel cell, the biochar acts as the unique separator which usually replaces the Nafion polymer membrane. Moreover, the biochar acts as a porous membrane having expanding sustainability than the other membrane and cost-effective eco-friendly catalyst material. The biochar-built PEM fuel cell was applicable for the Industrial and lab scale preparation.

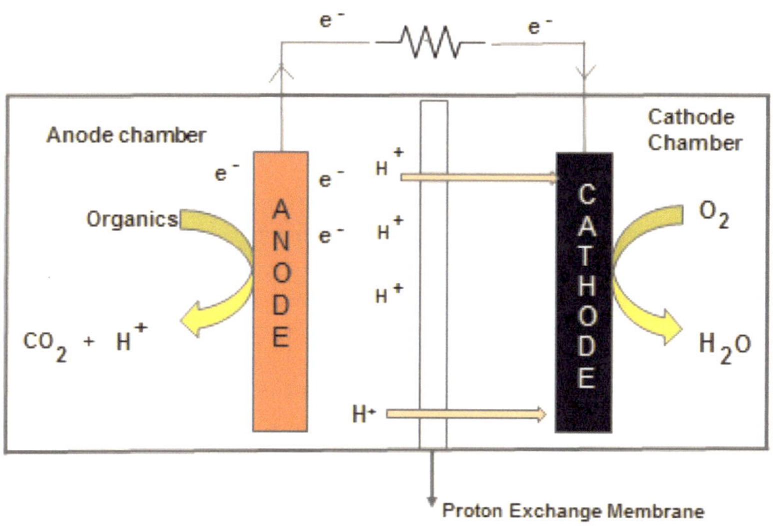

Figure 1.
Working of MFC.

S.No	Biomass	Electrode	Method of preparation	Processing temperature (°C)	Power Density (mW/m^2)	Ref
1	Balsa wood	Cathode	Pyrolysis	800	200.0	[22]
2	Banana	Cathode	Thermal treatment	550	393.7	[23]
				900	483.7	
			KOH activation	550	424.6	
				900	528.2	
3	Biodigester plant waste	Cathode	$KHCO_3$ activation	850	NA*	[24]
4	Pinus resinosa	Cathode	Pyrolysis	850	356.0	[25]
5	Coconut shell	Anode	Pyrolysis	400	283.4 ± 9.6	[26]
6	Pinewood chips	Anode	Carbonization	1000	457.0	[27]
	Pine sawdust				532.0	
7	Peanut shells	Anode	Pyrolysis	800	NA*	[28]
8	Maple wood	Anode	Pyrolysis	350–600	41.4	[29]
9	Rubber tree sawdust	Anode	Pyrolysis	500	326.0	[30]
10	Waste wood	Anode and cathode	Pyrolysis	1000	600.0	[31]
11	Wood	Cathode	KOH activation	1000	146.7	[32]
12	Corn straw	Anode and cathode	KOH activation	900	889.0	[33]
13	Giant cane stalk	Cathode	NA	900	NA*	[34]
14	Watermelon rind	Cathode	KOH activation	180	26.2	[35]

NA—not available.*

Table 1.
Sources of biochar and its utility in MFC.

2.2 Biochar based catalyst

The basic mechanism of oxygen reduction reaction (ORR) is the absorption of proton from the electrolyte by the oxygen molecule at the cathode. Followed by this, the transfer of electrons takes place from anode to metallic wire. Reaction requires more energy for the production of fuels. For the good performance of MFC, the anode and cathode play the catalyst role. For enhancing the sustainability, stability, and activity, the cathode fabrication is very important. The necessity of good cathodic material is to reduce the activation potential of ORR reaction and the cost-effective process. Initially, the fuel reactions were carried out using a platinum catalyst which serves as the catalyst for the reduction of oxygen and reduction. The economic preparation of the material was not affordable for the large-scale preparation as well as not suitable for the domestic purpose application. For the replacement of notable platinum catalyst, the non-transition metal, 2D material, carbon material, and porous

S.No	Biomass	Function	Method of preparation	Processing temperature (°C)	Power Density (mW/m^2)	Refs.
1	Giant cane-clay composite	Solid separators	Pyrolysis	350	40.0	[36]
			Nitrogen flow	900		
2	Coconut shell biochar blended with metal (20%)	Anode CS-$Si_{0.2}$	Pyrolysis	500	16.8	[37]
		CS-$Zn_{0.2}$			22.9	
		CS-$Cu_{0.2}$			38.7	
3	Sewage sludge	Catalyst (ORR)	Carbonization	900	500.0	[38]
4	Banana peel	PEM	H_2SO_4 activation	600	41.1	[39]

Table 2.
Sources of biochar, its preparation, functions in MFC, and the power obtained.

material are used for the fuel reaction. In recent research, the usage of biochar material from natural sources acts as the cathodic material for the ORR reduction and showed the good performance than the other catalyst materials. The mechanistic reactions are explained in various ways.

List of biochar derived from various biomass and its utility as separator, PEM, and catalyst for ORR are given in **Table 2**.

3. Direct carbon fuel cell

Global energy demand depends mainly on conventional sources such as coal, petroleum, and natural gas since earlier days. The excess use of coal as an energy source in the past is due to its low cost, abundance, and extensive distribution throughout the world [40]. As these sources are nonrenewable, its continuous usage leads to scarcity. Moreover, continuous usage of conventional sources leads to environmental damage, thereby exploration of clean energy source is the need of the hour. This put forward the steps to initiate energy generation from renewable sources like biomass. The direct carbon fuel cell (DCFC) employs carbon as anode operates on a high-temperature range of 700–900°C. It is superior over other fuel cells by attaining 80% efficiency (for power generation) [41, 42]. The carbon used as fuel in DCFC may be coal, biomass, and organic waste which are abundantly found in nature. DCFC transforms chemical energy trapped in the solid carbon fuel into electrical energy. DCFC accounts for green energy generation as it does not require any gasification processes and other conventional electric generators [43, 44]. The elemental carbon act as fuel contains high-energy density and is oxidized electrochemically at the electrodes [43, 44].

Based on the working electrolyte, there are three main categories of DCFCs, namely, molten hydroxide, molten carbonate, and solid oxide DCFCs. Molten hydroxide and molten carbonate DCFC utilizes NaOH/KOH and carbonates, respectively, as their electrolyte. The electrolyte is filled in a metal vessel, which acts as the cathode. The carbon materials function both as fuel and the anode and are dipped into the electrolyte. Solid oxide DCFC resembles the other two DCFCs except using an oxygen ion (O^{2-}) conducting ceramic electrolyte. The most commonly used electrolyte is Y_2O_3 stabilized zirconia due to high ionic conductivity, better stability, chemical and thermal compatibility, mechanical robustness, easy fabrication, and low cost [45–47]. Solid oxide DCFC though favorable for its simple design, but it has the limitation of low output [48]. This low output is explicitly due to the limited reaction zone at the carbon fuel and the electrolyte interface.

A DCFC consumes solid carbon and oxygen to produce electrical energy through electrochemical anodic and cathodic reactions [49, 50]. The overall reaction involved in DCFC is the simple combination of carbon and oxygen to form carbon dioxide. Electrooxidation of carbon occurs at the anode, whereas electroreduction of oxygen occurs at the cathode [40, 49, 50]. Working of different types of DCFC is shown in **Figure 2**.

Reaction at the anode (oxidation of carbon).

$$C + (\text{electrolyte})\ 4OH^- \rightarrow CO_2 + 2H_2O + 4e^-. \tag{3}$$

$$C + (\text{electrolyte})\ 2CO_3{}^{2-} \rightarrow 3CO_2 + 4e^-. \tag{4}$$

Overall anodic reaction.

$$C + 2O^{2-} \rightarrow CO_2 + 4e^-. \tag{5}$$

Reaction at the cathode(reduction of oxygen).

$$O_2 + 2H_2O + 4e^- \rightarrow 4OH^-. \tag{6}$$

$$2CO_2 + O_2 + 4e^- \rightarrow 2CO_3{}^{2-}. \tag{7}$$

$$O_2 + 4e^- \rightarrow 2O^{2-}. \tag{8}$$

If DCFC is supposed to operate at a high temperature (above 700°C), carbon electrooxidation is overlooked by the reverse Boudouard reaction (electrooxidation of CO). So direct electrooxidation of solid carbon is overlooked by direct electrooxidation of CO, termed as CO shuttling mechanism [51] which leads to low carbon fuel utilization [52].

$$C + CO_2 \leftrightarrow 2CO. \tag{9}$$

Being reverse Boudouard reaction is endothermic, decreasing the working temperature of DCFC will enhance the CO reduction in the anode exhaust, thereby increase carbon fuel usage [49, 53]. DCFC attracts the scholarly attraction owing to its low-maintenance cost, simple cell structure, and rich availability of carbon feedstocks, i.e., biomass. List of biochar derived from biomass which finds its utility in various types of DCFC is given in **Table 3**.

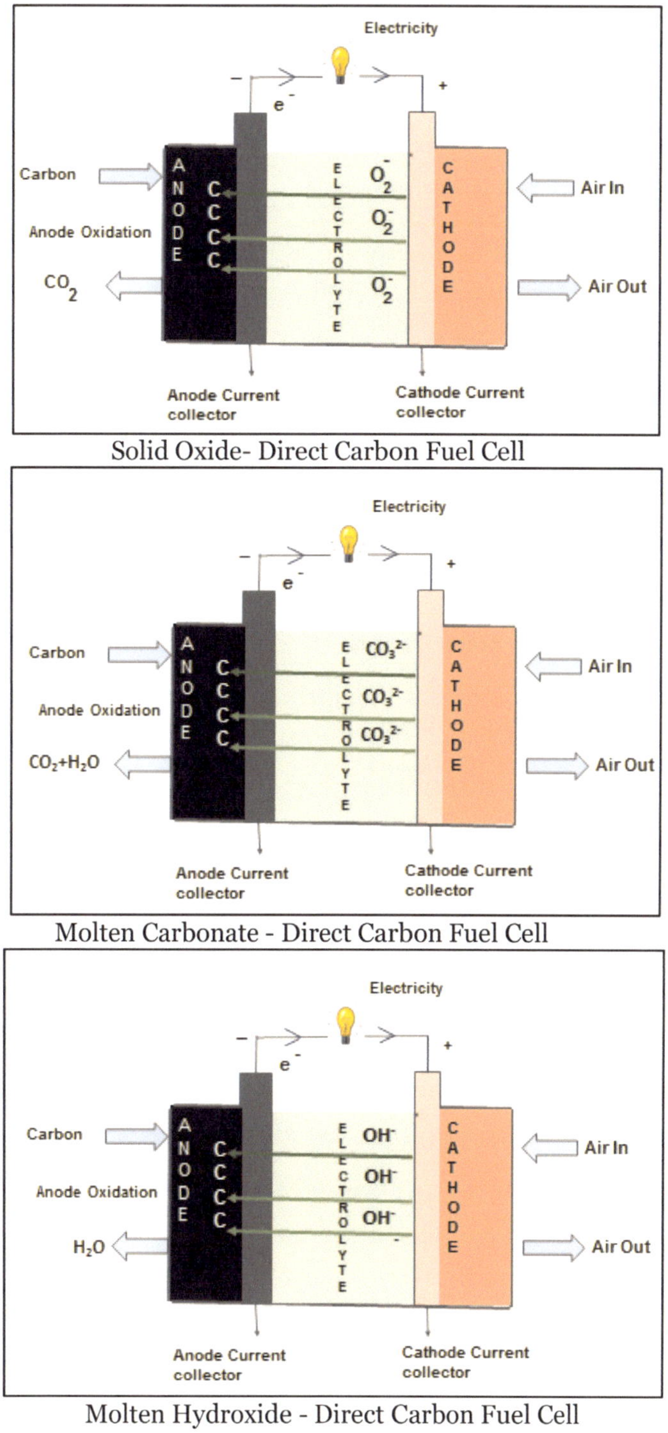

Figure 2.
Working of different types of DCFC.

S.No	Biomass	Electrolyte	Processing temperature (°C)	Open circuit voltage (V)	Power Density (mW/m^2)	Refs.
1	Corn cob	Samarium doped ceria (SDC) with eutectic carbonate phase	750	1.05	185.0	[54]
2	Almond shell	Ceria-carbonate composite	700	1.07	127.0	[55]
3	Olive wood	Ceria-doped samarium (SDC) combined with molten carbonate	700	1.02	105.0	[56]
4	Beech wood chips	Yttria-stabilized zirconia	800	1.00	100.0	[57]
5	Acacia wood chips	Yttria-stabilized zirconia	800	0.80	90.0	
6	Waste coffee grounds	Yttria-stabilized zirconia	900	1.10	87.2	[58]
7	Miscanthus straw	Yttria-stabilized zirconia	800	1.15	70.0	[59]
8		Molten Carbonate		0.88	12.0	
9	Coconut	Yttria-stabilized zirconia	800	0.87	60.0	[60]
10	Wheat	ScSZ electrolyte layer	800	1.18	67.0	[61]
11	Spruce	ScSZ	800	1.16	57	
12	Wood	Molten carbonate	700	1.00	25	[62]
13	Refuse plastic/ paper fuel	Molten carbonate	700	1.20	22	[63]
14	Refuse derived fuel	Molten carbonate	700	1.00	17	
15	Sunflower husks	Molten hydroxide	450	1.01	22	[64]
16	willow shavings			0.87	20	
17	Pine			1.05	18	
18	Black liquor	Yttria-stabilized zirconia	700–800	0.80	122	[65]
19	Eucalyptus leaves	Molten carbonate	700	0.84	NA*	[66]
20	Neem leaves			0.73		
21	Mast leaves			0.71		
22	Melon seed husk	Molten carbonate	450	0.71	550.0	[67]
23	Palm kernel shell	Yttria-stabilized zirconia	850	0.81	330.0	[68]
24	Pine pellets	Molten hydroxide	800	0.84	766.0	[69]
25	Pine bark pellets	Molten hydroxide	800	1.07	450.0	[70]
26	Pistachio shells	Yttria-stabilized zirconia	800	0.94	155.0	[71]
27	Pecan shells				140.0	
28	Sawdust				100.0	

S.No	Biomass	Electrolyte	Processing temperature (°C)	Open circuit voltage (V)	Power Density (mW/m²)	Refs.
29	Pomelo peel	Yttria-stabilized zirconia	850	1.02	3090.0	[72]
30	Reed	Ceria-doped samarium (SDC) combined with molten carbonate	750	0.96	3780.0	[73]
31	Rice husk	Yttria-stabilized zirconia	750	0.81	1790.0	[74]
32	Rubber wood	Yttria-stabilized zirconia	850	0.77	294.0	[75]
33	Walnut shell	YSZ	800	0.97	1470.0	[76]
34	Wheat straw	YSZ	800	1.0	1870.0	[77]
35	Corncob	Miscanthus straw biomass		0.98	2040.0	
36	Bagasse			0.99	2600.0	

NA—not available.*

Table 3.
List of Biochar served as fuel in DCFC.

4. Challenges and future perspective

To understand the mechanism behind the electrochemical reaction kinetics and carbon oxidation at the anode/electrolyte interface is still a challenge for us. In addition to that, metallic components in the biochar inhibit the electrochemical performance of the fuel cell. Determining the amount of ash accumulation is also a key factor and should be researched into as well to determine the lifetime of DCFC. Technological expertise in the cell design along with the clear understanding of kinetics will solve the issues in near future.

5. Conclusion

Biochar an inexhaustive renewable resource solves many environmental issues arised in recent decades, viz., pollution, remediation in soil, and water. Research studies in recent years advocate the multifarious utility of biochar as fuel in DCFC and as electrodes, separator, and catalyst for ORR in MFC. Moreover, biochar-based MFCs remove hazardous chemicals from wastewater, DCFC utilizes carbon from zero-cost sources as fuel along with the generation of electricity.

Author details

Ariharaputhiran Anitha* and Nagarajan Ramila Devi
Department of Chemistry, V.V. Vanniaperumal College for Women, Virudhunagar, India

*Address all correspondence to: anithaa@vvvcollege.org

References

[1] Kalyani P, Anitha A. Biomass carbon & its prospects in electrochemical energy systems. International Journal of Hydrogen Energy. 2013;**38**(10): 4034-4045. DOI: 10.1016/j.ijhydene. 2013.01.048

[2] Anitha A. Performance of activated carbon from Acalypha indica (Kuppaimeni) leaf as an adsorbent for copper (II) ions. Journal of Advanced Scientific Research. 2020;**11**(3): 214-219. Available from: https://www.sciensage.info/index.php/JASR/article/view/529

[3] Palanichamy K, Ariharaputhiran A. Areca leaves as a source of carbon: Preliminary investigation as catalyst support for electrolytic hydrogen evolution in acidic medium. International Journal of Hydrogen Energy. 2013;**38**(5):2263-2270. DOI: 10.1016/j.ijhydene.2012.12.003

[4] Kalyani P, Anitha A, Darchen A. Activated carbon from grass – A green alternative catalyst support for water electrolysis. International Journal of Hydrogen Energy. 2013;**38**(25): 10364-10372. DOI: 10.1016/j.ijhydene. 2013.06.022

[5] Fey GTK, Lin YY, Huang KP, Lin YC, Kumar TP, Cho YD, et al. Green energy anode materials: Pyrolytic carbons derived from peanut shells for lithium ion batteries. Advanced Materials Research. 2011;**415-417**:1572-1585. DOI: 10.4028/www.scientific.net/AMR.415-417.1572

[6] Kalyani P, Anitha A. Capacitor behaviour of activated carbon from used tea dust powder. Asian Journal of Chemistry. 2015;**27**(4): 1365-1370. DOI: 10.14233/ajchem.2015.17970

[7] Anitha A, Kalyani P. Refuse derived energy - Tea derived boric acid activated carbon as an electrode material for electrochemical capacitors. Portugaliae Electrochimica Acta. 2013;**31**(3):165-174. DOI: 10.4152/pea.201303165

[8] Jafri N, Wong WY, Doshi V, Yoon LW, Cheah K. A review on production and characterization of biochars for application in direct carbon fuel cells. Process Safety and Environment Protection. 2018;**118**: 152-166. DOI: 10.1016/j.psep.2018. 06.036

[9] Patwardhan SB, Pandit S, Gupta PK, Jha NK, Rawat J, Joshi HC, et al. Recent advances in the application of biochar in microbial electrochemical cells. Fuel. 2022;**311**:122501. DOI: 10.1016/j.fuel. 2021.122501

[10] Firdous S, Jin W, Shahid N, Bhatti Z, Iqbal A, Abbasi U, et al. The performance of microbial fuel cells treating vegetable oil industrial wastewater. Environmental Technology & Innovation. 2018;**10**: 143-151. DOI: 10.1016/j.eti.2018. 02.006

[11] Thomas YRJ, Picot M, Carer A, Berder O, Sentieys O, Barriere F. A single sediment-microbial fuel cell powering a wireless telecommunication system. Journal of Power Sources. 2013;**241**: 703-708. DOI: 10.1016/j.jpowsour. 2013.05.016

[12] Elshobary ME, Zabed HM, Yun J, Zhang G, Qi X. Recent insights into microalgae-assisted microbial fuel cells for generating sustainable bioelectricity. International Journal of Hydrogen Energy. 2021;**46**(4):3135-3159. DOI: 10.1016/j.ijhydene.2020.06.251

[13] Bhande R, Noori M, Ghangrekar M. Performance improvement of sediment microbial fuel cell by enriching the sediment with cellulose: Kinetics of cellulose degradation. Environmental Technology & Innovation. 2019;**13**:189-196. DOI: 10.1016/j.eti.2018.11.003

[14] Xiao Y, Zheng Y, Wu S, Zhang EH, Chen Z, Liang P, et al. Pyrosequencing reveals a core community of anodic bacterial biofilms in bioelectrochemical systems from China. Frontiers in Microbiology. 2015;**6**:1410. DOI: 10.3389/fmicb.2015.01410

[15] Oh S, Min B, Logan BE. Cathode performance as a factor in electricity generation in microbial fuel cells. Environmental Science & Technology. 2004;**38**(18):4900-4904. DOI: 10.1021/es049422p

[16] Rismani-Yazdi H, Carver SM, Christy AD, Tuovinen IH. Cathodic limitations in microbial fuel cells: An overview. Journal of Power Sources. 2008;**180**(2):683-694. DOI: 10.1016/j.jpowsour.2008.02.074

[17] Dong H, Yu H, Wang X, Zhou Q, Feng J. A novel structure of scalable air-cathode without Nafion and pt by rolling activated carbon and PTFE as catalyst layer in microbial fuel cells. Water Research. 2012;**46**(17):5777-5787. DOI: 10.1016/j.watres.2012.08.005

[18] Renslow R, Donovan C, Shim M, Babauta J, Nannapaneni S, Schenk J, et al. Oxygen reduction kinetics on graphite cathodes in sediment microbial fuel cells. Physical Chemistry Chemical Physics. 2011;**13**(48):21573-21584. DOI: 10.1039/C1CP23200B

[19] Wang Z, Mahadevan GD, Wu Y, Zhao F. Progress of air-breathing cathode in microbial fuel cells. Journal of Power Sources. 2017;**356**:245-255. DOI: 10.1016/j.jpowsour.2017.02.004

[20] Jadhav DA, Carmona-Martínez AA, Chendake AD, Pandit S, Pant D. Modeling and optimization strategies towards performance enhancement of microbial fuel cells. Bioresource Technology. 2021;**320**:124256. DOI: 10.1016/j.biortech.2020.124256

[21] Li S, Ho S-H, Hua T, Zhou Q, Li F, Tang J. Sustainable biochar as an electrocatalysts for the oxygen reduction reaction in microbial fuel cells. Green Energy Environment. 2021;**6**(5): 644-659. DOI: 10.1016/j.gee.2020.11.010

[22] Chang HC, Gustave W, Yuan ZF, Xiao Y, Chen Z. One-step fabrication of binder-free air cathode for microbial fuel cells by using balsa wood biochar. Environmental Technology & Innovation. 2020;**18**:100615. DOI: 10.1016/j.eti.2020.100615

[23] Yuan H, Deng L, Qi Y, Kobayashi N, Tang J. Nonactivated and activated biochar derived from bananas as alternative cathode catalyst in microbial fuel cells. The Scientific World Journal. 2014:1-8. Article No. 832850. DOI: 10.1155/2014/832850

[24] Mutuma BK, Sylla NF, Bubu A, Ndiaye NM, Santoro C, Brilloni A, et al. Valorization of biodigestor plant waste in electrodes for supercapacitors and microbial fuel cells. Electrochimica Acta. 2021;**391**:138960. DOI: 10.1016/j.electacta.2021.138960

[25] Nastro RA, Flagiello F, Silvestri N, Gambino E, Falcucci G, Chandrasekhar K. Use of Biochar-Based cathodes and increase in the electron flow by pseudomonas aeruginosa to improve waste treatment in microbial fuel cells. Processes. 2021;**9**:1941. DOI: 10.3390/pr9111941

[26] Thipraksa J, Chaijak P. Improved the coconut shell biochar properties for bio-electricity generation of microbial fuel cells from synthetic wastewater. Journal of Degraded and Mining Lands Management. 2022;**9**(4):3613-3619. DOI: 10.15243/jdmlm.2022.094.3613

[27] Huggins T, Wang H, Kearns J, Jenkins P, Ren ZJ. Biochar as a sustainable electrode material for electricity production in microbial fuel cells. Bioresource Technology. 2014;**157**: 114-119. DOI: 10.1016/j.biortech.2014.01.058

[28] Gajbhiye P, Maan KS, Malik N, Farque O, Rajabali K. Biochar production for microbial fuel cell electrode. Journal of Emerging Technologies and Innovative Research 2022;**9**(6):404885. Available from: https://www.jetir.org/view?paper=JETIR2206791

[29] Khudzari JM, Gariepy Y, Kurian J, Tartakovsky B, Raghavan GSV. Effects of biochar anodes in rice plant microbial fuel cells on the production of bioelectricity, biomass, and methane. Biochemical Engineering Journal. 2018; **141**:190-199. DOI: 10.1016/j.bej.2018.10.012

[30] Chaijak P, Sato C, Lertworapreecha M, Sukkasem C, Boonsawang P, Paucar N. Potential of biochar-anode in a ceramic-separator microbial fuel cell (CMFC) with a laccase-based air cathode. Polish Journal of Environmenal Studies. 2020;**29**(1): 499-503. DOI: 10.15244/pjoes/99099

[31] Huggins T, Latorre A, Biffinger J, Ren Z. Biochar based microbial fuel cell for enhanced wastewater treatment and nutrient recovery. Sustainability. 2016; **8**(2):169. DOI: 10.3390/su8020169

[32] Huggins TM, Pietron JJ, Wang H, Ren ZJ, Biffinger JC. Graphitic biochar as a cathode electrocatalyst support for microbial fuel cells. Bioresource Technology. 2015;**195**:147-153. DOI: 10.1016/j.biortech.2015.06.012

[33] Wang B, Wang Z, Jiang Y, Tan G, Xu N, Xu Y. Enhanced power generation and wastewater treatment in sustainable biochar electrodes based bioelectrochemical system. Bioresource Technology. 2017;**241**:841-848. DOI: 10.1016/j.biortech.2017.05.155

[34] Marzorati S, Goglio A, Fest-Santini S, Mombelli D, Villa F, Cristiani P, et al. Airbreathing bio-cathodes based on electro-active biochar from pyrolysis of giant cane stalks. International Journal of Hydrogen Energy. 2019;**44**(9): 4496-4507. DOI: 10.1016/j.ijhydene.2018.07.167

[35] Mo RJ, Zhao Y, Wu M, Xiao HM, Kuga S, Huang Y, et al. Activated carbon from nitrogen rich watermelon rind for high-performance supercapacitors. RSC Advances. 2016;**6**(64):59333-59342. DOI: 10.1039/C6RA10719B

[36] Cristiani P, Goglio A, Marzorati S, Fest-Santini S, Schievano A. Biochar-Terracotta Conductive composites: New design for bioelectrochemical systems. Frontiers in Energy Research. 2020;**8**: 1-12. DOI: 10.3389/fenrg.2020.581106

[37] Naveenkumar M, Senthilkumar K. Microbial fuel cell for harvesting bio-energy from tannery effluent using metal mixed biochar electrodes. Biomass and Bioenergy. 2021;**149**:106082. DOI: 10.1016/j.biombioe.2021.106082

[38] Yuan Y, Yuan T, Wang D, Tang J, Zhou S. Sewage sludge biochar as an efficient catalyst for oxygen reduction reaction in an microbial fuel cell. Bioresource Technology. 2013;**144**: 115-120. DOI: 10.1016/j.biortech.2013.06.075

[39] Chakraborty I, Sathe SM, Dubey BK, Ghangrekar MM. Waste-derived biochar: Applications and future perspective in microbial fuel cells. Bioresource Technology. 2020;**312**: 123587. DOI: 10.1016/j.biortech.2020. 123587

[40] Giddey S, Badwal SPS, Kulkarni A, Munnings C. A comprehensive review of direct carbon fuel cell technology. Progress in Energy and Combustion Science. 2012;**38**:360-399. DOI: 10.1016/ j.pecs.2012.01.003

[41] Cherepy NJ, Krueger R, Fiet KJ, Jankowski AJ, Cooper JF. Direct conversion of carbon fuels in a molten carbonate fuel cell. Journal of the Electrochemical Society. 2005;**152**:A80-A87. DOI: 10.1149/1.1836129

[42] Park SW, Jang CH, Baek KR, Yang JK. Torrefaction and low-temperature carbonization of woodybiomass: Evaluation of fuel characteristics of the products. Energy. 2012;**45**:676-685. DOI: 10.1016/j. energy.2012.07.024

[43] Hackett GA, Zondlo JW, Svensson R. Evaluation of carbon materials for use in a direct carbon fuel cell. Journal of Power Sources. 2006;**168**: 111-118. DOI: 10.1016/j. jpowsour.2007.02.021

[44] Li X, Zhu ZH, Chen JL, De Marco R, Dicks A, Bradley J, et al. Surface modification of carbon fuels for direct carbon fuel cells. Journal of Power Sources. 2009;**186**:1-9. DOI: 10.1016/j. jpowsour.2008.09.070

[45] Hagen A, Barfod R, Hendriksen P, Liu Y, Ramousse S. Degradation of anode supported SOFCs as a function of temperature and current load. Journal of the Electrochemical Society. 2006;**153**: A1165. DOI: 10.1149/1.2193400

[46] Holzer L, Iwanschitz B, Hocker T, Munch B, Prestat M, Wiedenmann D, et al. Microstructure degradation of cermet anodes for solid oxide fuel cells: Quantification of nickel grain growth in dry and in humid atmospheres. Journal of Power Sources. 2011;**196**:1279-1294. DOI: 10.1016/j.jpowsour.2010.08.017

[47] Nelson G, Grew K, Izzo J, Lombardo J, Harris W, Faes A, et al. Three-dimensional microstructural changes in the Ni–YSZ solid oxide fuel cell anode during operation. Acta Materialia. 2012;**60**(8):3491-3500. DOI: 10.1016/j.actamat.2012.02.041

[48] Li C, Shi Y, Cai N. Performance improvement of direct carbon fuel cell by introducing catalytic gasification process. Journal of Power Sources. 2010; **195**(15):4660-4666. DOI: 10.1016/ j.jpowsour.2010.01.083

[49] Jiang C, Ma J, Corre G, Jain SL, Irvine JTS. Challenges in developing direct carbon fuel cells. Chemical Society Reviews. 2017;**46**(10):2889-2912. DOI: 10.1039/C6CS00784H

[50] Kacprzak A. Hydroxide electrolyte direct carbon fuel cells-Technology review. International Journal of Energy Research. 2019;**43**:65-85. DOI: 10.1002/ er.4197

[51] Gur T. Critical review of carbon conversion in carbon fuel cells. Chemical Reviews. 2013;**113**:6179-6206. DOI: 10.1021/cr400072b

[52] Cao T, Huang K, Shi Y, Cai N. Recent advances in high-temperature carbon–air fuel cells. Energy & Environmental Science. 2017;**10**:460-490. DOI: 10.1039/ C6EE03462D

[53] Behling N, Managi S, Williams M. Updated look at the DCFC: The fuel cell technology using solid carbon as the fuel.

Mining, Metallurgy & Exploration. 2019; **36**:181-187. DOI: 10.1007/s42461-018-0022-x

[54] Yu J, Zhao Y, Li Y. Utilization of corn cob biochar in a direct carbon fuel cell. Journal of Power Sources. 2014;**270**: 312-317. DOI: 10.1016/j.jpowsour.2014.07.125

[55] Elleuch A, Boussetta A, Yu J, Halouani K, Li Y. Experimental investigation of direct carbon fuel cellfueled by almond shell biochar: Part I. Physico-chemical characterization of the biochar fuel and cell performance examination. International Journal of Hydrogen Energy. 2013;**38**:16590-16604. DOI: 10.1016/j.ijhydene.2013.08.090

[56] Elleuch A, Halouani K, Li Y. Investigation of chemical and electrochemical reactions mechanisms in a direct carbon fuel cell using olive wood charcoal as sustainable fuel. Journal of Power Sources. 2015;**281**:350-361. DOI: 10.1016/j.jpowsour.2015.01.171

[57] Dudek M, Socha R. Direct electrochemical conversion of the chemical energy of raw waste wood to electrical energy in tubular direct carbon solid oxide fuel cells. International of Electrochemical Science. 2014;**9**: 7414-7430. Available from: https://citeseerx.ist.psu.edu/viewdoc/summary?doi=10.1.1.660.7996

[58] Jang H, Ocon JD, Lee S, Lee JK, Lee J. Direct power generation from waste coffee grounds in a biomass fuel cell. Journal of Power Sources. 2015;**296**: 433-439. DOI: 10.1016/j.jpowsour.2015.07.059

[59] Adeniyi OD, Ewan BCR, Adeniyi MI, Abdulkadir M. The behaviour of biomass char in two direct carbon fuel cell designs. Journal of Energy Challenges and Mechanics. 2014;**1**:1-6. Available from: https://www.nscj.co.uk/JECM/PDF/1-4-6-Adeniyi.pdf

[60] Munnings C, Kulkarni A, Giddey S, Badwal SPS. Biomass to power conversion in a direct carbon fuel cell. International Journal of Hydrogen Energy. 2014;**39**:12377-12385. DOI: 10.1016/j.ijhydene.2014.03.255

[61] Adeniyi OD. Solid oxide direct carbon fuel cell electrochemical performance using wheat and spruce carbon fuels. Energy Sources, Part A: Recovery Utilization and Environmental effects. 2015;**37**(22):2401-2407. DOI: 10.1080/15567036.2012.715230

[62] Ahn SY, Eom SY, Rhie YH, Sung YM, Moon CE, Choi GM, et al. Utilization of wood biomass char in a direct carbon fuel cell (DCFC) system. Applied Energy. 2013;**105**:207-216. DOI: 10.1016/j.apenergy.2013.01.023

[63] Ahn SY, Eom SY, Rhie YH, Sung YM, Moon CM, Choi GM, et al. Application of refuse fuels in a direct carbon fuel cell system. Energy. 2013;**51**: 447-456. DOI: 10.1016/j.energy.2012.12.025

[64] Kacprzak A, Kobylecki R, Wlodarczyk R, Bis Z. The effect of fuel type on the performance of a direct carbon fuel cell with molten alkaline electrolyte. Journal of Power Sources. 2014;**255**:179-186. DOI: 10.1016/j.jpowsour.2014.01.012

[65] Teoh SW, Yoon LW, Wong WY, Low JH. Application of black liquor derived carbonaceous compound as fuel in direct carbon fuel cell. International Journal of Integrated Engineering. 2022; **14**(2):146-156. Available from: https://publisher.uthm.edu.my/ojs/index.php/ijie/article/view/10949

[66] Adeniyi OD, Idemudia GE, Usman AA, Adeniyi MI, Paul SH,

Olutoye MA, et al. Electrical power generation from direct carbon duel fuel using biochar from eucalyptus, neem and mast leaves. International Journal of Scientific & Engineering Research. 2018; **9**(4):406

[67] Adeniyi OD, Ngozichukwu B, Adeniyi MI, Olutoye MA, Musa U, Ibrahim MA. Power generation from melon seed husk biochar using fuel cell. Ghana Journal of Science. 2020;**61**(2): 38-44. DOI: 10.4314/gjs.v61i2.4

[68] Jafri N, Yoon LW, Wong WY, Cheah KHSN. Power generation from palm kernel shell biochar in a direct carbon fuel cell. Applied Sciences. 2020; **2**:386. DOI: 10.1007/s42452-020-2189-2

[69] Hao W, Luo P, Wu Z, Mi Y, Gao Z. The effect of biomass pyrolysis temperature on the performance of biochar-fed molten hydroxide direct carbon fuel cells. Biomass and Bioenergy. 2021;**150**:106122. DOI: 10.1016/j.biombioe.2021.106122

[70] Hao W, Luo P, Wu Z, Sun G, Mi Y. Feasibility of pine bark pellets and their pyrolyzed biochar pellets as fuel sources in molten hydroxide direct carbon fuel cells. Energy & Fuels. 2020;**34**: 16756-16764. DOI: 10.1021/acs.energyfuels.0c03171

[71] Konsolakis M, Kaklidi N, Marnello GE, Zaharaki D, Komnitsa K. Assessment of biochar as feedstock in a direct carbon solid oxide fuel cell. RSC Advances. 2015;**5**:73399-73409. DOI: 10.1039/C5RA13409A

[72] An W, Sun X, Jiao Y, Juliao PSB, Wang W, Wang S, et al. A solid oxide carbon fuel cell operating on pomelo peel char with high power output. International Journal of Energy Research. 2018;**43**(7):2514-2526. DOI: 10.1002/er.4097

[73] Wang J, Fan L, Yao T, Gan J, Zhi X, Hou N, et al. A High-Performance direct carbon fuel cell with reed rod biochar as fuel. Journal of the Electrochemical Society. 2019;**166**(4):F175-F179. DOI: 10.1149/2.0321904jes

[74] Palniyandy LK, Wong WY, Yap JJ, Doshi V, Yoon LW. Effect of alkaline pre-treatment on rice husk-derived biochar for direct carbon fuel cell. Journal of Engineering Science and Technology 6th EURECA 2016 Special Issue May. 2017: 84 – 100. Available from: https://jestec.taylors.edu.my/Special%20Issue%20eureca%202016_1/eureca6_paper%207.pdf

[75] Palniyandy LK, Yoon LW, Wong WY, Yong ST, Pang MM. Application of biochar derived from different types of biomass and treatment methods as a fuel source for direct carbon fuel cells. Energies. 2019;**12**(13): 2477. DOI: 10.3390/en12132477

[76] Xie Y, Xiao J, Liu Q, Wang X, Liu J, Wu P, et al. Highly efficient utilization of walnut shell biochar through a facile designed portable direct carbon solid oxide fuel cell stack. Energy. 2021;**227**: 120456. DOI: 10.1016/j.energy.2021.120456

[77] Qiu Q, Zhou M, Cai W, Zhou Q, Zhang Y, Wang W, et al. A comparative investigation on direct carbon solid oxide fuel cells operated with fuels of biochar derived from wheat straw, corncob, and bagasse. Biomass and Bioenergy. 2019;**121**:56-63. DOI: 10.1016/j.biombioe.2018.12.016

Chapter 10

Biochar Synergistic New Ammonia Capture of CO_2 and High-Value Utilization of Intermediate Products

Yu Zhang, Yalong Zhang, Dongdong Feng, Jiabo Wu, Jianmin Gao, Qian Du and Yudong Huang

Abstract

In the face of global warming and the urgent need for CO_2 reduction, carbon capture, utilization, and storage, technology plays an important role. Based on the traditional liquid-phase and solid-phase CO_2 capture technologies, the liquid-phase ammonia and biochar CO_2 capture technologies are reviewed with emphasis. A multiphase carbon capture technology that uses biochar to enhance the mass transfer-crystallization process of the new ammonia CO_2 capture technology is proposed. High CO_2 capture efficiency, limited ammonia escape, and low system energy consumption can be achieved through the orderly construction of three-dimensional graded pore channels and the directional functionalization of biochar. The intermediate products of CO_2 captured by the ammonia process and the special agricultural waste rice husk components were considered. The use of rice husk-based biochar for CO_2 capture by synergistic new ammonia method and the process regulation of intermediate products to prepare nano-silica to achieve high-value utilization of interstitial products of carbon capture. This technology may be important to promote the development of CO_2 capture technology and CO_2 reduction.

Keywords: CO_2 capture, biochar, new ammonia, rice husk, nano-silica

1. Introduction

1.1 Current status of CO_2 emissions and CCUS technology

Carbon is cycled between different sources (atmosphere, ocean, terrestrial biota, and marine biota) in the form of carbon dioxide, carbonates, and organic compounds. Human activities have disrupted the balance of this cycle, and a large amount of CO_2 emissions has led to an increasingly serious greenhouse effect. Global climate change has caused widespread concern in the international community. According to the report of the International Energy Agency (IEA) [1], to achieve the target of global average temperature increase within 2°C above the pre-industrial level by 2100 and to try to limit it to 1.5°C, direct CO_2 emissions from industrial production need to be reduced by about 30%, and CO_2 emissions per unit of GDP need to be reduced by about 60% by 2050 compared with the current level. However, as things stand today, global CO_2 emissions from energy combustion and industrial processes will rebound in 2021 to the highest annual level ever recorded (**Figure 1(a)**). Emissions increased by 6% compared with 2020 (**Figure 1(b)**). The largest increase in CO_2 emissions by sector in 2021 is from electricity and heat production, accounting for 46% of global emissions (**Figure 1(d)**). Coal accounts for more than 40% of the increase in total global CO_2 emissions, a record high (**Figure 1(c)**). As the most important coal-consuming industry, coal-fired power plants are the most important source of CO_2 emissions. Hence, the research on CO_2 reduction technology for coal-fired power plants has profound significance.

Carbon Capture, Utilization, and Storage (CCUS) technology are considered the most economical and feasible way to reduce greenhouse gas emissions and mitigate global warming on a large scale in a short period. CCUS technology captures CO_2 from large point sources such as power plants or directly from the atmosphere. The captured CO_2 will be compressed and transported for various applications or injected into deep geological layers for permanent storage. As early as 2005, the Intergovernmental Panel on Climate Change (IPCC) identified CCUS as a key technology in mitigating the greenhouse effect [2]. Today, strengthened climate goals and new investment incentives have created unprecedented momentum for CCUS, and many countries have taken steps to develop CCUS technologies [3–7]. Projections indicate [8] that the least-cost pathway to "≤2°C" is to capture and sequester about 4 billion tons of CO_2 per year by 2040 and that the current CO_2 capture capacity is still

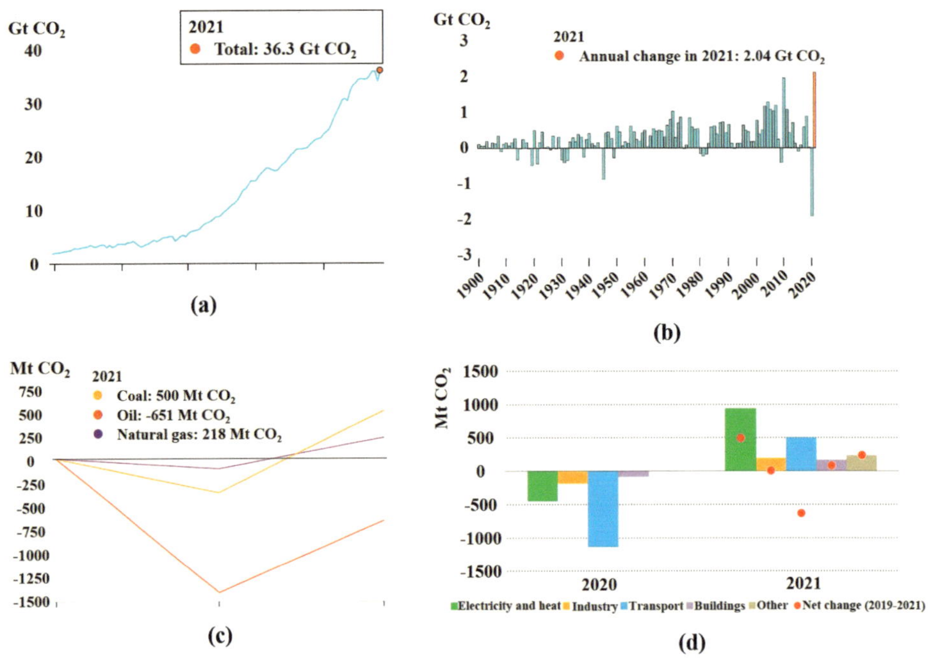

Figure 1.
(a) CO_2 emissions from energy combustion and industrial processes, 1900–2021, (b) annual change in CO_2 emissions from energy combustion and industrial processes, 1900–2021, (c) change in CO_2 emissions from fossil fuels, 2019–2021, relative to 2019 levels, (d) annual change in CO_2 emissions by sector, 2020–2021 [1].

far from the required amount, making CO_2 capture technology critical in the overall carbon reduction and CCUS system.

1.2 CO_2 capture

1.2.1 CO_2 capture technology

There are four main CO_2 capture technology routes: pre-combustion capture, oxygen-enriched combustion, post-combustion capture, and chemical loop combustion. In pre-combustion capture technology, fossil fuels are converted to a syngas of carbon dioxide and hydrogen before combustion using gasification or reforming technology so that the "carbon" in the fuel does not participate in the combustion process [9]. Oxyfuel combustion uses oxygen instead of air for combustion and can be used without considering the separation of nitrogen and carbon dioxide, a technically feasible process [10]. Post-combustion capture technologies remove CO_2 from the flue gas after combustion has occurred. In recent years, chemical loop combustion (CLC) has also been developed. It uses metal oxides to transport the oxygen required for combustion to prevent direct contact between fuel and air, with its inherent CO_2 capture capability [11]. Of the above capture technologies, post-combustion CO_2 capture is the most mature and most thorough and is the preferred option for retrofitting existing power plants.

1.2.2 Post-combustion CO_2 capture

Post-combustion CO_2 capture technologies mainly include adsorption, absorption, membrane separation, and low-temperature distillation. Low-temperature distillation is a method of separation using the difference in boiling point or volatility of each component gas in the gas mixture. This method has high CO_2 separation efficiency and purity and can directly produce liquid CO_2 for storage and transportation [12]. The absorption method includes chemical absorption and physical absorption. Physical absorption involves using a physical solvent to dissolve a component gas. The solubility increases with increasing pressure and decreasing temperature; therefore, the optimal conditions for the CO_2 absorption process are high pressure and low temperature [13]. The chemical absorption method uses an alkaline absorber to contact and react with CO_2 in the flue gas to remove CO_2. The salts generated by the reaction will decompose and release CO_2 under certain conditions, thus removing and enriching CO_2 from the flue gas [14]. The principle of membrane separation is that different components pass through the membrane with different selectivity. The membrane allows only specific gases to pass through, thus achieving separation and enrichment. The performance of the membrane system is influenced by the flue gas conditions [15], the enriched CO_2 concentration is low, and the separation conditions are demanding. Adsorption can be divided into physical adsorption and chemisorption, with physical adsorption having a weak binding force, a relatively small heat of adsorption, and easy desorption. On the other hand, chemisorption is caused by chemical bonding between the adsorbent and the adsorbent, the adsorption is often irreversible, and the heat of adsorption is usually larger [16]. Adsorption differs from the absorption process in that the adsorption efficiency is mainly influenced by the specific surface area, selectivity, and regeneration characteristics. **Table 1** compares the above four post-combustion CO_2 capture technologies, and all of these methods inevitably have various problems. Therefore, the development of new ammonia decarbonization technology will become the main

Technology	Advantage	Disadvantages	Reference
Low temperature	Technology maturity	Only for high CO_2 concentration, low temperature, high energy consumption	[17, 18]
Adsorption	Reversible process, recyclable adsorbent, high adsorption efficiency	Requires high-temperature adsorbent and high energy for desorption	[19, 20]
Membrane separation	High separation efficiency	Operational problems include low flux and scaling	[21]
Absorbent	High absorption efficiency, renewable absorbent, mature process	Absorption efficiency depends on CO_2 concentration, high energy consumption for absorber regeneration	[22]

Table 1.
Comparison of different CO_2 capture technologies.

theme of CO_2 capture technology. However, its ammonia escape problem also needs to be further strengthened. The study of solid-phase adsorption combined with ammonia liquid-phase absorption to achieve two-phase synergistic CO_2 capture will have far-reaching significance in the future.

2. Ammonia-based liquid-phase CO_2 capture technology

2.1 Liquid-phase chemical absorption of CO_2

The commonly used absorbents for chemical absorption targeting CO_2 capture are monoethanolamine (MEA), ammonia, and potassium carbonate. The CO_2 capture efficiency of MEA is very high, but it has high regeneration energy, a high corrosion rate, and is susceptible to oxidative degradation. The high energy consumption of CO_2 capture using aqueous amines is also one of the main drawbacks that limit its wide application. Non-aqueous absorbents have an absorption capacity comparable to aqueous MEAs and higher desorption efficiency, leading to a larger cycle capacity and nearly half the energy consumption (**Figure 2(a)**). Bougie et al. [27] investigated new non-aqueous MEA absorbers that greatly reduced energy consumption and improved CO_2 absorption kinetics. The regeneration of MEAs is also a major challenge, with approximately 80% of the total energy consumption in the CO_2 capture process occurring in the solvent regeneration process [28]. Many studies have shown that carbonate solutions can be used for CO_2 uptake, and K_2CO_3 solutions have higher capture capacity than other carbonate solutions and are more commonly used in industry [29]. Although carbonate solutions have been extensively studied, the kinetics and thermodynamics of their absorption solutions still need to be investigated, and K_2CO_3 solvents may be subject to corrosion due to flue gas contaminants and solvent degradation.

2.2 Absorption of CO_2 by ammonia

2.2.1 Ammonia CO_2 capture technology

Figure 2(c) shows that reliable absorbents for low concentration CO_2 capture without pressurization are amine-based and ammonia-based CO_2 capture technologies. Ammonia-based CO_2 capture is considered a viable carbon capture technology

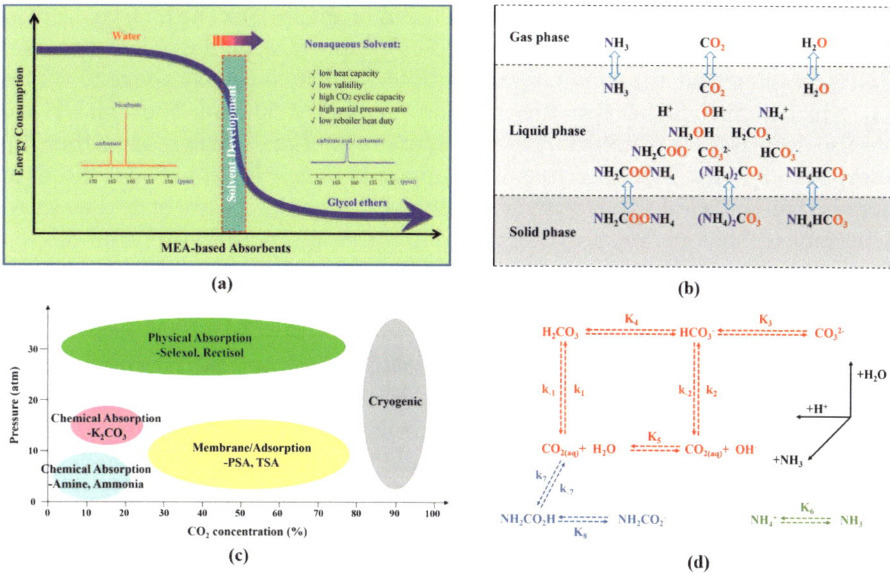

Figure 2.
(a) Comparison of aqueous and non-aqueous MEA absorbents [23], (b) Schematic diagram of NH_3-CO_2-H_2O three-phase system [24], (c) Applicable range of different CO_2 capture technologies (based on operating pressure and CO_2 concentration) [25], (d) Reaction mechanism of CO_2 absorption by ammonia method [26].

due to the high corrosiveness of MEA and regeneration problems over conventional amine-based CO_2 capture technologies in terms of technical and economic advantages. The CO_2-NH_3-H_2O system (**Figure 2(b)**) thermochemical properties have been reasonably well explained in recent studies. Although ammonia is the simplest amine, its interaction with CO_2 is quite complex, involving gas-liquid-solid three-phase reactions, making the application of CO_2-NH_3-H_2O systems in CO_2 capture poses some challenges. Thomsen and Rasmussen [30] developed a thermodynamic model with a temperature. The model can be used not only for gas-liquid systems but also for gas-liquid-solid equilibria, including forming NH_4HCO_3, $(NH_4)_2CO_3$-H_2O, and NH_2COONH_4. Que and Chen [31] developed an electrolyte NRTL activity coefficient model that can well represent the thermodynamic properties of the NH_3-CO_2-H_2O system when the CO_2 loading reaches a consistent level. The availability of these models allows to reliably calculate the thermochemical properties of the CO_2-NH_3-H_2O system under various conditions and to assess the energy performance of the capture process [32]. The uptake of CO_2 by ammonia is a relatively slow process; therefore, it is important to understand the reaction mechanisms/kinetics involved in the uptake chemistry. The most important reaction in the presence of free ammonia is the reaction of NH_3 with CO_2, and the reaction scheme is shown in **Figure 2(d)**. The equilibrium constant of carbamate of MEA is much higher than that of ammonia, and the yield of ammonia-derived carbamate is lower than that of the equivalent monoethanolic ammonium carbamate, indicating that ammonia possesses a higher CO_2 capture capacity.

2.2.2 Ammonia escape

Ammonia CO_2 capture technology has many advantages, but it also has drawbacks in current applications: (1) low CO_2 absorption rate; (2) serious ammonia escape; and

(3) high energy consumption for desorption and regeneration. The high volatility of ammonia is the main drawback of ammonia CO_2 capture technology. The concentration of NH_3 escaping from the emission gas of this technology is usually above 10,000 ppm [33], which is much higher than the emission standard of 50 ppm. The high NH_3 escape rate also decreases the concentration of NH_3 in the solution, which reduces the CO_2 absorption capacity [34]. Therefore, it is imperative to develop effective methods to suppress ammonia leakage or recover the leaked ammonia. The use of acid washing, membrane technology, and additives are common strategies to control ammonia escape.

2.2.3 Ammonia-ethanol mixture absorber

To better solve the above problems, many scholars have proposed the modification of CO_2 absorption by ammonia solution using additives, which can inhibit not only NH_3 escape but also improve CO_2 absorption performance. Many scholars have studied the CO_2 capture performance of ammonia with additives [35–41], among which Gao and Zhang et al. [39, 41] have shown significant advantages in various aspects of using ethanol as an additive. Ammonia and additives can, to some extent, promote each other to improve the CO_2 uptake rate of ammonia [42]. However, a slight contradiction emerged between the hybrid absorber improving the absorption rate and inhibiting ammonia escape [43]. The additive mainly binds the free ammonia in the ammonia solution by hydrogen bonding and thus inhibits ammonia escape. However, the additive cannot achieve effective ammonia release when this hybrid absorber absorbs CO_2, which will reduce the liquid-phase partial pressure of free ammonia and adversely affect the absorption process. The advantages of an "ammonia-ethanol adsorbent mixture" for CO_2 absorption and capture are significant [41]. However, many aspects still need to be improved. It is urgent to develop a new ammonia carbon capture technology based on this idea to maintain its advantages and avoid its shortcomings.

3. Biochar-new ammonia synergistic carbon capture

3.1 Solid-phase adsorption CO_2 capture

The adsorption of CO_2 by porous carbon materials is an exothermic process, with the heat of adsorption of physical adsorption processes ranging from −25 to −40 kJ/mol [44], and the amount of adsorption is directly related to the porous structure of the adsorbent and the active functional groups on the surface. The molecular kinetic diameter of CO_2 is 0.33 nm, so micropores (<1 nm) are the main sites for CO_2 adsorption (**Figure 3(a)**). Still, only micropores cannot achieve high adsorption capacity, and a suitable pore structure is required [45]. Macropores and mesopores act as channels for diffusive CO_2 transport and can facilitate CO_2 adsorption in micropores. CO_2 being polar and acidic molecules, basic and polar functional groups (e.g., pyridine, pyrrole nitrogen) also plays an important role in adsorption [45]. Therefore, when selecting CO_2 adsorbent, the economy and reliability should be satisfied. The adsorbent's pore structure and surface functional groups should be considered to ensure that the distribution of the two reaches a certain balance. Too much pursuit of one side will lead to the deterioration of the other side, resulting in a worse adsorption effect. (**Figure 3(b–e)**) [47].

Non-carbon-based solid adsorbents, mainly MOF and zeolite, are well studied and widely used. Almost all metals and a large amount of organic matter can make MOF, which is widely used in adsorption due to its extremely high porosity and specific

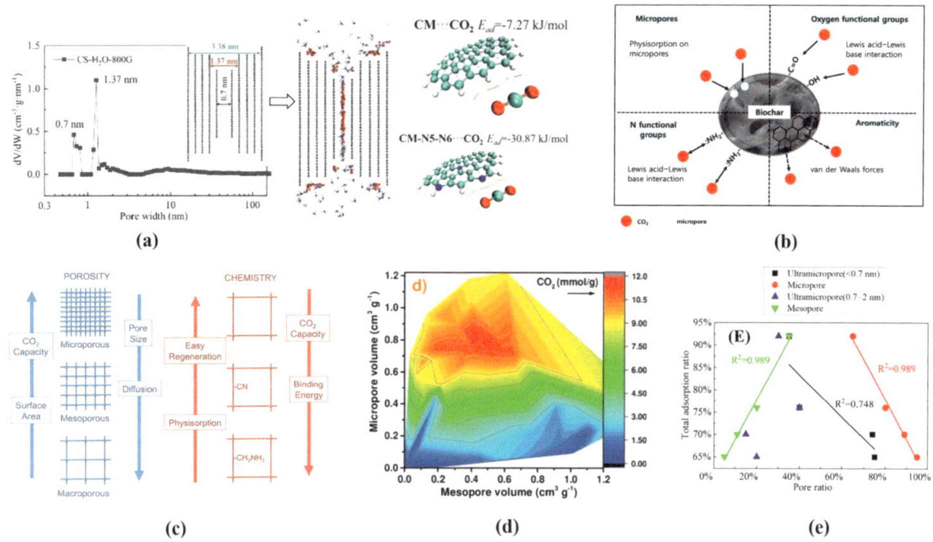

Figure 3.
(a) Molecular dynamics and quantum chemical simulation of CO_2 adsorption by porous carbon materials [45], (b) various structures in biochar micropores that effectively enhance CO_2 adsorption [46], (c) effect of adsorbent porosity and chemical properties on CO_2 adsorption performance [47], (d) correlation between CO_2 adsorption, micropore, and mesopore volumes at 25°C and 5 bar [48], (e) correlation between different pore ratios and relative CO_2 adsorption [45].

surface area. When the partial pressure of CO_2 is low (<0.2 bar), the adsorption capacity of MOF is poor [49], and impurity gases replace the skeletal ligands during the CO_2 capture process, leading to degradation of MOF and a decrease in the capture capacity. Zeolites have a regular pore size of 0.5–1.2 nm [50] and have been widely investigated for CO_2 capture due to the strong electrostatic interaction between CO_2 and alkali metal cations in the zeolite skeleton [51]. Siriwardane et al. [52] showed that natural zeolites with high sodium content exhibited high CO_2 adsorption capacity. However, the electrostatic interaction between CO_2 and alkali metal cations in the zeolite skeleton is reduced by water [53], and therefore only in dry gas streams is CO_2 separation effective. Among the carbon-based materials, activated carbon is one of the most commonly used adsorbents in industry. It is less costly than other adsorbents [54], but its adsorption capacity is only comparable to that of zeolites at high CO_2 pressure [55], and the heat of adsorption is lower than that of zeolites. By introducing impurity atoms or acid-base sites, activated carbon can appropriately improve adsorption selectivity and adsorption capacity. As a new carbon-based material, carbon nanotubes have also received attention in gas adsorption [56, 57].

3.2 Biochar adsorbent

The raw materials of biochar are widely sourced, and the cost is lower than other adsorbents. The biochar prepared from different raw materials is different due to their intrinsic elemental composition ratio and structure. Biochar prepared from raw materials with high strength and carbon content, such as wood chips, coconut shells, date kernels, and rice husks, has a more desirable CO_2 adsorption capacity [58]. During preparation, the carbonization temperature affects the structure, surface functional groups, and elemental composition of the final material and 500–800°C is considered

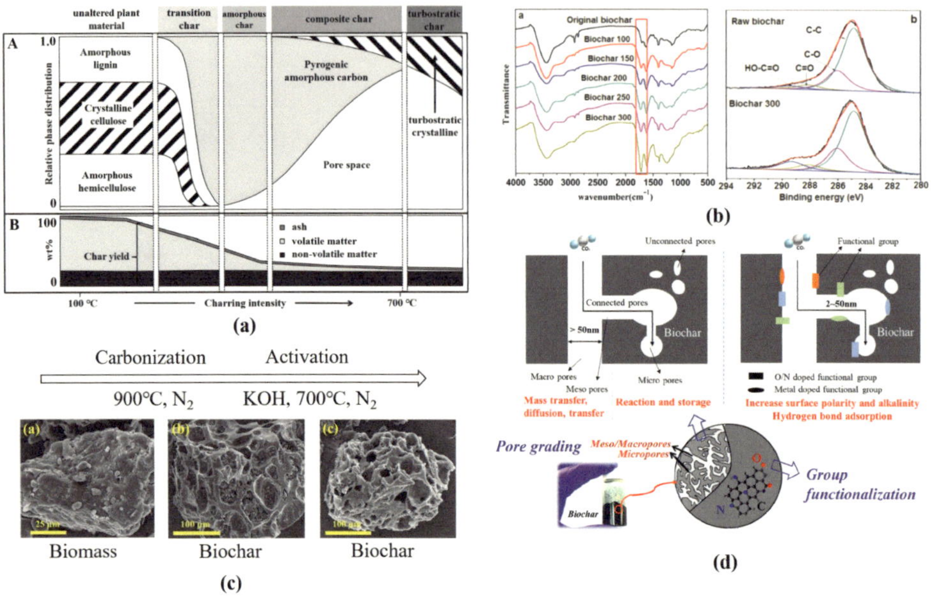

Figure 4.
(a) Dynamic molecular structure of biochar derived from plant biomass [60], (b) Infrared spectra of biochar after heat treatment at different temperatures and comparison of XPS spectra of raw biochar and biochar after heat treatment at 300°C [61], (c) SEM images of different stages of biochar preparation [62], (d) Mechanism of biochar pore classification and group functionalization [2].

the optimal temperature range for carbonization [59]. Thermal degradation of biomass at high temperatures in limited or complete anoxia is central to biomass conversion into porous carbon. Most biomass consists of lignin, cellulose, and hemicellulose, prepared under different pyrolysis and activation conditions to obtain different pore structures, group ratios, and surface chemistry (**Figure 4(a** and **b)**).

3.3 Biochar modification

The adsorption of CO_2 by biochar is highly dependent on the pore structure and surface physicochemical properties, and the optimal pore size is about twice the kinetic diameter of CO_2 molecules. However, the CO_2 adsorption capacity of directly carbonized biochar is low. The authors' previous studies [45] have been conducted to enhance the CO_2 adsorption capacity of biochar by sequential construction of pore channels and surface functionalization modification through activation (physical and chemical activation). Molecular dynamics simulations were also performed by clearly modeling the hierarchical pore channels to explain the experimental phenomena from a microscopic perspective. The mechanism is shown in **Figure 4(d)**. On the other hand, the carbonization temperature plays a key role in controlling activated porous biochar's functional groups and specific surface area. Therefore, the reasonable selection of the amount of activator and carbonization temperature becomes a necessary part of preparing activated porous biochar materials.

3.3.1 Biochar pore hierarchy construction

According to the International Union of Pure and Applied Chemistry (IUPAC) standards, biochar is classified as macroporous, mesoporous, and microporous.

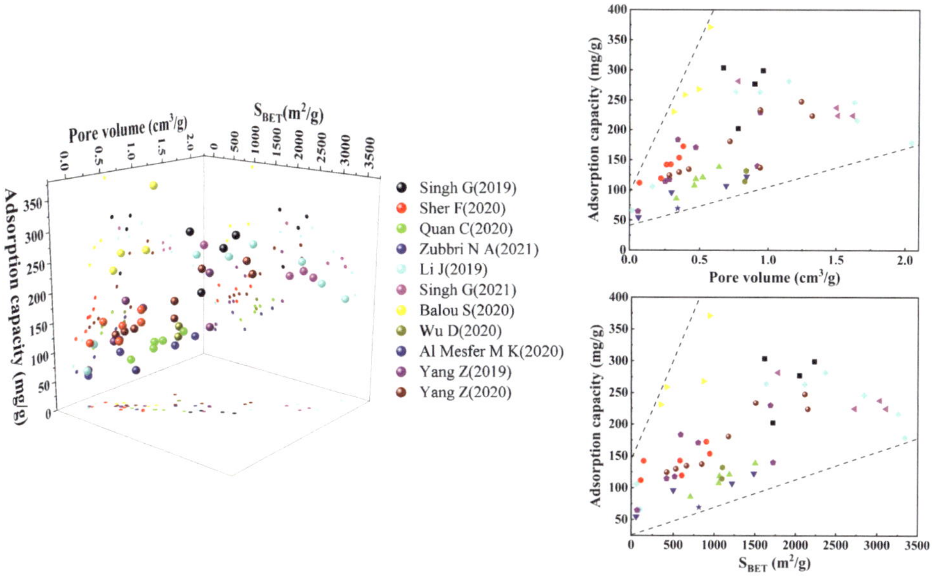

Figure 5.
Relationship between specific surface area and pore volume of biochar and CO_2 adsorption capacity [64–74].

Usually, the pore size of macroporous exceeds 50 nm, mesoporous is 2–50 nm, and microporous is less than 2 nm. For the CO_2 adsorption process, macropores and mesopores help diffusive transport of CO_2 molecules, while micropores provide adsorption sites as direct storage sites for CO_2. Therefore, a reasonable construction of graded pores can effectively enhance the CO_2 capture performance of biochar. Lingyu et al. [46] prepared biochar from seven types of straw and wood biomass to study their CO_2 adsorption performance and found that wood biochar has better pore structure than straw biochar with 2.73–4.40 times larger specific surface area, and biochar with super pore structure has higher CO_2 adsorption capacity. Capacity was higher, and good pore structure played a crucial role in the CO_2 adsorption. Avanthi et al. [63] prepared biochar using pine sawdust and steam activated it at the same temperature for 45 min after completion of pyrolysis. Due to the high surface area and microporosity, pine sawdust biochar showed significantly higher CO_2 adsorption capacity than paper mill sludge biochar, which may be due to the Steam activation increased the microporosity, surface area, and oxygen-containing basic functional groups. In this paper, we summarized the literature that studied the CO_2 adsorption capacity of biochar with different pore structures in recent years, and the relationship between their specific surface area, pore-volume, and biochar CO_2 capture capacity is shown in **Figure 5**.

3.3.2 Biochar functionalization construction

The adsorption of CO_2 on the biochar surface is influenced by the chemical properties of the biochar surface. Many studies have shown that the introduction of basic nitrogen functional groups can increase the alkaline sites on biochar and enhance the adsorption of acidic CO_2 [75]. Nitrogen-containing functional groups are the main contributors to the alkalinity of the biochar surface, and activation in different nitrogen-containing reagents was performed to introduce nitrogen-containing functional groups to the biochar surface. The commonly used activation reagents are

KOH, NaOH, CO_2, and K_2CO_3. Activation of biochar with KOH or NaOH can dissolve compounds such as ash, lignin, and cellulose, thus increasing the O content and surface alkalinity of biochar. Some new activation reagents such as $NaNH_2$, CH_2COOK, and H_2SO_4 have been gradually investigated. He et al. [76] prepared activated carbon by KOH activation using rice husk as raw material and modified biochar with chitosan as a nitrogen source. They found that the modified AC exhibited better CO_2 adsorption performance in comparison. Yang et al. [77] prepared N doped porous carbon, and the CO_2 adsorption capacity could reach 6.33 mmol/g at 273.15 K and 100 kPa, which was significantly higher than most of the carbon-based adsorbents reported in the literature due to the introduction of nitrogen-containing functional groups that increased the CO_2 adsorption sites. In addition, unlike the acid-base interactions between CO_2 and biochar surfaces, it has been shown that the presence of oxygen-containing acidic functional groups such as hydroxyl and carboxyl groups also promotes hydrogen bonding between CO_2 molecules and carbon surfaces, thus increasing CO_2 adsorption on carbon-containing surfaces [78]. Ma et al. [79] synthesized a series of carbon materials with different functional group contents. The experimental results showed that introducing oxygen functional groups into the carbon framework can again improve CO_2 capture efficiency in N-doped porous carbon. According to the theoretical calculations (**Figure 6**), the carbon framework with high oxygen content further enhanced the hydrogen bonding and electrostatic interaction for CO_2 adsorption. Wu et al. [80] prepared biochar from corn kernels by KOH activation, and the samples possessed a very high number of oxygen functional groups (45.5%) and exhibited a large CO_2 adsorption capacity. The presence of alkali and alkaline earth metal (AAEM) elements such as Na, K, Ca, and Mg can also promote the formation of basic sites, which have a strong affinity for CO_2 with acidic properties [81]. Therefore, the presence of biochar's AAEM elements may enhance the CO_2 adsorption capacity of biochar, and the introduction of alkaline metal sites in the biochar skeleton may also enhance the CO_2 adsorption of biochar in the order of Mg > Al > Fe > Ni > Ca > Raw biochar > Na [75].

3.4 "Biochar-new ammonia" CO_2 capture system

The excellent CO_2 adsorption performance and low regeneration energy consumption of biochar are closely related to its well-developed specific surface area,

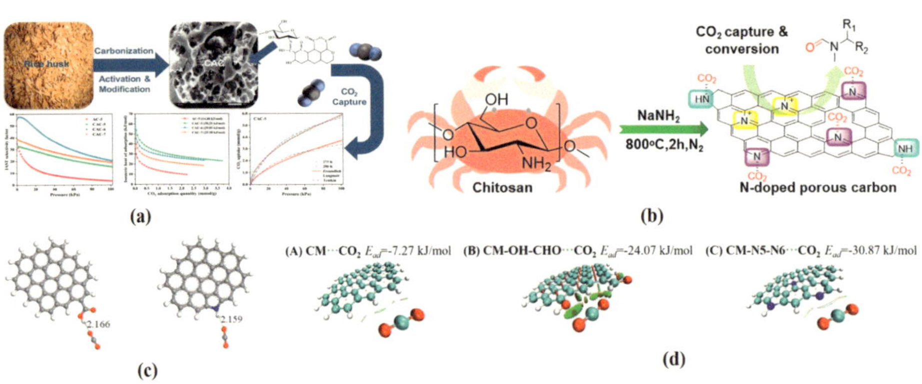

Figure 6.
(a) modification of rice husk-based biochar and CO_2 adsorption capacity [76], (b) preparation of N-doped porous carbon from chitosan and $NaNH_2$ for CO_2 adsorption [77], (c) hydrogen bonding between CO_2 and functionalized biochar surface [79], (d) adsorption energy of different functional groups of biochar for CO_2 [45].

three-dimensional through-gradient pore structure, and unique oxygen/nitrogen surface chemistry. Dagaonkar et al. [82] estimated the effective diffusion coefficient of CO_2 within biochar to be 9.645×10^{-7} m^2 s^{-1}. Suppose biochar particles are used as a modified material to enhance the mass transfer properties of the liquid phase. In that case, their stronger CO_2 diffusion properties can be fully utilized to improve the overall reaction rate of the carbon capture system. Biochar has also been used to adsorb ammonia nitrogen, and more than half of the mass of ammonia nitrogen adsorbed was completed within 2 h [83]. In view of this, biochar can be applied in ammonia water CO_2 absorption systems to achieve effective inhibition of ammonia escape through the sequestration of free ammonia by its active surface groups.

By combining the respective development potentials of biochar adsorbent and ammonia-ethanol absorber, biochar adsorbent was cross-linked with ammonia-ethanol absorber to realize the functionalized cross-linking of the biochar-enhanced new ammonia carbon capture mass transfer-crystallization process (**Figure 7(a)**).

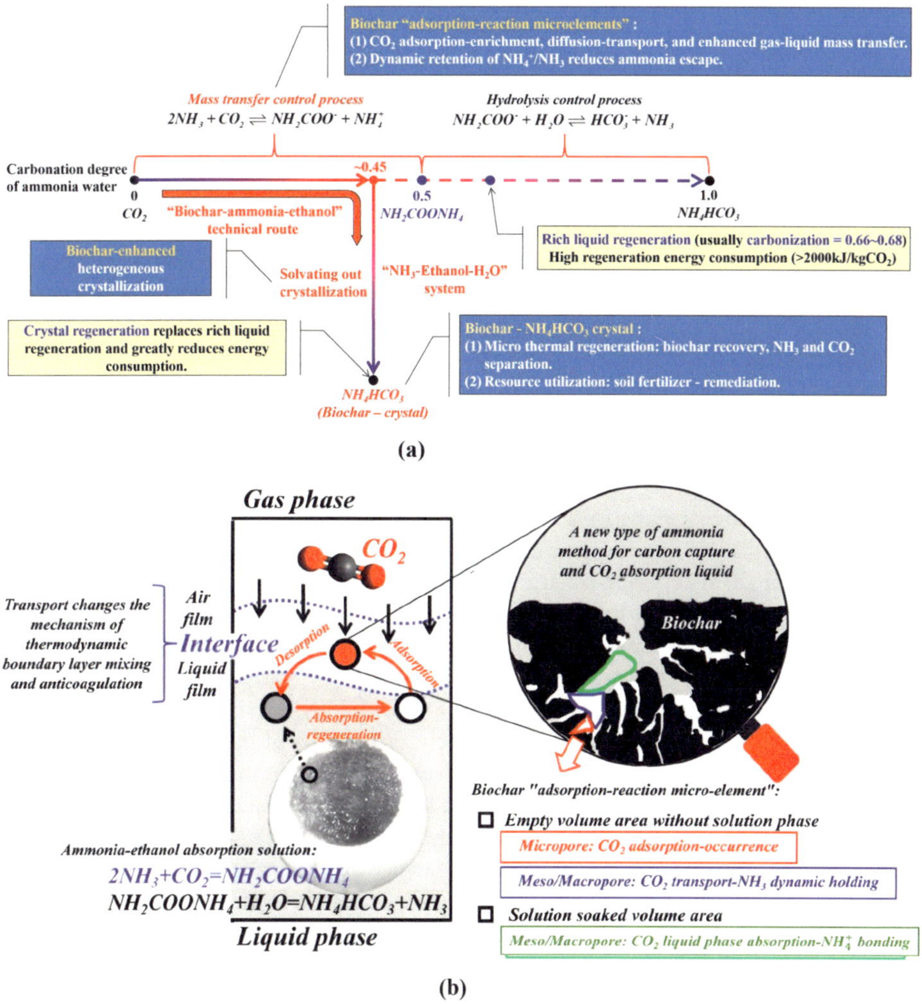

Figure 7.
(a) Functionalized cross-linking of biochar-enhanced novel ammonia-based carbon capture mass transfer-crystallization processes, (b) Enhanced mass transfer mechanism of biochar in ammonia-ethanol mixed absorbent [2].

This system transforms the carbon capture process from the traditional ammonia carbon capture gas-liquid two-phase reaction to a gas-liquid-solid three-phase process. The system can achieve CO_2 adsorption and enrichment in micropores, CO_2 diffusion and transport in mesopores/macropores, and dynamic sequestration of free ammonia by regulating the hierarchical structure of biochar nanopores and the orderly grouping of active functional groups on the surface. The cross-scale multiphase system processes, such as functionalization, pore gradation, biochar/NH_4HCO_3 dissolution and crystallization, and adsorption/absorption coupling, are cross-linked by crystal regeneration instead of liquid-rich regeneration. The synergistic effect of "graded adsorption—efficient absorption—dissolution crystallization—crystal regeneration" in the system is accomplished. The multiple goals of ammonia carbon capture, such as increasing absorption rate, suppressing ammonia escape, and reducing system energy consumption, are achieved. This process can improve a series of shortcomings of ammonia CO_2 capture and overcome the shortcomings of biochar adsorbents. The synergy of the CO_2 capture process in the solid-liquid system of "biochar-ammonia-ethanol" is achieved by "taking the advantages of each and avoiding the shortcomings."

3.4.1 Biochar efficiency transfer

The absorption of CO_2 by ammonia is a typical non-homogeneous reaction process in which CO_2 in the gas phase is first dissolved in the absorption solution and then reacts with NH_3 in the liquid phase. Therefore, the absorption rate is controlled by the "chemical reaction in the liquid phase" and the "mass transfer characteristics between gas and liquid." The generation and hydrolysis of carbamate in the reaction process is the most important factor affecting the chemical reaction rate, roughly divided by the carbonation degree of ammonia absorption CO_2 solution ≈ 0.5, as shown in **Figure 7(a)**. The liquid membrane mainly controls the mass transfer resistance of ammonia absorption CO_2 reaction process, when the hydrolysis of ammonium carbamate mainly controls the carbonation degree >0.5, ammonia absorption CO_2, so that the liquid phase carbon capture rate is significantly reduced, and this process has been the bottleneck to improve the absorption rate in the later stage of the reaction in the traditional process. This process has been the bottleneck to improving the absorption rate in the later stage of the reaction in the traditional process. The key to reducing the liquid film mass transfer resistance in the process of CO_2 adsorption by ammonia and improving the low CO_2 absorption rate is to get rid of the influence of carbonation degree on the regeneration energy consumption and to control the CO_2 absorption reaction by ammonia only in the rapid generation phase of ammonium carbamate with carbonation degree <0.5. The mechanism of mass transfer characteristics of the new ammonia carbon capture process with biochar efficiency enhancement is shown in **Figure 7(b)**. Using the highly efficient adsorption performance of biochar hierarchical pore channels, the initial rapid CO_2 sequestration is completed, and the biochar is used as a carrier to bring CO_2 into the ammonia absorption system. Subsequently, the transfer of CO_2 from solid particles' adsorption space to the ammonia liquid phase's absorption space is further realized. The release of CO_2 from biochar and the absorption of CO_2 by ammonia is completed, which greatly increases the contact time between CO_2 and ammonia liquid phase, thus realizing the reduction of liquid film resistance and prolonging the residence time of CO_2 in the solid-liquid phase system to increase the material transfer and chemical absorption rate in the ammonia liquid phase system. The liquid-liquid phase ammonia system can be used to increase the rate of material transfer and chemical absorption.

3.4.2 Limiting ammonia escape

The ammonia escape process is shown in **Figure 8**. Among many parameters affecting ammonia escape, the temperature is one of the most sensitive [84]. From the ammonia escape point of view, the absorption temperature should be as low as possible, requiring a large amount of energy to maintain cold ammonia. For the solid-liquid two-phase CO_2 capture system, the pore surface functional groups in the solution permeable region of biochar/macropore can undergo cation exchange with NH^{4+} in solution [85], which promotes the reverse migration of the hydrolyzing process of ammonia monohydrate. At the same time, the free ammonia in the liquid phase was held by the van der Waals force and chemical hybrid force [86] so that the production of free ammonia in the liquid phase could be effectively controlled. Therefore, the hierarchical functionalized construction of biochar pore structure ensures the hierarchical adsorption of CO_2/NH_3 by biochar particle pore, improves the material transfer and chemical absorption rate in the ammonia liquid phase system, and makes the dynamic balance of NH_3 adsorption and fixation in unsaturated solution impregnation space present in biochar pore. To a great extent, the effective concentration of free ammonia that can participate in the reaction in the liquid phase system is ensured, the dynamic partial pressure of free ammonia in the liquid phase is maintained, and the ammonia escape is limited.

3.4.3 Dissolution crystallization instead of rich liquid regeneration

The traditional ammonia-rich liquid thermal regeneration process is the largest energy-consuming part of the whole ammonia carbon capture process. The regeneration energy consumption is mainly composed of three parts: the sensible heat of rich liquid warming, the latent heat of vaporization, and the heat absorption of

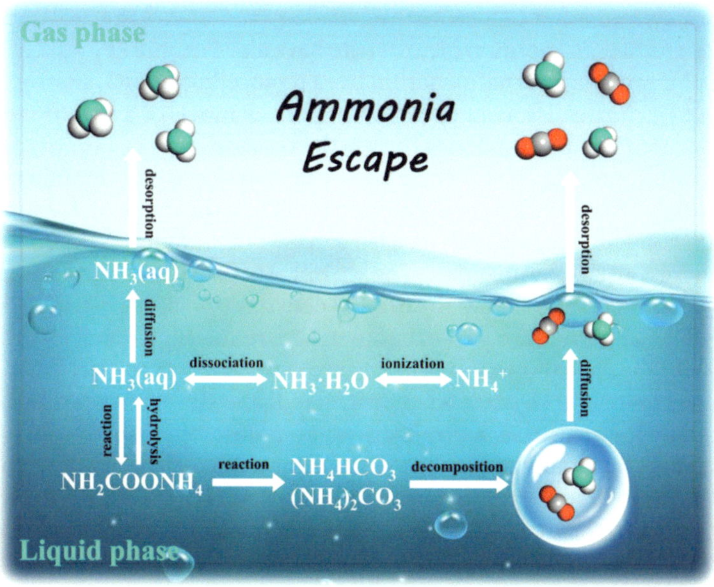

Figure 8.
Ammonia escape mechanism.

regeneration reaction, of which 50–70% of the energy is consumed in the warming and vaporization of rich liquid solvent [87]. The solubility of the product of the reaction process of CO_2 absorption by ammonia is known: ammonium carbamate is soluble in water and ethanol; ammonium bicarbonate is soluble in water-insoluble in ethanol. The main mechanism of solvation crystallization is to use the different chemical structures of the main solvent molecules and solvating agent molecules to make a difference in the microscopic forces between the ions of the substances to be separated and to change the macroscopic properties of the mixed solvent by changing the microscopic forces of the particles in the solution, thus greatly reducing the solubility of the solute, and using the solubility difference as the driving force to make the solute continuously precipitate out of the liquid phase in the form of crystals, so that the solvent and the solute are separated. The solvent and solute are separated. In the "biochar-ammonia-ethanol" carbon capture system, the crystallization process is strengthened by the solvation and precipitation method, and the regeneration of crystals replaces the regeneration of carbon-rich liquid, which greatly reduces the energy consumption of regeneration. The biochar functionalized Meso-/macropore pores ensure the NH_3/NH^{4+} concentration in the liquid phase. The pores' active surface structure provides nucleation sites for the crystallization process, which accelerates the formation and growth of carbonated liquid solvation crystals in the liquid phase system. The dynamic balance between the crystallization process's residence time and the biochar's saturation time for efficient adsorption can provide a stable CO_2 adsorption-absorption-crystallization series process.

3.5 High-value utilization of carbon capture products

The main crystallization product of the novel ammonia CO_2 capture technology described above is ammonium bicarbonate, which is widely used in agriculture, food, pharmacy, and ecological management, but its utilization process's complexity and economics have prevented its use widespread development. Therefore, further optimization of the new ammonia CO_2 capture technology and high-value utilization of the intermediate product ammonium bicarbonate have also become key issues. Rice husk is widely available, and its internal structure has a lignocellulose-SiO_2 crossover network, and the SiO_2 in it can be dissolved to construct pore channels of a specific size. With this unique structural advantage, rice husk is the best raw material for preparing graded porous carbon [88]. The particle size of SiO_2 in rice husk is mainly concentrated in the range of 8–22 nm, with a small fraction of SiO_2 in the range of 1–7 nm [89], indicating that SiO_2 in rice husk can be used as a natural template to induce mesopore generation in situ after solubilization. The chemical activation of agricultural waste rice husk as a raw material enables the orderly construction of high-quality rice husk-based biochar with a hierarchical pore structure and high specific surface area. Combined with the new ammonia carbon capture technology, the rice husk-based biochar-ammonia-ethanol system was constructed, and the nano-silica carbon black was produced by the acid-base neutralization and redecomposition reaction between NH_4HCO_3, an intermediate product of the new ammonia carbon capture, and silicate, an intermediate product of the rice husk-based biochar (**Figure 9**). This route greatly solved the problems of carbon capture product consumption and agricultural waste pollution and produced high-value products of rice husk-based biochar carbon and nano-silica at the same time.

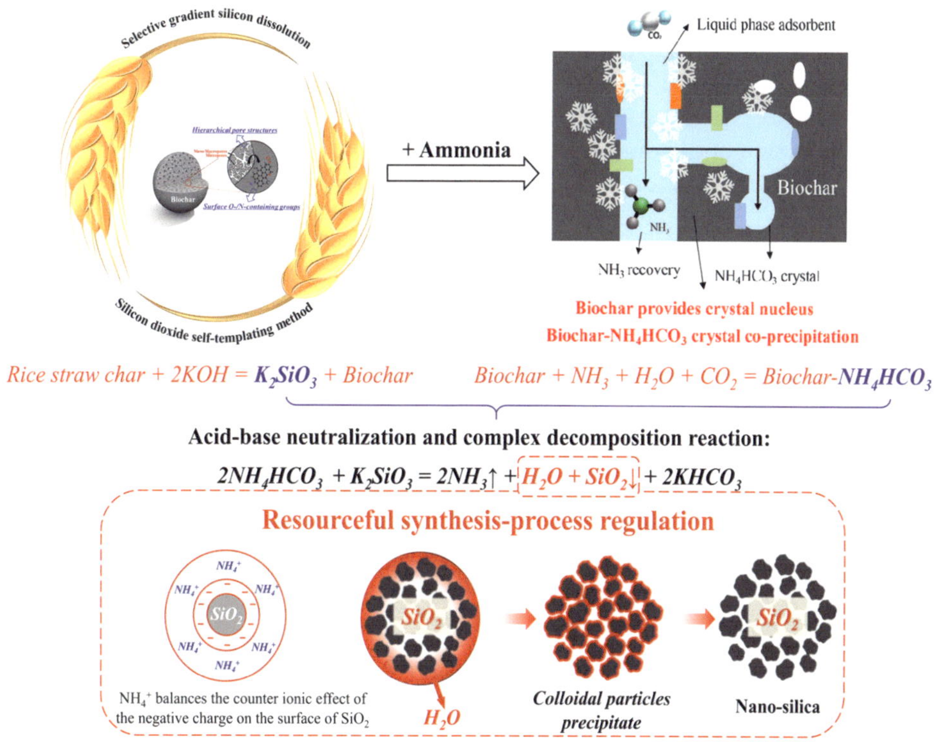

Figure 9.
Technology roadmap for high-value utilization of process products from rice husk-based biochar-new ammonia-based carbon capture technology.

4. Summary and outlook

CO_2 capture is a crucial part of CCUS technology, and absorption and adsorption have been widely studied as the main means of CO_2 capture. The mainstream CO_2 liquid-phase chemical absorption method is difficult to avoid high regeneration energy consumption, degradation problems, and high corrosiveness. In contrast, the "ammonia-ethanol" system effectively avoids these problems but still has serious ammonia escape problems and crystallization control difficulties, and the technology needs to be improved. Biochar has excellent CO_2 adsorption performance due to its specific surface area, three-dimensional through-gradient pore structure, and unique oxygen/nitrogen surface chemistry. However, it still has many problems, such as poor CO_2 selectivity, limited adsorption capacity, high cost, and short service life. Combining the above-mentioned new ammonia carbon capture technology, the carbon capture process is transformed from the traditional ammonia carbon capture gas-liquid two-phase reaction to a gas-liquid-solid three-phase process, which maximizes the efficiency of CO_2 capture by graded adsorption of biochar and efficient absorption of ammonia-ethanol solution:

1. Enhancement of mass transfer between solid and liquid phases to improve the carbon capture rate;

2. Biochar hierarchical pore channel fixation of CO_2/NH_3 to achieve a "win-win" situation of enhancing liquid phase absorption and suppressing ammonia escape;

3. Enhancement of dissolution and crystallization of carbonized liquid to replace rich liquid regeneration with crystal regeneration to reduce energy consumption.

In order to realize the high-value utilization of intermediate products, we propose a system of rice husk-based biochar—new ammonia method—process product resource synthesis—process regulation, which provides new ideas and directions for the CO_2 capture industry.

Author details

Yu Zhang, Yalong Zhang, Dongdong Feng*, Jiabo Wu, Jianmin Gao,
Qian Du and Yudong Huang
School of Energy Science and Engineering, Harbin Institute of Technology,
Harbin, Heilongjiang, China

*Address all correspondence to: 08031175@163.com

References

[1] IEA. Global Energy Review: CO_2 Emissions in 2021. Paris: IEA; 2022. Available online: https://www.iea.org/reports/global-energy-review-co2-emissions-in-2021-2

[2] Zhang Y, Wang S, Feng D, Gao J, Dong L, Zhao Y, et al. Functional biochar synergistic solid/liquid-phase CO2 capture: A review. Energy & Fuels. 2022;**36**(6):2945-2970

[3] Jiang K, Ashworth P, Zhang S, Liang X, Sun Y, Angus D. China's carbon capture, utilization and storage (CCUS) policy: A critical review. Renewable and Sustainable Energy Reviews. 2020;**119**:109601

[4] Liu H, Were P, Li Q, Gou Y, Hou Z. Worldwide status of CCUS technologies and their development and challenges in China. Geofluids. 2017;**2017**

[5] Bachu S. Identification of oil reservoirs suitable for CO2-EOR and CO2 storage (CCUS) using reserves databases, with application to Alberta, Canada. International Journal of Greenhouse Gas Control. 2016;**44**:152-165

[6] Leonzio G, Bogle D, Foscolo PU, Zondervan E. Optimization of CCUS supply chains in the UK: A strategic role for emissions reduction. Chemical Engineering Research and Design. 2020;**155**:211-228

[7] Leonzio G, Foscolo PU, Zondervan E. An outlook towards 2030: Optimization and design of a CCUS supply chain in Germany. Computers & Chemical Engineering. 2019;**125**:499-513

[8] Jefferson M. IPCC Fifth Assessment Synthesis Report:"Climate Change 2014: Longer Report": Critical Analysis. Technological Forecasting and Social Change. 2015;**92**(mar.):362-363

[9] Olajire AA. CO2 capture and separation technologies for end-of-pipe applications–A review. Energy. 2010;**35**(6):2610-2628

[10] Buhre BJ, Elliott LK, Sheng C, Gupta RP, Wall TF. Oxy-fuel combustion technology for coal-fired power generation. Progress in Energy and Combustion Science. 2005;**31**(4):283-307

[11] Adánez J, Abad A. Chemical-looping combustion: Status and research needs. Proceedings of the Combustion Institute. 2019;**37**(4):4303-4317

[12] Xu G, Liang F, Yang Y, Hu Y, Zhang K, Liu W. An improved CO2 separation and purification system based on cryogenic separation and distillation theory. Energies. 2014;**7**(5):3484-3502

[13] Ban ZH, Keong LK, Mohd Shariff A, editors. Physical absorption of CO2 capture: A review. In: Advanced Materials Research. Trans Tech Publ; 2014

[14] Fang M, Yi N, Di W, Wang T, Wang Q. Emission and control of flue gas pollutants in CO2 chemical absorption system–A review. International Journal of Greenhouse Gas Control. 2020;**93**:102904

[15] Brunetti A, Scura F, Barbieri G, Drioli E. Membrane technologies for CO2 separation. Journal of Membrane Science. 2010;**359**(1-2):115-125

[16] Klein G. Principles of adsorption and adsorption processes. Reactive Polymers Ion Exchangers Sorbents. 1985;**4**(1):62-62

[17] Göttlicher G, Pruschek R. Comparison of CO2 removal systems for fossil-fuelled power plant processes.

Energy Conversion and Management. 1997;**38**:S173-S1S8

[18] Tuinier M, van Sint AM, Kramer GJ, Kuipers J. Cryogenic CO2 capture using dynamically operated packed beds. Chemical Engineering Science. 2010;**65**(1):114-119

[19] Webley PA. Adsorption technology for CO_2 separation and capture: a perspective. Adsorption. 2014;**20**(2):225-231

[20] Clausse M, Merel J, Meunier F. Numerical parametric study on CO2 capture by indirect thermal swing adsorption. International Journal of Greenhouse Gas Control. 2011;**5**(5): 1206-1213

[21] Gibbins J, Chalmers H. Carbon capture and storage. Energy Policy. 2008;**36**(12):4317-4322

[22] Bhown AS, Freeman BC. Analysis and status of post-combustion carbon dioxide capture technologies. Environmental Science & Technology. 2011;**45**(20):8624-8632

[23] Guo H, Li C, Shi X, Li H, Shen S. Nonaqueous amine-based absorbents for energy efficient CO2 capture. Applied Energy. 2019;**239**:725-734

[24] Zhao B, Su Y, Tao W, Li L, Peng Y. Post-combustion CO2 capture by aqueous ammonia: A state-of-the-art review. International Journal of Greenhouse Gas Control. 2012;**9**:355-371

[25] Han K, Ahn CK, Lee MS. Performance of an ammonia-based CO2 capture pilot facility in iron and steel industry. International Journal of Greenhouse Gas Control. 2014;**27**:239-246

[26] Qin F, Wang S, Hartono A, Svendsen HF, Chen C. Kinetics of CO2 absorption in aqueous ammonia solution. International Journal of Greenhouse Gas Control. 2010;**4**(5):729-738

[27] Bougie F, Pokras D, Fan X. Novel non-aqueous MEA solutions for CO2 capture. International Journal of Greenhouse Gas Control. 2019;**86**:34-42

[28] Oyenekan BA, Rochelle GT. Alternative stripper configurations for CO2 capture by aqueous amines. AIChE Journal. 2007;**53**(12):3144-3154

[29] Ochedi FO, Yu J, Yu H, Liu Y, Hussain A. Carbon dioxide capture using liquid absorption methods: A review. Environmental Chemistry Letters. 2021;**19**(1):77-109

[30] Thomsen K, Rasmussen P. Modeling of vapor–liquid–solid equilibrium in gas–aqueous electrolyte systems. Chemical Engineering Science. 1999;**54**(12):1787-1802

[31] Que H, Chen C-C. Thermodynamic modeling of the NH3–CO2–H2O system with electrolyte NRTL model. Industrial & Engineering Chemistry Research. 2011;**50**(19):11406-11421

[32] Yu H, Qi G, Wang S, Morgan S, Allport A, Cottrell A, et al. Results from trialling aqueous ammonia-based post-combustion capture in a pilot plant at Munmorah Power Station: Gas purity and solid precipitation in the stripper. International Journal of Greenhouse Gas Control. 2012;**10**:15-25

[33] Zhang M, Guo Y. Rate based modeling of absorption and regeneration for CO2 capture by aqueous ammonia solution. Applied Energy. 2013;**111**:142-152

[34] Wang F, Zhao J, Miao H, Zhao J, Zhang H, Yuan J, et al. Current status and challenges of the ammonia escape inhibition technologies in

ammonia-based CO2 capture process. Applied Energy. 2018;**230**:734-749

[35] Lu R, Li K, Chen J, Yu H, Tade M. CO2 capture using piperazine-promoted, aqueous ammonia solution: Rate-based modelling and process simulation. International Journal of Greenhouse Gas Control. 2017;**65**:65-75

[36] Pachitsas S, Bonalumi D. Parametric investigation of CO2 capture from industrial flue gases using aqueous mixtures of ammonia (NH3) and potassium carbonate (K2CO3). International Journal of Greenhouse Gas Control. 2022;**114**:103567

[37] Jiang K, Li K, Yu H, Feron PHM. Piperazine-promoted aqueous-ammonia-based CO2 capture: Process optimisation and modification. Chemical Engineering Journal. 2018;**347**:334-342

[38] Lillia S, Bonalumi D, Fosbøl PL, Thomsen K, Jayaweera I, Valenti G. Thermodynamic and kinetic properties of NH3-K2CO3-CO2-H2O system for carbon capture applications. International Journal of Greenhouse Gas Control. 2019;**85**:121-131

[39] Gao J, Zhang Y, Feng D, Du Q, Yu M, Xie M, et al. A new technique of carbon capture by ammonia with the reinforced crystallization at low carbonized ratio and initial experimental research. Fuel Processing Technology. 2015;**135**:207-211

[40] Xu Y, Jin B, Chen X, Zhao Y. Performance of CO2 absorption in a spray tower using blended ammonia and piperazine solution: Experimental studies and comparisons. International Journal of Greenhouse Gas Control. 2019;**82**:152-161

[41] Zhang Y, Feng D, Gao J, Du Q, Wu S. Thermodynamic properties in ternary system of NH4HCO3–H2O–ethanol based on antisolvent method to strengthen crystallization of carbonized ammonia. Adsorption Science & Technology. 2019;**37**(1-2):127-138

[42] Yang N, Yu H, Dy X, Conway W, Maeder M, Feron P. Amino acids/NH3 mixtures for CO2 capture: Effect of neutralization methods on CO2 mass transfer and NH3 vapour loss. Energy Procedia. 2014;**63**:773-780

[43] Ma S, Chen G, Zhu S, Wen J, Gao R, Ma L, et al. Experimental study of mixed additive of Ni (II) and piperazine on ammonia escape in CO2 capture using ammonia solution. Applied Energy. 2016;**169**:597-606

[44] Wang J, Yuan X, Deng S, Zeng X, Yu Z, Li S, et al. Waste polyethylene terephthalate (PET) plastics-derived activated carbon for CO2 capture: A route to a closed carbon loop. Green Chemistry. 2020;**22**(20):6836-6845

[45] Feng D, Guo D, Zhang Y, Sun S, Zhao Y, Chang G, et al. Adsorption-enrichment characterization of CO2 and dynamic retention of free NH3 in functionalized biochar with H2O/NH3·H2O activation for promotion of new ammonia-based carbon capture. Chemical Engineering Journal. 2021;**409**:128193

[46] Cao L, Zhang X, Xu Y, Xiang W, Wang R, Ding F, et al. Straw and wood based biochar for CO2 capture: Adsorption performance and governing mechanisms. Separation and Purification Technology. 2022;**287**:120592

[47] Patel HA, Byun J, Yavuz CT. Carbon dioxide capture adsorbents: Chemistry and methods. ChemSusChem. 2017;**10**(7):1303-1317

[48] Zhang Z, Schott JA, Liu M, Chen H, Lu X, Sumpter BG, et al. Prediction of carbon dioxide adsorption via deep learning. Angewandte Chemie. 2019;**131**(1):265-269

[49] Dasgupta S, Biswas N, Aarti GNG, Divekar S, Nanoti A, et al. CO2 recovery from mixtures with nitrogen in a vacuum swing adsorber using metal organic framework adsorbent: A comparative study. International Journal of Greenhouse Gas Control. 2012;7: 225-229

[50] Vedrine JC. Zeolite Chemistry and Catalysis. Studies in Surface Science and Catalysis. 1976;**69**:25-42

[51] Zhang J, Singh R, Webley PA. Alkali and alkaline-earth cation exchanged chabazite zeolites for adsorption based CO2 capture. Microporous and Mesoporous Materials. 2008;**111**(1-3): 478-487

[52] Siriwardane RV, Shen M-S, Fisher EP. Adsorption of CO2, N2, and O2 on natural zeolites. Energy & Fuels. 2003;**17**(3):571-576

[53] Mason JA, McDonald TM, Bae T-H, Bachman JE, Sumida K, Dutton JJ, et al. Application of a high-throughput analyzer in evaluating solid adsorbents for post-combustion carbon capture via multicomponent adsorption of CO2, N2, and H2O. Journal of the American Chemical society. 2015;**137**(14):4787-4803

[54] Balsamo M, Budinova T, Erto A, Lancia A, Petrova B, Petrov N, et al. CO2 adsorption onto synthetic activated carbon: Kinetic, thermodynamic and regeneration studies. Separation and Purification Technology. 2013;**116**:214-221

[55] Wahby A, Ramos-Fernández JM, Martínez-Escandell M, Sepúlveda-Escribano A, Silvestre-Albero J, Rodríguez-Reinoso F. High-surface-area carbon molecular sieves for selective CO2 adsorption. ChemSusChem. 2010;**3**(8):974-981

[56] Peng X, Vicent-Luna JM, Jin Q. Water–gas shift reaction to capture carbon dioxide and separate hydrogen on single-walled carbon nanotubes. ACS Applied Materials & Interfaces. 2021;**13**(9):11026-11038

[57] Xiong Y, Wang Y, Jiang H, Yuan S. MWCNT decorated rich N-doped porous carbon with tunable porosity for CO2 capture. Molecules. 2021;**26**(11):3451

[58] Hidayu AR, Muda N. Preparation and characterization of impregnated activated carbon from palm kernel shell and coconut shell for CO2 capture. Procedia Engineering. 2016;**148**:106-113

[59] Chatterjee R, Sajjadi B, Chen W-Y, Mattern DL, Hammer N, Raman V, et al. Effect of pyrolysis temperature on physicochemical properties and acoustic-based amination of biochar for efficient CO2 adsorption. Frontiers in Energy Research. 2020;**8**

[60] Keiluweit M, Nico PS, Johnson MG, Kleber M. Dynamic molecular structure of plant biomass-derived black carbon (Biochar). Environmental Science & Technology. 2010;**44**(4):1247-1253

[61] Chen Z, Ma L, Li S, Geng J, Song Q, Liu J, et al. Simple approach to carboxyl-rich materials through low-temperature heat treatment of hydrothermal carbon in air. Applied Surface Science. 2011;**257**(20):8686-8691

[62] Hong S-M, Jang E, Dysart AD, Pol VG, Lee KB. CO2 capture in the sustainable wheat-derived activated microporous carbon compartments. Scientific Reports. 2016;**6**(1):34590

[63] Igalavithana AD, Choi SW, Shang J, Hanif A, Dissanayake PD, Tsang DCW, et al. Carbon dioxide capture in biochar produced from pine sawdust and paper mill sludge: Effect of porous structure

and surface chemistry. Science of the Total Environment. 2020;**739**:139845

[64] Singh G, Lakhi KS, Ramadass K, Sathish C, Vinu A. High-performance biomass-derived activated porous biocarbons for combined pre-and post-combustion CO2 capture. ACS Sustainable Chemistry & Engineering. 2019;**7**(7):7412-7420

[65] Sher F, Iqbal SZ, Albazzaz S, Ali U, Mortari DA, Rashid T. Development of biomass derived highly porous fast adsorbents for post-combustion CO2 capture. Fuel. 2020;**282**:118506

[66] Quan C, Jia X, Gao N. Nitrogen-doping activated biomass carbon from tea seed shell for CO2 capture and supercapacitor. International Journal of Energy Research. 2020;**44**(2):1218-1232

[67] Zubbri NA, Mohamed AR, Lahijani P, Mohammadi M. Low temperature CO2 capture on biomass-derived KOH-activated hydrochar established through hydrothermal carbonization with water-soaking pre-treatment. Journal of Environmental Chemical Engineering. 2021;**9**(2):105074

[68] Li J, Michalkiewicz B, Min J, Ma C, Chen X, Gong J, et al. Selective preparation of biomass-derived porous carbon with controllable pore sizes toward highly efficient CO2 capture. Chemical Engineering Journal. 2019;**360**:250-259

[69] Singh G, Bahadur R, Mee Lee J, Young Kim I, Ruban AM, Davidraj JM, et al. Nanoporous activated biocarbons with high surface areas from alligator weed and their excellent performance for CO2 capture at both low and high pressures. Chemical Engineering Journal. 2021;**406**:126787

[70] Balou S, Babak SE, Priye A. Synergistic effect of nitrogen doping and ultra-microporosity on the performance of biomass and microalgae-derived activated carbons for CO2 capture. ACS Applied Materials & Interfaces. 2020;**12**(38):42711-42722

[71] Wu D, Liu J, Yang Y, Zheng Y. Nitrogen/oxygen Co-doped porous carbon derived from biomass for low-pressure CO2 capture. Industrial & Engineering Chemistry Research. 2020;**59**(31):14055-14063

[72] Al Mesfer MK. Synthesis and characterization of high-performance activated carbon from walnut shell biomass for CO2 capture. Environmental Science and Pollution Research. 2020;**27**(13):15020-15028

[73] Yang Z, Zhang G, Xu Y, Zhao P. One step N-doping and activation of biomass carbon at low temperature through NaNH2: An effective approach to CO2 adsorbents. Journal of CO2 Utilization. 2019;**33**:320-329

[74] Yang Z, Guo X, Zhang G, Xu Y. One-pot synthesis of high N-doped porous carbons derived from a N-rich oil palm biomass residue in low temperature for CO2 capture. International Journal of Energy Research. 2020;**44**(6):4875-4887

[75] Lahijani P, Mohammadi M, Mohamed AR. Metal incorporated biochar as a potential adsorbent for high capacity CO2 capture at ambient condition. Journal of CO2 Utilization. 2018;**26**:281-293

[76] He S, Chen G, Xiao H, Shi G, Ruan C, Ma Y, et al. Facile preparation of N-doped activated carbon produced from rice husk for CO2 capture. Journal of Colloid and Interface Science. 2021;**582**:90-101

[77] Yang C, Zhao T, Pan H, Liu F, Cao J, Lin Q. Facile preparation of N-doped

porous carbon from chitosan and NaNH2 for CO2 adsorption and conversion. Chemical Engineering Journal. 2022;**432**:134347

[78] Xing W, Liu C, Zhou Z, Zhou J, Wang G, Zhuo S, et al. Oxygen-containing functional group-facilitated CO2 capture by carbide-derived carbons. Nanoscale Research Letters. 2014;**9**(1):189

[79] Ma X, Li L, Zeng Z, Chen R, Wang C, Zhou K, et al. Experimental and theoretical demonstration of the relative effects of O-doping and N-doping in porous carbons for CO2 capture. Applied Surface Science. 2019;**481**:1139-1147

[80] Wu R, Ye Q, Wu K, Wang L, Dai H. Highly efficient CO2 adsorption of corn kernel-derived porous carbon with abundant oxygen functional groups. Journal of CO2 Utilization. 2021;**51**:101620

[81] Xu X, Kan Y, Zhao L, Cao X. Chemical transformation of CO2 during its capture by waste biomass derived biochars. Environmental Pollution. 2016;**213**:533-540

[82] Dagaonkar M, Heeres H, Beenackers A, Pangarkar V. The application of fine TiO2 particles for enhanced gas absorption. Chemical Engineering Journal. 2003;**92**(1-3):151-159

[83] Park MH, Jeong S, Kim JY. Adsorption of NH3-N onto rice straw-derived biochar. Journal of Environmental Chemical Engineering. 2019;**7**(2):103039

[84] Mathias PM, Reddy S, O'Connell JP. Quantitative evaluation of the chilled-ammonia process for CO2 capture using thermodynamic analysis and process simulation. International Journal of Greenhouse Gas Control. 2010;**4**(2):174-179

[85] Gai X, Wang H, Liu J, Zhai L, Liu S, Ren T, et al. Effects of feedstock and pyrolysis temperature on biochar adsorption of ammonium and nitrate. PLoS One. 2014;**9**(12):e113888

[86] Liu N, Sun ZT, Wu ZC, Zhan XM, Zhang K, Zhao EF, et al., editors. Adsorption characteristics of ammonium nitrogen by biochar from diverse origins in water. In: Advanced Materials Research. Trans Tech Publ; 2013

[87] Li K, Yu H, Feron P, Tade M, Wardhaugh L. Technical and energy performance of an advanced, aqueous ammonia-based CO2 capture technology for a 500 MW coal-fired power station. Environmental Science & Technology. 2015;**49**(16):10243-10252

[88] Wang Z, Smith AT, Wang W, Sun L. Versatile nanostructures from rice husk biomass for energy applications. Angewandte Chemie International Edition. 2018;**57**(42):13722-13734

[89] Larichev YV, Yeletsky P, Yakovlev V. Study of silica templates in the rice husk and the carbon–silica nanocomposites produced from rice husk. Journal of Physics and Chemistry of Solids. 2015;**87**:58-63

Chapter 11

Microwaved Flux Matter- Char Sand Production of Waste Coal Char/Biochar/Gypsium Ash and Fly Ash Mixtures for Mortar- Fire Retardent Composite

Yıldırım İsmail Tosun

Abstract

Fire inhibiting materials as cement filler are used in mortar constructions especially using gypsium board, similar isolator mortars. The mortar covered char and ash sand mixtures insulate heat and reduce fire blazing activity. Ytong, or porous briquettes and clay is the world's most popular insulating construction material retarding blaze due to its porous durability, processability, and cost. However, producing concrete or mortar with high isolation with HD styrene panels is insulating the structure, protecting the cement board against flammable fire risk. Slag-type masonry requires high heavier fire inhibiting matter in construction. Styrene type isolation provides fire inhibiting at lightweight masonry or mortar generation with the use of waste gypsium fines and waste coal slimes and high ash char "char sands" and ash fines. The growing environmental concerns motivated researchers to search for char waste slag-type inhibiting materials using gypsium fines and biomass waste char fines leading to alternative routes of fire-retardant mortar construction. In this way, several alternative materials of isolation mortar have prompted.

Keywords: microwave, fire retardent, composite mortar, waste fire retardant, plaster, analysis of gradations, porous structure, light weight retardent, heat absorbance, composite plaster durability

1. Introduction

Molten plastic extruded belts or strips may easily be produced through the nozzle hole of pressed waste plastic fluids by microwave radiation till 300°C for recycling waste materials as granule compost [1, 2]. The use of waste concrete debris and broken glass or plastic slags cause an important cost decrease in masonry stone production [3, 4]. Even the use of waste materials as aggregate and sand size make them beneficial in concrete mixture evaluation in most light weight constructions [5, 6]. The melted plastics and bitumen asphalt may be replaced by cement in masonry brick and roof tile production as binder compound while providing impermeable and high resistive durability to thaw and freezing in cold climates. Plastic extrusion may need suitable fluidization quality and antifouling powders use such as clay at a certain amount. Presently, around 70% of the construction is produced through the conventional slag-type masonry as inhibiting masonry constructions [3, 4].

In the region, the municipal bottom ash wastes of asphaltite combustion in boilers as wastes, containing the high porous content. 70–80% content of bottom ash is over 25 mm size suitable as lightweight aggregate discarded and collected. The villagers for heating house collect agricultural oak tree and bush waste, municipal waste and agricultural, manure waste products such as forest waste at 21% of total waste [5, 6]. The biomass waste collected in the region is combusted and bottom ash mixed with asphaltite bottom ash at the density of 0.7 kg/l is about 450 thousand tons for wet production [7–9]. The wood char fine in Siirt and Hakkari is evaluated for fire inhibitor. The waste plastics are collected as sludge waste and shredded wet and converted plastic pellet noodles. The plastic noddle products and belts both should be evaluated by pyrolysis oil content below 350°C and the other slag plastics is becoming hard porous slag such as, fine matters gradation of aggregate subjected to mixing and melted asphalt briquetting of the sludge waste and subsequently briquetted products for concrete compost aggregate below 25-50 mm [10–12].

In this study; the effect on the physical parameters of briquetting, shear model patterns making preliminary tests to determine the briquetting and processing conditions, indention and sawing shear rate were investigated for rock and waste plastic or asphalt compost aggregate concrete in comparison with cemented aggregate [13]. This assay has been determined to be advantageous in the plastic and asphalt bound aggregate briquette production from sludge content solution with the waste plastic and their mixture rate with porous local stone [14–16].

1.1 Carbon source-biomass potential of Turkey

In the cement and retardant material consumption, use of waste materials as a carbon source from agricultural biomass waste and forest biomass waste depending on crop production in the market and waste straw used for various purposes, such as other waste cotton stalks, corn stalks, sunflower stalks, nut leafs are evaluated in production of retardend carbon source at finer sizes as filler material. The total amount of different wastes are given in the **Table 1** [14]. The total waste field crops in Turkey and waste quantities are given in **Table 2** [17, 18].

Biomass wastes are evaluated in char carbon production as active carbon or fire-retardant carbon even in the low-quality high ash containing matter as waste source. The biochar carbon resources may be produced from country oil resources, or crop oil, oily wastes as composted sources as given in **Table 2** in Turkey [17].

The asphaltite coal type is widely deposited in the Şırnak province with high amount of shale content. The shale ash content is illustrated in **Figure 1**. The combustion is retardent act over 45% ash content leaving about 20% unburned carbon in the ash [19, 20]. The ash content change of Şırnak asphaltite coal and char used as fire-retardant in this study in terms of density is illustrated in **Figure 1**.

1.2 Gradation

Aggregate size distribution is changing by ASTM standards of soil classification over the foundation stability research in detailed [14] is given with Sieve analysis

Waste Type	Waste Statistics		
	Heat Value, kJ/kg	Country, Actual million ton/year	Eastern Anatolian Region Actual, 1000tons/year
Textile,Rubber,Plastics	18,200	0.6	2.1
Wood, Cardboard, Paper	17,600	2.4	1.6
Organic Municipal Waste	13,500	2.2	29
Animal Waste	13,500	1.9	21
Forestry and Agricultural Biomass	18,500	2.8	63
Total	18,000	9.9	116.7

Table 1.
The total amount of municipal waste divided into actual values in Turkey and eastern Anatolian region in 2019 [17].

Waste Type	Waste Statistics		
	Heat Value, kJ/kg	Eastern Anatolian Region Actual 1000 ton/year	Şırnak Actual, 1000 ton/year
Plastic	17,200	2.1	1.3
Agricultural waste	17,600	2.8	1.6
Cow, Sheep Poultry Wastes	13,400	21	11
Forest Waste	18,600	60	33
Total	17,000	85.9	46.9

Table 2.
The total annual production of biomass waste in Şırnak and eastern Anatolian region [17].

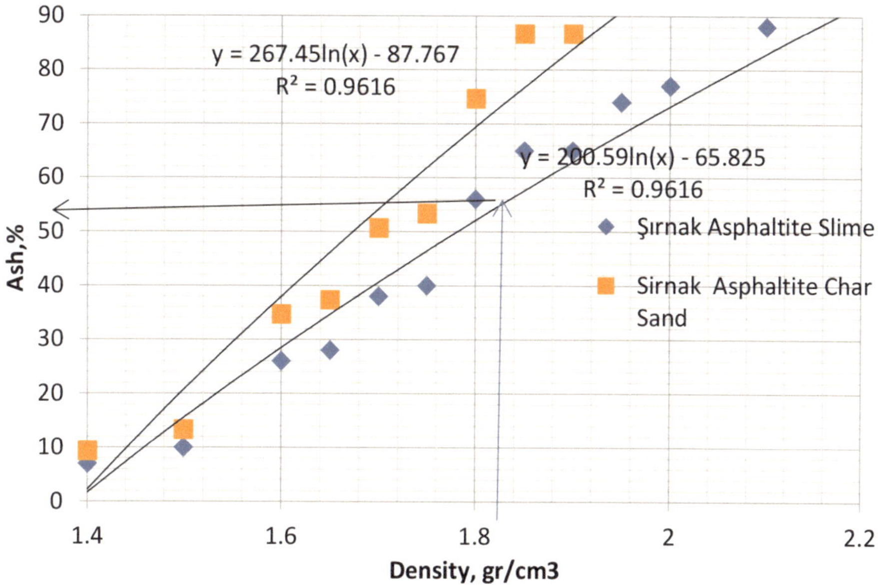

Figure 1.
The ash content change of Şırnak asphaltite coal and char in terms of density.

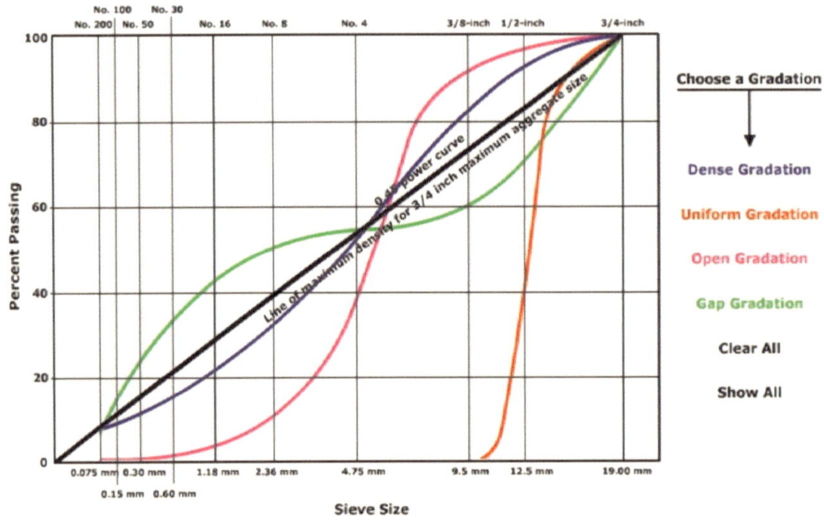

Figure 2.
The soil -aggregate classification in ASTM standard [14].

results as shown in **Figure 2**. The permeability of soil is also determined regarding the chart illustrated in Standard as **Figure 3**. Physical properties of the clay material [21, 22] Parameter Value Color Dark Brown Specific Weight 2.69 Sand Content (%) 17.33 Silt Content (%) 6.22 Clay Content (%) 76.44 Liquid Limit (%) 43.9 Plastic Limit (%) 21.8 Ground Class (USCS) C_L 10%, 30%, 50% liquid limit values of the waste Şırnak asphaltite slime and clay material mixed into the clay sample were calculated. The liquid limit value of the clay sample containing 10% waste slime clay corresponding to the sinking of 20 mm was determined as 28%. The textural and strength properties of the Şırnak shale clay showed that the water absorption of the texture is high and the chlorite mineral is suitable for volume changes. Due to this structure of the clay, the plasticity and strength of the material changes and the swelling and shrinkage activities of the clay can lead to different behaviors and cause structural problems. For this reason, it is of great importance to perform volumetric shrinkage tests of asphaltite slime or ash slime.

1.3 Fire retardent slags

The recycling needs of waste plastics in housing in cities forced to energy use and construction use of polymer wastes and many other filler areas such as fly ash composted ornaments and masonry areas are increasing. The large-scale reconstruction projects offer the use of demolished buildings concrete, transportation of those debris materials and crushing and compacted with water and cement cause a high amount of cement and water even increase cost elements. The dams, factories and the construction sector, which aims to protect the stability of concretre structures, gradually need much cheap aggregate production. To meet the cost reductions of all masonry and mortar construction, waste materials are evaluated similarly to masonry bricks regarding strength and durability [23, 24]. Although, the aggregate materials obtained from the quarries can be widely used in the construction

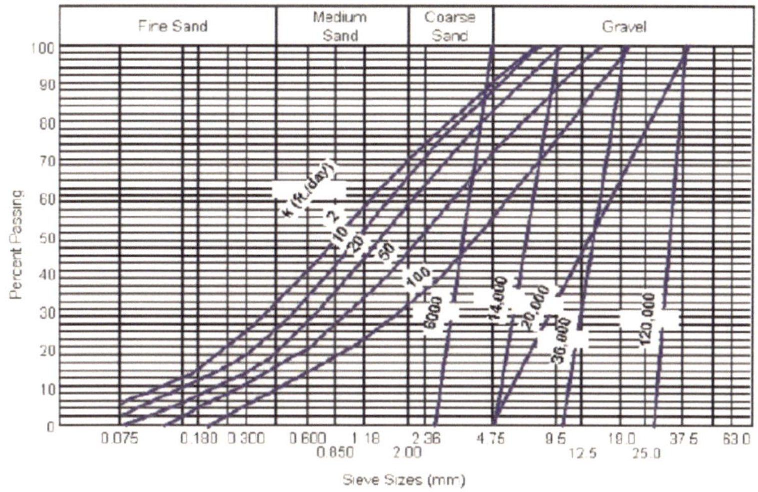

Figure 3.
The soil permeability regarding void in soil classification in ASTM standard [13].

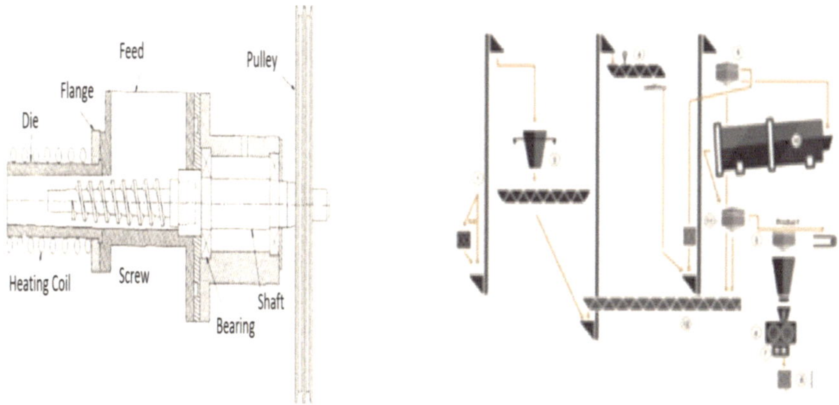

Figure 4.
(a) extruding ball die, (b) plastic waste and asphaltite slag and slime asphaltite char mixing briquetting [22].

industry. Lightweight materials are waiting in high house constructions as surplus stock. Crusher residue fine-grained materials are scattered around by 10–15% at depending on the crusher type. As a result, the waste plastic nodules or belts may compost as slag waste sand, fine material that remain in the dust collectors. Therefore, the general standard provisions stated in the construction materials fire resistivity for plastic contents not over 30% volume regarding bitumen asphalt or other masonry mixing fines [25–28]. In the mortar tests, the mixed waste briquettes are aimed to prevent fire reaction that occurs as a result in the blazing fire contact, degradation of stability of residents. The importance of fire inhibiting or control practices in the evaluation, the regulation was emphasized on that way. Disposal of plastic waste heaps is in the form of shredded waste and can be used by extruding.

Many plastic waste recycling articles dealing with many issues such as such are included in the literature [29–32]. As an industrial raw material, lightweight volcanic cinder instead of broken glass is used as the main raw material and additive material in lightweight brick sectors depending on the masonry use. The aggregates obtained by crushing, sieving and sizing according to the sector in which porous stone will be used are evaluated in appropriate gradation sizes according to the geotechnical strength purpose in briquetted brick use. However, the fine-grained material remaining under the sieve during the sizing process is awaiting stock surplus in the local dumps' areas. Şırnak asphaltite bottom ash slag with high porosity is in search of new areas of use to utilize the organic soils they obtain as waste other than dumping activities. In this respect, the light weight mixture with the recycled waste plastic product that occurs in retort furnace is searched detailed for lightweight briquette production without causing environmental pollution. In addition, shale fine in soil environments, which are quite commonly layered in certain regions of Şırnak is used for gradation mixing encountered for high strength. In this study, it was aimed to examine the behavior of these two different types of materials by mixing them in variable ratios because of the optimum gradation amount of these materials in the region and the specific characteristics of briquetted materials without cement are evaluated. It is stated that pumice is a suitable additive for the stabilization of high plasticity clay. It is emphasized the usability of plastic waste materials in improving the engineering properties of briquette with shale powder and porous limestone added to briquette blocks cemented in certain proportions. It was also determined that in the asphalt-based mixtures prepared by using fly ash and limestone fine in the improvement of the fine-grained ground sample, the limestone aggregate decreased the shrear strength by 35% and the volcanic cinder increased by 22% [33–36]. Indentation and shear properties of the briquetted materials by plastic waste melted and asphalt melted to be used in the experiment were carried out in the Şırnak University.

1.4 Fire retardant chemical materials

Fire retardent salts such as the construction materials used in the environments we live, the building materials are non-combustible and are produced from salt hydrates, chlorides as chemical materials. Since slag chars or melted/foam salts ignite and shine more quickly than natural materials, a possible fire spreads quickly. The heat from the flame source destroys the oxygen in the environment very quickly and starts to pose life risks in 90 seconds. It is not possible to prevent a fire in a closed environment after 3 minutes without external intervention. In a fire, blazing hazard and toxic gases hazard, chemical gas hazards, explosion and sustainable fire hazard, structure collapse occur. Again, in a possible fire, the temperature rises to 550°C in the first 5 minutes and to 720°C after half an hour. The temperature can reach 950°C degrees after 90 minutes and 1100°C degrees after 3 hours. In some large fires, it is claimed that a temperature of 1500–1700°C occurs [35]. From dripping bricks in buildings fire, which is an exothermic chemical reaction, continuously generates heat and enlarges and spreads the adjacent materials in a chain way by reaching their ignition temperature. In order to eliminate the devastating effects of risk factors in an indoor fire prompt the use of fire retardent construction materials that slow down and stop the progression and spread of fire; Various substances were Flame Retarder. The commonly used salt materials are

- Calcium Potassium salts,
- Phosphoric Acid,
- Hydro Chloric acid-based substances,
- Nitrogen systems containing Phenol and Formaldehyde,
- Products containing Ammonia and Antimony tri oxide,
- Strong basic products, Boron and Derivatives regarding the EU provisions [36].

Regarded the fires encountered the retardant material need have revealed the importance. The necessity of fire-proof materials in the world, especially the wood industry, cable (plastic) industry, and the textile industry have started to produce fire-resistant materials [36–38].

Fire retardant coating and mortars are needed for insulater's coverings and construction boards and plates. The gypsium is providing a good retardant protection however the strength and heat insulation change the strength stability of boards as given in **Table 3** [38].

1.5 Şırnak fly ash, waste ash slag and Şırnak asphaltite char with ash materials

The geological petrographic, geochemical and physical properties of the fine-grained slag and cinder material are found in the local quarry in Tatvan and Şırnak region. The volcanic cinder such as pumice stone, iso foam stone, ash slag stone, are two types of porous texture and contain at least 70–80% porous formation as a result of basic volcanic gaseous activities. The Tatvan basic volcanic cinder is similar to acidic pumice, which is the most widely found and used in the world, has a white dirty appearance and a grayish-white color. The silica ratio is higher in acidic pumices, and it can be widely used in the construction industry [12–15]. A volcanic cinder is a browny reddish porous, glassy volcanic rock that is formed as a result of volcanic gaseous eruptions highly sponge and resistant to chemical reactions at high abrasion strength. It contains pores from macro to micro scale due to the sudden release of the gases in the body during its formation and its sudden cooling. Volcanic cinders have high permeability and high heat and sound insulation. Its hardness is 5–6 according to the Mohs scale. In Eastern Anatolia, severe volcanic events have occurred in very wide areas since the Middle Miocene. Tectonic activity is covering wide areas near Van Lake as volcanic craters lake,

Organic compost	Inorganic salts
Fire retardant wire coating polymeric salts	Phosphoric acid and boric acid salts
Char, Carbon compounds	Calcium Potassium salts
Ammonium Poly Phosphate Resin binders	Gypsium, Anhydrite Magnesia Ferric oxide Fly Ash

Table 3.
Fire retardent chemical materials classification.

craters heel, disseminated tuff covers and tuff debris lava remnants carried by water-floods. It has been active starting from the Mid Miocene period until the end of the Quaternary [36]. Tatvan unit consists of volcanic cinders with 78–83% porous cinders as block flows, debris of flow tuffs, and andesitic, basaltic and rhyolitic lavas [37]. Pumices are light browny macroscopically and dark gray colored in certain places. It has a vesicular texture formed by the cavities left by the gases that expand as a result of sudden pressure decrease under atmospheric conditions. The gray acidic cinder contains coarse plagioclase minerals showing feldspat, biotite minerals and chlorite minerals as accessories are observed in the rock [38–44].

2. Method

The method of compaction for retardant wet material at 15% optimum fluid weight rate as water muddy content pushed to nozzles of the extruder for board plaque production as illustrated in **Figure 4a**. The aggregate mixing the retort mixer is used in laboratory-scale a meter and 30 cm diameter retort used in 10 minutes for mortar homogenized wetting at optimum retardant compositions as showed in flowsheet procedure followed in **Figure 4b**.

In the fire-retardant mixture, preparation used volcanic cinder prepared as slag based on the main element iron, manganese oxide and trace elements given in **Table 4**, the analysis results of the waste Şırnak Asphaltite bottom ash slag material of ultrabasic magma (**Table 5**).

2.1 Porous char slag asphalt sand production

Material is located in Şırnak Province, Southeastern Anatolia Region, are located in the south part of Tatvan and chlorite shale formations limestone formations

Element	Tatvan volcanic cinder %	Asphaltite ash slag
SiO2	27.359	27.359
Al2O3	8.668	8.668
K2O	955	955
Fe2O3	22.42	22.42
Na2O	1.730	1.730
CaO	1.342	1.342
MgO	1.064	1.064
TiO2	0.273	0.273
MnO0,074	0.073	0.073
P2O50,042	0.033	0.033
Cr2O3	1.001	1.001
Loss in Fire3	1, 7	1, 7
Total	98	98

Table 4.
Composition of the waste cinder and Şırnak bottom ash slag material.

Component and parameters	Tatvan volcanic cinder %	Asphaltite bottom ash slag
Gravel (%) 0	—	—
Sand content (%)	94	94
Clay and silt content (%)	4.8	4.8
Effective diameter (d67,mm)	0.12	0.12
Specific gravity	0.81	0.93
Uniformity coefficient (u)	3.1	3.1
Curvature coefficient (n)	0.72	0.72
Classification (USCS)	SP	SP
Plastic limit	NP	NP

Table 5.
Physical properties of waste volcanic cinder.

contained quartz, feldspar, calcite, dolomite and limonite, hematite minerals and asphaltite slag is red color due to the hematite mineral in its composition at 17%. It can be found in light yellow colors depending on the ratio of limonite in the gray shale ground [21–23]. The porous limestones shale texture, marl shows a heterogeneous texture (**Figure 2**).

2.2 Particle size distribution- gradation

M mass of aggregate is, the void is affected by compaction of briquetting and binder distribution. Especially melted asphalt and ash distribution are controlled by volume % of compaction. The bulk sand eating by microwave will also be controlled by the amount of little as 1% binder ash bound as a volume.

where, γg = _density of aggregate, g/cm3; $V(r)$ and $dN(r)$ are the volume and particle amount of aggregate in the size region of integration of cumulative pile from r, $to\ r + dr$), respectively. *V*e volumetric equation is calculated as,

$$dMr = \gamma g\ \mathrm{V(r)dN(r)} \tag{1}$$

$$V(\mathrm{r}) = k\, r^3 \tag{2}$$

where, k is the shape factor. Substitute Eqs. (1) and (2) using.

2.2.1 Aggregate particle size distribution

Particle size distribution is defined by aggregate crushing matter, the type of milling affects the size distribution and the fineness matter ranged below 20 microns determined as given the Eq. (3) below; and RRS logarithmic size distribution is defined as given I Eq. (4) below [24].

$$\mathrm{u(x, d_f, c)} = (\chi/\mathrm{d_f})\ (1 + \mathrm{k/d_f}\ \ (\mathrm{x} - \chi))^{-1/\mathrm{k}} \tag{3}$$

$$Rss(n) = f(n)W(n)\sum_{m=1}^{n} x'/(x - r)^m \tag{4}$$

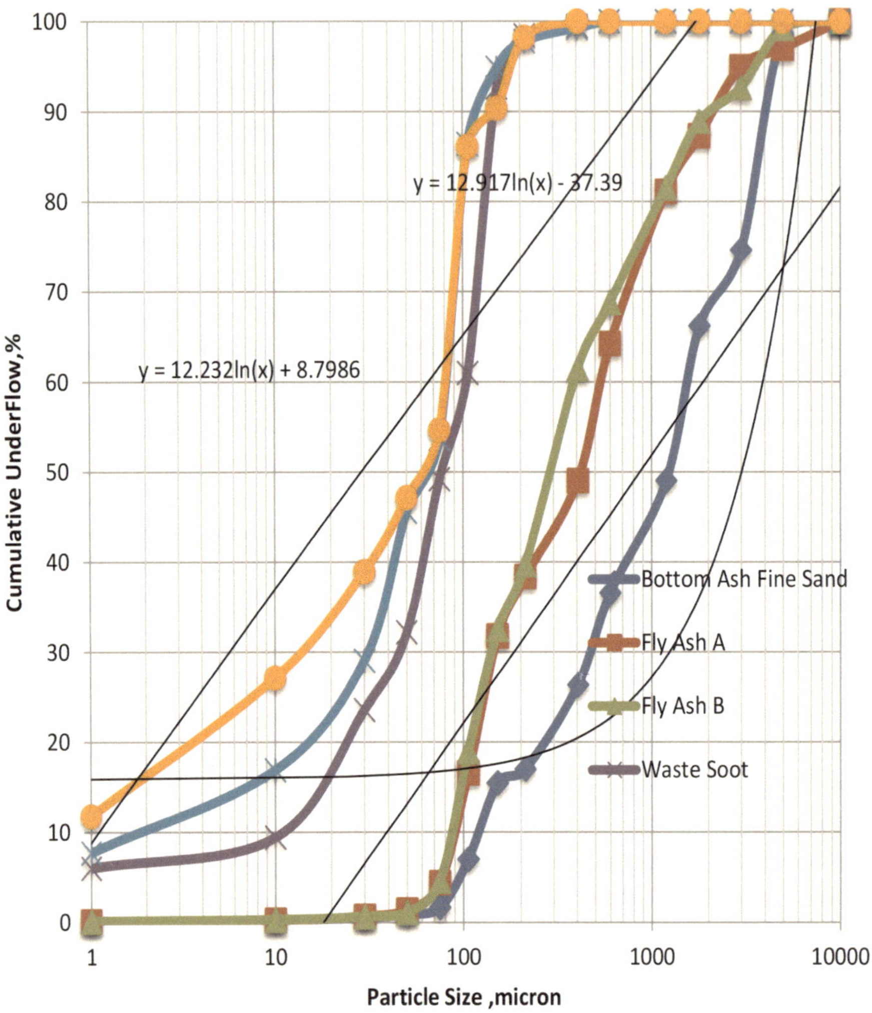

Figure 5.
The Şırnak Fly ash and Şırnak asphaltite slime particle size distribution in gradation in ASTM standard.

The weight of fineness below 100 microns is determined by hydrolic settling analysis. The rate of material used in the experimentation is illustrated in **Figure 5**. The d60 values of the particle distribution of Şırnak Fly ash, Şırnak char slime and waste slime are below 100-micron fine size.

2.3 Fire tests

Flame gas brulor is blazed on the thick 10 mm board and the resistance to fracture and bubbling on a 5 minutes time flash burning at a distance far from 10 mm. The depth of disturbed face of board in the fire resistivity test is determined as an opened hole or as weight rate of burning natter weight. The time of burning of fire contact according to ASTM D-635 was investigated over the extent of depth measured by extensometer of mortar boards reported if the specimen does not burn on the board of 10 mm thickness. An average burning depth rate was also determined.

3. Results and discussion

3.1 Gradation of retardent mixture- *asphalt ash/char amount and briquette porosity*

3.1.1 Particle İndex

The coarse particle distribution avoids the heat conduction so that fineness of particle size distribution in the construction gradation provides optimum fire retardent heat activity on the surface without breaking the mortar face.

$$Ia = 1,25\, V_{10} - 0,25V_{50} - 32 \quad (5)$$

where Ia is particle index, V10 is voided in aggregate compacted at 10 drops per layer, V50 is voided in aggregate compacted at 50 drops per layer. Especially fly ash content in the retardent mixture was decreasing compaction ability. The amount of reaching 30% fly ash addition reduced the permeability of texture compacted in the mortar briquettes at 27% volume rate decrease.

The sand matters are thought as rounded and smooth particles as an ideal form. This may have a low particle index of around 6 or 7, while silty sands composed of angular, rough particles may have a high particle index of between 15 and 20 or more.

3.1.2 Fineness modulus

The fire retardent mortar sands may contain optimum gradation with very fine clays or fly ash on standard content description as happening in ASTM C 125 with a gradation curve as illustrated in **Figure 2**. In this study, the fly ash fineness is determined by the RRS diagram and n distribution coefficient as in Eq. (6) below and illustrated in **Figure 6** for the samples studied as fire retardent.

$$\mathrm{F}(\mathrm{d};\mathrm{t};i) = \sum_{i=0}^{n} \mathrm{u}(\mathrm{x},\mathrm{t}) + \phi(\mathrm{x};\mathrm{t};i).e^{-tin} \quad (6)$$

3.2 Compressive strength test

It is based on the determination of the compressive strength from the indenting of the briquette sample in seconds as drilling bit penetration on rock sample [21]. Then, the indenting depth is determined using the extensometer dipping measure by the pattern is obtained for rock samples used in Şırnak. The fire retardent additives show stable porosity and strength suitable for mortar mixture while cement is locking the fixed coverage over wood in the fire flame tests (**Figures 6** and 7).

3.2.1 Mortarcompost - porous texture strength

The massive mortar mixture of rock sands show different porous structures and strengths. The compaction indentation depth for porous rock stones and fire retardent materials are depended on particle size and fines amount as given below Equations;

$$Elasticity(0) = -af \sum_{m=1}^{M} Cm(1+Xr)^{m} \quad (7)$$

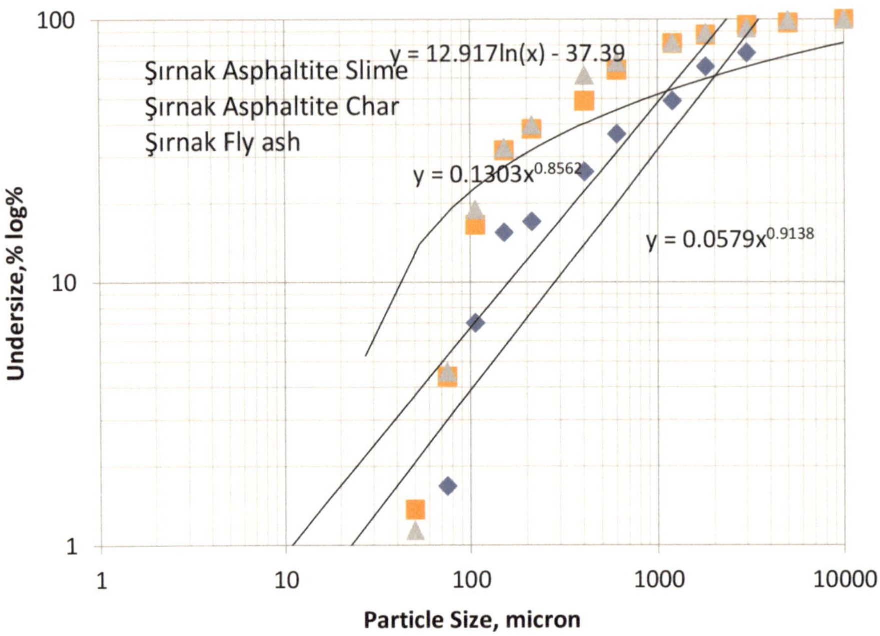

Figure 6.
The fineness of Şırnak fly ash and Şırnak asphaltite slime and char slime regarding gradation, RRS distribution factor of 0.45.

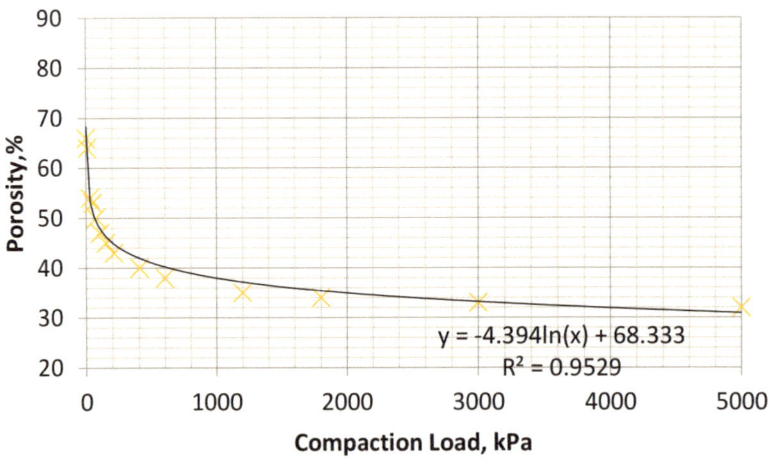

Figure 7.
The Şırnak Fly ash and Şırnak asphaltite slime compost compaction regarding gradation factor below 100 microns solid.

$$E\,deformation(0) = f\sum_{m=1}^{M} Cm(1+Xr)^{m} \tag{8}$$

After this process, the sinking amount of the cone was determined from the electronic measuring stick on the device. Some samples taken from the submerged part of the cone were dried in the oven and the water content corresponding to the determined sinking was found.

Samples with volume 10%, 20%, 30% and 40% waste phospahate salt and char/ salt flux composts blazed on the depth-averaged from three different points for fire retardent manner. The sample used is thick at 5.425 mm for 10 mm wood. The advantage of this experiment is that it minimizes the errors of the candle fire flame over 50 mm experiment according to the standard gradation.

Considering inferences, extreme deformations can be observed under fire load on wood-covered fire retardent mortar that is saturated with a dried binder depending on firing time. Due to these negative weight effects, various chemical burning weight changes on the wood are required for the unflammable ground environment to reduce fire weight decrease, reduce cracking and prevent the negative consequences of bubbling melting in the ground structure. The plastic slag and char regulate the air mixing and permeability on the wood substrates where it is criticized in **Figure 8**, while Şırnak asphaltite char addition reduces air diffusion and reduces heat conduction to wood (**Figure 9**).

The optimum mixing fineness content of the ash sample containing 3% plastic was determined as 30.5% and the maximum dry unit volume weight was 17.52 kN /m^3. The dry unit weight graph of the plastic slag sample containing 50% waste plastic, the optimum mixing binder content of 25.25% and the maximum dry unit weight of 16.13 kN /m^3 for Şırnak asphaltite char.

Accordingly, the results revealed that the asphalt mixing values decrease with the increase of the ratio of plastic slag chars because plastic char slag is a binding material. As a result of the indentation experiments, the optimum Optimum mixing fineness content as below 5-micron content and dry unit weight reduction in fire tests as given in **Figure 10**.

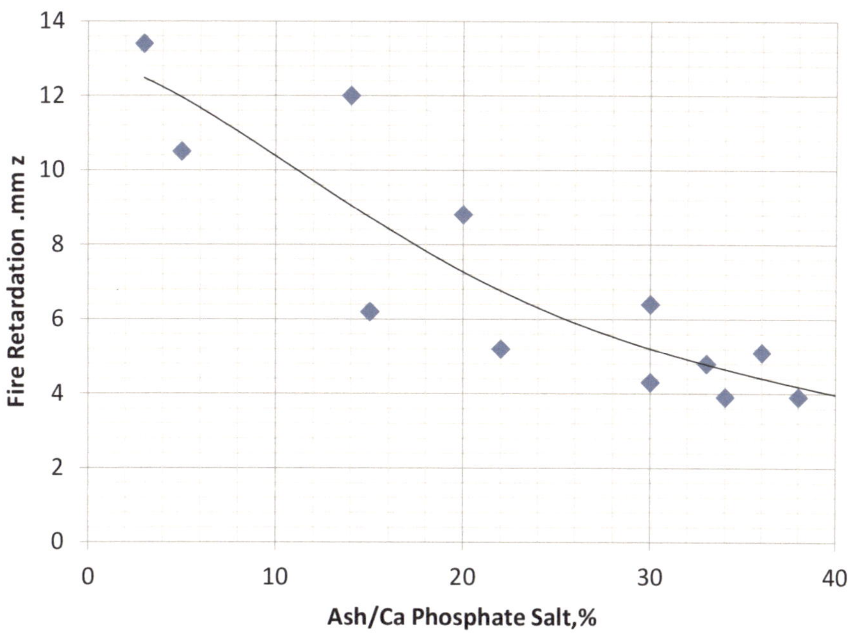

Figure 8.
The char/ ash and phosphate salt slime with retardation to board depth.

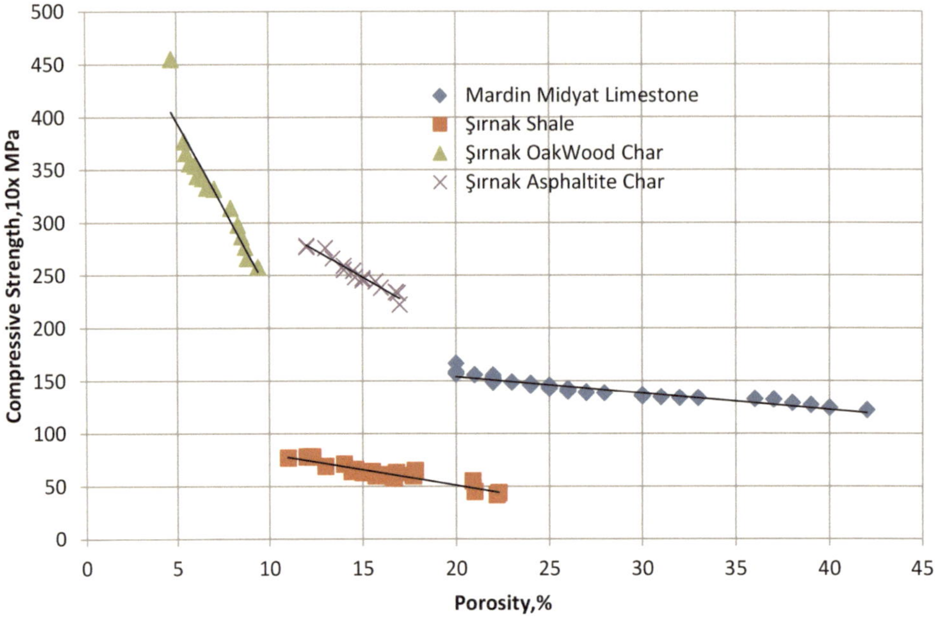

Figure 9.
The Şırnak fire-retardant mortar sand types regarding strength vs. porosity change.

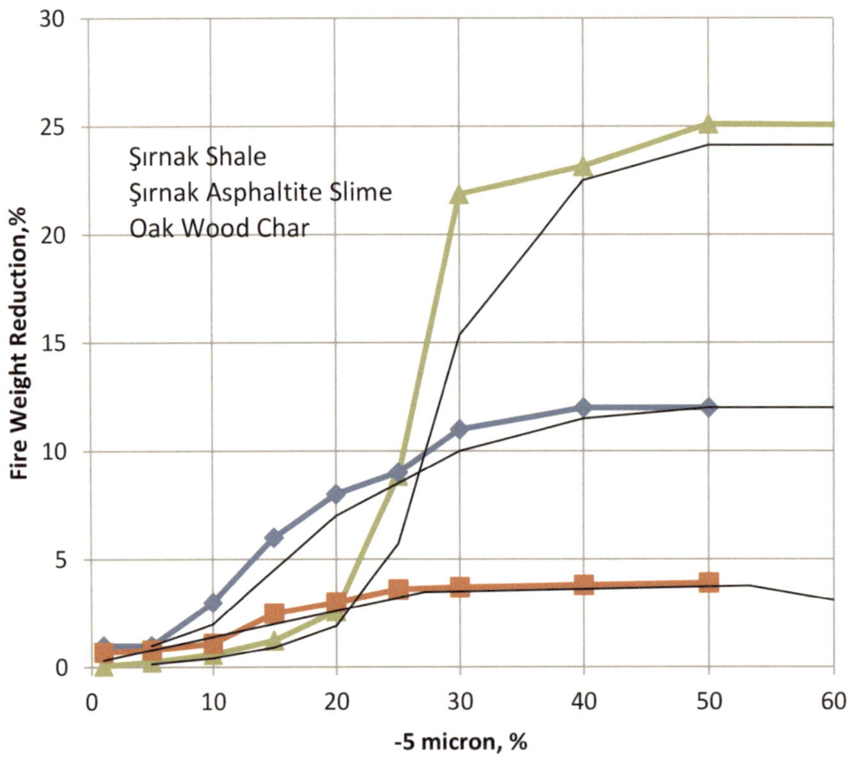

Figure 10.
Optimum mixing fineness content – Effect of dry unit −5 micron weight values on fire resistance

4. Conclusions

All materials undergo deformation when the load is applied. It is predicted that soils are also compact without shear deformation together with the decrease in volume under stress. However, this decreases in the volume of the soil mass, the compression of the plastergrains, the type of voids, the structure and its continuity reveal different types of behavior depending on the way and duration of removal of mixing light weight plaster and air in the cavities. In this context, the study emphasizes the importance of positively improving engineering properties such as compaction and fire inhibiting mortar mixing, char, plastic slag binder content by using different types of materials together.

Salt content over 20% with char and fly ash fire retardent sands the depth of deterioration decreased 200%.

As the amount of Şırnak asphaltite char and plastic slag in the briquette sample increases fineness weight rate with fire weight rate reduction rate, the optimum mixing binder content increases and the maximum dry unit volume weight decreases. This behavior is an expected situation by adding certain proportions of aggregate high- sand rate mixtures because of the plastic unit weight value of the used sand to low ash. The unit weight value of briquette with cinder decreased the bulk density of the mixture. The utility of briquette as lightweight concrete takes attention to the low cost of the material. In this text, the gradation of the much finer potential of fly ash is reducing density with char being critical in fire retardation for oxygen uptake as construction materials.

Author details

Yıldırım İsmail Tosun
Engineering Faculty, Mining Engineering Department, Şirnak University, Şirnak, Turkey

*Address all correspondence to: yildirimismailtosun@gmail.com

References

[1] Anonymous. EPA, 2015, Facts and Figures of Waste Management USA. 2017. Available from: https://www.epa.gov/sites/production/files/2016-11/panelizersidemswimage2014_0.jpg

[2] Anonymous, TKI. The Turkish Ministry of Energy. Lignite Coal Report Ankara: Energy, Dept; 2015

[3] Tosun YI. Foam Concrete Production with Şırnak Mine Waste Claystone and Fly ash, Construction, Materials and Structures. Ekolu SO, Dundu M, Gao X, eidtors. Amsterdam: IOS Press; 2014. pp 1364-1369. 5 Aralık 2014 - 1548 sayfa, ISBN: 978-1-61499-465-7, ISBNe: 978-1-61499-466-4

[4] Tosun Yİ. In: Ekolu SO, Dundu M, Gao X, editors. Foam Concrete Landfill Use in Landslide Hazardous Area in West Şırnak Road, Construction, Materials and Structures. Amsterdam: IOS Press; 2014. pp. 226-231. 5 Aralık 2014 - 1548 sayfa. ISBN: 978-1-61499-465-7, ISBNe: 978-1-61499-466-4

[5] Tosun YI. In: Eraslan M, Demirkaya H, Direkci B, Aslan F, Kilinç M, Kitapliği G, editors. Zayıf yapıların mikro fiber, ko-polimer -uçucu kül ile çimentolanması ve güçlendirilmesi, Bölüm: Mühendislik. Ankara: DİSİPLİNLERARASI AKADEMİK ÇALIŞMALAR- I; 2018. ISBN: 978-605-288-362-4, Çankaya/ ANKARA www. gecekitapligi.com

[6] Tosun Yİ. In: Eraslan M, Demirkaya H, Direkci B, Aslan F, Kilinç M, Kitapliği G, editors. Use of Coal Boiler Bottom Ash, Fly Ash and Geosynthethics for Bottom Seal Layer for Municipal Waste Landfill, Bölüm: Mühendislik. Ankara: DİSİPLİNLERARASI AKADEMİK ÇALIŞMALAR- I; 2018. ISBN: 978-605-288-362-4, Çankaya/ ANKARA. www. gecekitapligi.com

[7] ASTM C 330. Standart Specifications For Lightweight Aggregates for Structural Concrete. Philadelphia: ASTM; 2013

[8] ASTM C 136. Standard test method for Sieve Analysis of Fine and Coarse Aggregates. Pennsylvania; 2013

[9] ASTM D6024-07. Standard test method for Ball Drop on Controlled Low Strength Materials. Pennsylvania: ASTM; 2007

[10] Bieniawski ZT. Mechanism of brittle failure of rock Part I - Theory of fracture process. I. J. of Rock Mech. and Min. Sc. and Geomech. Abstr. 1967;**4**(4):s. 395-406

[11] Campione G, MendolaL L. Behavior in compression of lightweight fiber reinforced concrete confined with transverse steel reinforcement. Cem ConcrCompos. 2004;**26**:645-656

[12] Cavaleri L, Miraglia N, Papia M. Pumice concrete for structural wall panels. Engineering Structures. 2003;**25**: s.115-s.125

[13] Chen B, Liu J. Experimental Application Of Mineral Admixtures in Lightweight Concrete With High Strength And Workability. Construction and Building Materials. 2008;**22**:s. 655-s. 659

[14] ASTM D 3398. Index of Aggregate Particle Shape and Texture. Philedelphia: ASTM;

[15] ASTM D 4791. Flat Particles, Elongated Particles, or Flat and

Elongated Particles in Coarse Aggregates. Philedelphia: ASTM;

[16] ASTM C 1252. Uncompacted Void Content of Fine Aggregate. Philedelphia: ASTM;

[17] Demir H. Synergistic effect of natural zeolites on flame retardant additives [MScThesis]. İzmir: Izmir Technology Institute; 2004

[18] Kalabokidis KD. Effects of wildfire suppression chemicals on people and the environment - a review. Global Nest: the Int. J. 2000;**2**(2):129-137

[19] Blahm TH, Snyder GR. Effect of chemical fire retardants on the survival of juvenile Salmonids. Prescott, Oregon, USA: National Marine Fisheries Service; 1974

[20] Blakely AD. Combustion recovery of flaming pine needle fuel beds sprayed with water/MAP mixtures. In: USDA For. Serv. Res. Pap. INT-421, Ogden, Utah, USA. 1990

[21] Bradstock R, Sanders J, Tegart A. Short-term effects on the foliage of a eucalypt forest after an aerial application of a chemical fire retardant. Australian Forestry. 1987;**50**:71-80

[22] George CW. Fire retardant ground distribution patterns from the CL-215 air tanker. In: USDA For. Serv. Res. Pap. INT-165, Ogden, Utah, USA. 1975

[23] Hardy CE. Chemicals for forest fire fighting. 3rd ed. Boston, Massachusetts, USA: National Fire Protection Association; 1977

[24] Johnson CW, George CW. Relative corrosivity of currently approved wildland fire chemicals. In: USDA For. Serv. Res. Pap. INT-437, Ogden, Utah, USA. 1990

[25] Larson JR, Duncan DA. Annual grassland response to fire retardant and wildfire. Journal of Range Management. 1982;**35**:700-703

[26] NWCG Fire Equipment Working Team. Lot acceptance, quality assurance, and field quality control for fire retardant chemicals, 5th edition, National Wildfire Coordinating Group, NFES 1245. California, USA: San Dimas; 1995b

[27] Pyne SJ, Andrews PL, Laven RD. Introduction to wildland fire. 2nd ed. New York: John Wiley & Sons, Inc; 1996. ISBN: 9780471549130

[28] Coelho A. and De Brito, J. , 2013, Conventional demolition versus deconstruction techniques in managing construction and demolition waste (CDW), 141-185, Handbook of Recycled Concrete and Demolition Waste, A volume in Woodhead Publishing Series in Civil and Structural Engineering, Editors F. Pacheco-Torgal, V.W.Y. Tam, J.A. Labrincha, Y. Ding, J. de Brito, p 672. ISBN: 978-0-85709-682-1

[29] A.E. Richardson, 2013, Strength and durability of concrete using recycled aggregates (RAs), Handbook of Recycled Concrete and Demolition Waste, A volume in Woodhead Publishing Series in Civil and Structural Engineering, 2013, Editors F. Pacheco-Torgal, V.W.Y. Tam, J.A. Labrincha, Y. Ding, J. de Brito, Woodhead Publishing, p 672. ISBN: 978-0-85709-682-1. pp 330-348

[30] B. Gómez-Meijide and I. Pérez, 2013, Recycled aggregates (RAs) for asphalt materials p 378-393, Handbook of Recycled Concrete and Demolition Waste, A volume in Woodhead Publishing Series in Civil and Structural Engineering, 2013, Editors F. Pacheco-Torgal, V.W.Y. Tam, J.A. Labrincha, Y. Ding, J. de Brito, Woodhead Publishing, p 672. ISBN: 978-0-85709-682-1

[31] Gündüz LA. Use of quartet blends containing flyash, scoria, perlitic pumice and cement to produce cellular hollow lightweight masonry blocks for non-load bearing walls. Construction Building Materials. 2008;**22**:s. 747-s. 754

[32] Gündüz LB. The effects of pumice aggregate/ cement ratios on the low-strength concrete properties. Construction and Building Materials. 2008;**22**(5):s721-s728

[33] Gündüz L, Bekar M, Şapcı N. Influence of a new type of additive on the performance of polymer-lightweight mortar composites. Cem ConcrCompos. 2007;**29:s**:594-602

[34] Gündüz L, Uğur İ. The effects of different fine and coarse pumice aggregate/cementratios on the structural concrete properties without using any admixtures. Cement and Concrete Research. 2005;**35, 9**:s. 1859-1864

[35] Gündüz L, Sariisik A, Tozaçan B, Davraz M, Uğur İ, Çankıran O. Pumice technology. Vol. 1. Isparta: Süleyman Demirel University; 1998. pp. 275-285

[36] Howard PH, Datta RS. In: Wheelock T, editor. Chemical Cominution: A Process for Liberating the Mineral matter from Coal, Coal Desulfurization. Washington: ACS Series 64; 1976. pp. 58-64

[37] Tosun Yİ. Gypsum/Desulfurization Fly Ash/Activated Shale Char/Claystone of Şırnak with Popped Biochar Composite Granules as Fire Inhibitor for Fire Hazard Risk in Forest Management, Chapter 6. In: Zhang L, editor. 208 pagesAdvances in Forest Management under Global Change. ISBN: 978-1-83968-307-7. Rijeka: InTech; 2020. pp. 141-170. DOI: 10.5772/ intechopen.69927. Chapter 6, published September 9th, 2020, DOI: 10.5772/ intechopen.87525, ISBN: 978-1-83968-307-7, Print ISBN: 978-1-83968-306-0, eBook (PDF) ISBN: 978-1-83968-308-4, pp 95-122

[38] Moulia M, Khelafib H. Performance Characteristics Of Lightweight Aggregate Concrete Containing Natural Pozzolan. Building and Environment. 2008;**43**:s. 31–36

[39] Park CK, Noh MH, Park TH. Rheological Properties Of Cementitious Materials Containing Mineral Admixtures. Cement And Concrete Research. 2005;**35**(5):s 842-s 849

[40] Piora LS, Piora IL. Production of expanded-clay aggregate for lightweight concretefromnon-self bloating clays. Cem Concr Compos. 2004;**26**:s.639-s.643

[41] Szilagyi H, Terec L, Bricks recycled aggregates for structural green lightweight concrete, 13th SGEM GeoConference on Nano, Bio And Green Technologies For A SustainableFuture, www.sgem.org, SGEM2013 Conference Proceedings,Varna. ISBN: 978-619-7105-06-3 / ISSN: 1314-2704, s. 375 - s. 380 doı:10.5593/sgem2013/bf6/s26.004

[42] Tosun YI. Chemical Activated Grinding of Cement RawMaterials. In: Proceedings of XIVth International Mineral Processing Symposium, Kuşadası, Turkey. 2014

[43] ASTM C 150. AASHTO M 85, Standard specifications for Portland Cement and testing shall comply with Specification 3101. Philedelphia: ASTM;

[44] ASTM C 989. Ground granulated blast furnace slag (GGBFS) shall comply with Specification 3102, AASHTO M 302, and ASTM C 989. Philedelphia: ASTM;

Chapter 12

Biofuel and Biorefinery Technologies

Abdulkareem Ghassan Alsultan, Nurul Asikin-Mijan, Laith Kareem Obeas, Aminul Isalam, Nasar Mansir, Maadh Fawzi Nassar, Siti Zulaika Razali, Robiah Yunus and Yun Hin Taufiq-Yap

Abstract

The global demand for energy is expected to rise up to 59% by the year 2035. This is due to the increasing technology developments and contemporary industrialization. Continues trends of these simultaneously will affects the crude fossil oil reserves progressively. Therefore, biofuels that are predominantly produced from the biomass based feedstocks such as plant, algae material and animal waste. Liquid or gaseous biofuels are the most simple to ship, deliver, and burn since they are easier to transport, deliver, and burn cleanly. The key contributor to the elevated green house gaseous concentration is carbon dioxide (CO2). Two-thirds of global anthropogenic CO2 emissions are due to fossil fuel combustion, with the remaining third attributed to land-use changes. Interestingly, recent literature has announced that the utilization of liquid biofuels capable of reducing the CO and CO2 emissions. Other positive impacts of the liquid biofuels are; (1) reduce the external energy dependence, (2) promote the regional engineering, (3) increase the Research & Development activities, (4) reduce the environmental effects of electricity generation and transformation, (5) improve the quality of services for rural residents and (6) provide job opportunities.

Keywords: catalysis, bioenergy, biofuel, hydrogen energy, green fuel

1. Introduction to biofuel

The global demand for energy is expected to rise up to 59% by the year 2035 [1–7]. This is due to the increasing technology developments and contemporary industrialization. Continues trends of these simultaneously will affects the crude fossil oil reserves progressively. Therefore, biofuels that are predominantly produced from the biomass based feedstocks such as plant, algae material and animal waste [2, 3]. Liquid or gaseous biofuels are the most simple to ship, deliver, and burn since they are easier to transport, deliver, and burn cleanly [4]. The key contributor to the elevated green house gaseous concentration is carbon dioxide (CO_2). Two-thirds of global anthropogenic CO_2 emissions are due to fossil fuel combustion, with the remaining third attributed to land-use changes. Interestingly, recent literature has announced that the utilization of liquid biofuels capable of reducing the CO and CO_2 emissions [5, 6]. Other positive impacts of the liquid biofuels are; (1) reduce the external energy dependence, (2) promote the regional engineering, (3) increase the Research & Development activities, (4) reduce the environmental effects of electricity generation and transformation, (5) improve the quality of services for rural residents and (6) provide job opportunities [7].

1.1 Type of biofuels

The oxygen content is the most important difference between biofuels and petroleum based fuels [8]. The biofuels produced from different renewable resources are typically non-toxic, accessible and abundant. Biogas, syngas, biobuthanol, bioethanol, biodiesel, bio-ether, and green fuel are various forms of biofuels. Gaseous biofuels are commonly used for heat and energy production purposes.

Biogas, is a gas fuel that burns much like fossil fuels, and for this reason, it gradually gains its position. Biogas consists mainly of methane gas, although it is produced from the degradation of anaerobic biomass. Many agricultural businesses use biogas, and the fuel is currently being packaged for domestic use in gas cylinders. The fuel is extracted from a combination of flora and fauna since each provides a specific ingredient (animals and plants). Plants have significant carbon and hydrogen in them, while they have nitrogen in them for animals. The components above are necessary and required for the production of biofuels.

Liquid biofuels, on the other hand, are widely used in the automotive industry. Biogas, obtained by anaerobic fermentation from organic materials, has a 40–70% CH_4 composition, 30–60% CO_2 and other gases such as H_2S, H_2, N_2 and CO. Biobutanol is capable of replacing both petrol and diesel. Using a bacterium to ferment biomass appears to be the most promising approach for processing biobutanol at the moment. The acetone-butanol-ethanol process is the name of the method. Acetone (propanone), butanol and ethanol are produced here. Clostridium species such as *Clostridium beijerinckii, Clostridium acetobutylicum*, *Clostridium saccharoperbutylacetonicum* and *Clostridium saccharobutylicum,* are used for the fermentation process of acetone-butanol-ethanol to produce biobased butanol [9]. Algae, sugar beet, sugar cane, maize, sorghum and cassava are the feedstocks used successfully so far. To make pure butanol, the materials are fractionally distilled. Bioethanol is a transparent liquid that is biodegradable, non-toxic, and environmentally safe. It's chemically known as ethyl alcohol, and it's made from the plant's fermentable sugars (such as glucose, sucrose, and other sugars) through microorganisms. It is possible to mix bioethanol with gasoline as well. Fermentation of bioethanol is a biological process in which microorganisms convert sugars to produce bioethanol and CO_2. Yeasts are the most widely used microorganism in the fermentation process and Saccharomyces cerevisiae is the preferred option for bioethanol fermentation among yeasts [10]. For example, fermentation bacteria used are *Clostridium acetobutylicum, Lactobacillus fermentum* etc. The basic schematic fermentation process is as follow (**Figure 1a**).

Green fuels (green-diesel and green-gasoline) are an oxygen-free hydrocarbon comprised of short chain and long chain carbon fractions within a range of C_8–C_{12} and C_{13}–C_{20}, respectively. The green fuels also free from sulfur and aromatic compounds. They contain *n*-alkanes and *n*-alkenes, which similar to those found in the petroleum-based gasoline and diesel. As the green fuel is entirely compatible as the petroleum-derived fuels, their fuel properties are vastly similar to each other.

DOI: http://dx.doi.org/10.5772/intechopen.104984

$$C_6H_{12}O_6 \dashrightarrow 2C_2H_5OH + 2CO_2 + \text{heat}$$

Glucose/fructose bio-ethanol carbon dioxide

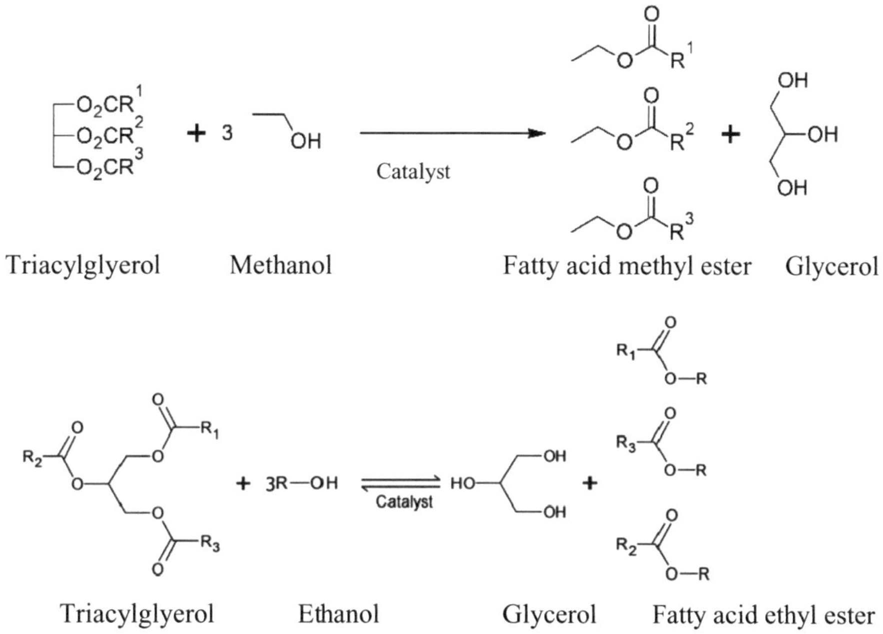

Figure 1.
(a) Fermentation process, (b) Transesterifcation reaction of triacylglycerol using methanol and (c) Transesterification of triacylglycerol using ethanol.

However, green fuel is completely different in chemical structure as compared to the well-established commercialized biodiesel [11–13]. To produce green diesel, biological oil feedstocks such as algae, vegetable or plant oils, and animal fats are thermally catalytically hydrocracked. Hydrocracking is a refinery process that uses elevated temperatures and pressure to break down these larger molecules (natural oils) into a shorter mixture of hydrocarbon chains in the presence of strong chemical heterogeneous catalysts. Green diesel also known as renewable diesel [14]. One of the most well-established liquid biofuels is biodiesel or fatty acid methyl ester (FAME), not only because of its lower environmental effect, but also because it provides the benefits of being renewable, biodegradable and non-toxic. By transesterification reaction with alcohol, biodiesel will be fabricated from either vegetable oils or animal fats using required catalyst [15]. It can be used as a diesel engine fuel in its pure form (B100) or in mixtures (B10, B15), but it is typically used as a diesel additive to reduce the levels of diesel-powered particulates, CO and hydrocarbons. The following is a diagram of the transesterification process (**Figure 1b**). Biodiesel fuel used in car and lorry engines typically comprised of higher boiling FAME fraction and its alkanes contain 14–22 carbon atoms [16]. Biodiesel also can be a mixture between their FAME and fatty acid ethyl ester (FAEE) and FAEE is prepared as following reaction (**Figure 1c**).

Table 1 depicts the most recent global developments on modern transportation fuels in 2019, published by NS Energy [17]. There are three liquid biofuels – bioethanol, biogasoline and biodiesel account for the vast majority of global biofuel production and use today. The United States is on the first world rank of biofuel

Country	Production (barrel/day)	Type of biofuel	Feedstock
US	1190.2	Bio-gasoline Bioethanol	Soybean
Brazil	693.2	Bio-gasoline Biodiesel Bioethanol	Sugarcane Soybean
Indonesia (Indexmundi2020)	>63.0	Biodiesel	Palm oil
Germany	75.8	Biodiesel	Rapeseed WCO
Argentina	70.6	Bioethanol Biodiesel	Sugarcane Corn
China	68.0	Ethanol Ethanol-gasoline blend	Import

Table 1.
Top countries for biofuel production across the globe in 2019 [17].

manufacturer followed by Brazil. In Europe, Germany is the largest producer. Argentina produced almost similar production amount with the Germany. As been expected, China is the leading country for biofuel production in Asia. The majority of biofuels were produced from soybean, sugarcane, rapeseed, corn and waste cooking oil (WCO). In the case of China, bioethanol and ethanol-blended gasoline are primary products. In the United States and Brazil, soybean oil is widely used. Many European countries, primarily Germany, use rapeseed oil for biofuel production. Note that Germany also utilized the WCO as a biofuel feedstock. It should be noted that the WCO derived biofuels initiative is also used in many nations, including Australia, China, Italy, Portugal, the United Kingdom, the United States, Austria and Spain. Brazil also produced WCO-based biodiesel, but it only accounted for 0.5% of total biodiesel production. In Brazil, only 2.5% of the WCO produced in Brazil is estimated to be reused for biofuel production, while the rest is improperly discarded [18]. Sugarcane is widely used in Brazil and Argentina. Brazil is the global leader in producing bioethanol from sugarcane.

Overall, all liquid biofuels are made mainly from agricultural commodities, such as grain and sugar (bioethanol) and vegetable oil, based on the above results (biodiesel, bio-gasoline). It can be observed that each nation concentrated on distinct feedstock. Lowest oil price, high oil content, required fatty acid composition (saturated or unsaturated acid), low cultivation maintenance and expense, controllable growth and harvesting season, consistent seed maturity rates, and potential demand for agricultural by-products are all desirable characteristics when choosing the best biofuel feedstock [19].

1.2 Biofuel uses

There are applications of biofuel other than an alternative to diesel fuel. Most claim that the material is used for transportation only. But hydrogen, cleaning oil, cooking oil, and more can be provided by biofuel. As an alternative to substitute energy needs from vehicle fuel to core home heating, biofuels can work.

Here are the top ten biofuel applications.

1.2.1 Transportation services

In the United States, about 30% of the energy consumed is used for moving cars. Transport accounts for 24% of electricity and more than 60% of the absorbed oil worldwide. This means that more than a third of the oil is used for vehicle operations.

The key issue with alternatives is that it is not feasible for transport to use solar, wind, and other renewable energies. Experts think that successful breakthroughs in developments in practical technology are still decades away.

In short, biofuel can be converted into steam of hydrogen that is intended to be used in the fuel cell adjacent to it. More important automotive brands have already invested in biodiesel vehicle stations.

1.2.2 Generating electricity

Fuel cells provide a power-generating application that is used for electricity and providing fuel for transport. In backup systems where pollutants matter the most, biofuels could be used to generate electricity. This involves facilities located in suburban areas, such as schools, hospitals, and other styles. In reality, the greatest biofuel market in the United Kingdom will turn over 350,000 homes from landfill gas into power generation.

1.2.3 Provide heat

Over the last few years, bioheat has developed. The heat coming from hydraulic fracking would contribute to natural gas development as the primary use of natural gas that comes from fossil fuels. Although there is no need for natural gas to come from fossil materials, it can also derive from newly grown materials.

There is a large amount of biofuel that is used for heating. Since wood is the most practical heating process, houses that use wood-burning stoves instead of gas or electricity are used. A biodiesel blend would reduce the production of both nitrogen and sulfur dioxide.

1.2.4 Electronics charging

According to scientists from Saint Luis University, a fuel cell was built with cooking oil and sugar to produce electricity; customers would be able to use these cells instead of generating electricity. Instead of batteries, customers will be able to use fuel cells to charge everything from laptops to mobile phones. Cells have the ability to become a ready source of power when they are still in the process of growth.

1.2.5 Spills and grease from clean oil

Biofuel is considered to be environmentally friendly and can also help clean up oil and grease spills. For areas where crude oil polluted the waters, it was checked to act as a possible cleaning agent.

It has also been found that the results improve the areas of recovery and allow it to be extracted from the water. Biofuel can also be used for metal cleaning as an industrial solvent, which is also useful because of its lack of harmful effects.

1.2.6 Cooking

Although the most common ingredient to be used for stoves and non-wick lanterns is kerosene, biodiesel works equally well.

1.2.7 As a lubricant

In order to decrease the Sulphur concentration, diesel fuel is needed as Sulphur offers the most fuel lubricity. When it comes to maintaining the engine running correctly and preventing infection's premature failure, this is critical.

1.2.8 Remove paint and adhesive

Biofuels can replace toxic materials in order to eliminate paint and adhesives. The best approach for eliminating non-critical applications is often known to be biofuel.

1.2.9 Create energy when fossil fuel runs out

As the supply of oil is beginning to run out. This has led us to ask how, without damaging the ecosystem, fuel can be extracted. Biofuel would assist the government in forming a sustainable, cost-effective method of generating energy.

1.2.10 Reduce cost and need for imported oil

In the United States, over 84% of the world's petroleum is used. The U.S. has recently begun to reduce the need since 2006, despite the rise in fuel requirements. This makes it possible for biofuels to become the strongest emission reduction factor.

Analysts claim that when oil is disrupted, substituting imported oil with biofuel would help to balance the Economy. It does not matter how much the Americans Spend on oil imports, but how to balance the overall Economy (Top 10 Uses for Biofuel, 2016).

1.3 Biofuel feedstock

Biomass feedstocks for energy production can be produced from plants directly grown for energy use or parts of plants, waste, residues, and materials extracted from humans and animals' activities. In 2005, the U.S. Department of Energy evaluated these feedstocks' results and found that it was possible to sustainably harvest and deliver more than 1 billion tons of agricultural and forestry-related biofuels to bio-refineries. Feedstocks may be defined by plant or residue types, the energy products they make, or any other way. The following categories of feedstocks will be used for discussion purposes.

1.3.1 Sugar and starch crops

Many of the sugar and starch crops that are contenders for biofuel production are already being used for agricultural and food grains or sweetener sugars. Root and tuber starches are usually used across the globe as food staples. Via conventional fermentation methods, these crops and their particular products can easily be transferred to ethanol and related alcohols for transport and other uses.

DOI: http://dx.doi.org/10.5772/intechopen.104984

Competition for resources and the need for genetic, development, and manufacturing modifications to increase energy production sustainably would be the unique challenges facing most of these crops. Some examples are:

- Corn Grain
- Sweet Sorghum

1.3.2 Fibre and grass cellulosic crops

Many of the grass and related crops that have been cultivated for decades as pasture and grazing for feeding livestock or for soil conservation can be used as an energy resource. In general, these crops are higher in fiber (cellulose, hemicellulose, lignin) and poorer in carbohydrates, proteins, and oils. A variety of methods may turn these crops into energy, including direct heat and/or power combustion, cellulosic conversion into ethanol, thermochemical processing for fuel supplements, or anaerobic methane digestion. Some examples are:

- Miscanthus
- Energy cane

1.3.3 Oil crops

While several crops generate at least a small amount of vegetable oil, 15–50% of oil is provided by various crops. By grinding the seed and squeezing the oil out, oil can be extracted. To produce biodiesel, the oil is transesterified. Oil crops may also be transformed as alternatives to fossil fuel materials into high-value biochemicals and biomaterials, thus reducing the use of fossil fuels in turn. Some examples are:

- Soybean
- Canola/Rapeseed
- Mustard
- Camelina
- Warm Climate Feedstocks for Biodiesel
- Hazelnut/Filbert

1.3.4 Crop residues, manures, and organic wastes

Critical biomass residues remain after corn, sugar, starch, or oil plants are harvested for feed and food components. Abundant crop residues that can be transformed into renewable fuels are corn stover, corn cobs, wheat, and small grain straw. Sustainable maintenance of the agricultural production system is a crucial obstacle when removing crop residues. To increase soil organic matter quality as well as soil and water conservation objectives, crop residues are usually incorporated into the soil. In order to evaluate the influence of stover removal on the sustainability of crop production, a great deal of research is performed to study the effects on ecosystem services and the diversity of insects, vertebrates, and microbes. Some examples are:

- Corn Stover
- Corn Cobs

Manures are a result of livestock's digestion of plants. Anaerobic digestion techniques have been used for years to transform these and other organic waste to methane and related gases in addition to their usual land application for nutrient content. In exchange, methane can be used explicitly for combustion heat, fueling diesel generators, or supplementing natural gas with further processing and cleaning.

In metropolitan areas, food production and industrial waste, including restaurant grease, leaves, grass cuttings, and other garden waste, are contained in large quantities and can be processed and converted to electricity through a number of methods.

1.3.5 Wood products

Trees and their associated products were used as a direct source of ignition and combustion for heating and cooking for decades. The thermal efficiency of these wood products, when dry, is about two-thirds that of coal and about 10% greater than that of deciduous plant biomass. As the fuel source for gasification and cellulosic conversion to ethanol, wood and its derivatives have also been used. Although it is generally possible to use any wood supply, different research projects have been ongoing to create so-called "energy forests" or "wood energy farms." Woody Crops [20]. Some examples are:

- Hybrid Poplar
- Willow Shrub

1.4 Advantages of biofuels

Biofuels offer a wide range of benefits.

1.4.1 Renewable energy sources

Globally, there is strong energy demand. Nonetheless, most power sources are non-renewable, lead to the greenhouse effect, or, as is the case with nuclear energy, may lead to major ecological problems. Biofuels, which are renewable fuel sources and environmentally friendly, are derived from plant and animal manure.

The majority of fossil fuels will expire and one day wind up in flames. Since most sources, such as manure, maize, switchgrass, soybeans, crop, and plant waste, are renewable and are not likely to be running out any time soon, the use of biofuels in nature is effective. These crops can also be replanted again and again, as well.

1.4.2 Sovereignty

Unlike fossil fuels, whose deposits aren't really found in all countries, any country can undertake biofuels' development without interference with all other countries' energy sources. By impacting or determining the world's fuel prices and petroleum-based goods, countries with fossil fuel reserves have always taken full advantage of their economic resources. If a nation can manufacture its own biofuel, it can easily set its own prices for goods without many regional and global constraints.

Although local crops have decreased the nation's reliance on fossil fuels, many experts agree that addressing our energy needs will take a very long time. We need more renewable energy options to reduce our dependence on fossil fuels as crude oil prices are hitting sky high.

1.4.3 Ensure economy's sustainability

The sustainable quality of biofuels has contributed to states worldwide adopting them and supporting a decrease in fossil fuel use. Instead of high-cost imports of fossil fuels from Middle Eastern countries, policymakers should reduce this reliance and instead fund biofuel plants that are cheaper in the long term.

Locally generated biofuels can minimize reliance on other fuels and thus increase the security of energy and economic prosperity. Fewer imports imply more exports and, therefore, greater self-dependence.

1.4.4 Low expenses

The majority of biofuels are easy to manufacture and cheaper than fossil fuels. Therefore, their use will make life easier for ordinary citizens and help boost people's living standards by reducing the increasing cost of living globally due to reliance on fossil fuels. As of now, as gasoline does, biofuels cost the same as the market. However, the net cost-benefit of using them is much more significant. They are safer fuels, which implies that they generate lower burning pollutants. They also have the ability to become cheaper in the future with the growing demands for biofuels.

1.4.5 Clean fuel

A lot of carbon is emitted by fossil fuels, which results in large levels of air pollution. This carbon also mixes with other greenhouse gases, such as methane, which contributes to unfavorable weather conditions. On the other hand, since they are clean fuels, biofuels do not release this amount of carbon into the environment.

1.4.6 Efficient fuel

Biofuel is made from renewable resources and, compared to fossil diesel, is relatively less combustible. It has considerably stronger hydrating characteristics. Compared to standard diesel, this produces less toxic carbon emissions. It is possible to manufacture biofuels from an extensive range of materials. The net cost-benefit of someone using them is considerably greater.

1.4.7 Extensive durability of vehicles' engine

In most conditions, biofuels are able to adapt to existing engine designs and perform very well. It has higher levels of cetane and more robust lubricating properties. The longevity of the engine improves when biodiesel is being used as a flammable fuel.

Engine conversion is not required. This allows the engine to run for longer, needs less maintenance, and reduces the cost of pollution control overall. Engines intended to run on biofuels generate fewer emissions than other diesel engines.

1.4.8 Less smoke generation

Automobiles and factories using fossil fuels such as petroleum and diesel commonly create a lot of atmospheric smoke. As biodiesels have oxygen atoms in their chemical structure, they burn better and contain less carbon deposits. Biodiesels emit less smoke as a byproduct and are more environmentally friendly.

1.4.9 Minimize monopoly

Fossil fuels are more likely to be favored by biofuels due to their widespread use. This has created a monopoly over the years, contributing to price increases and the ever-increasing standard of living. Since biofuels are equivalent replacements for fossil fuels, they can be used to help minimize the fossil-fuel monopoly.

Biogas could be used in the same way as fossil fuels, for instance. Consequently, people have the option of converting to Biogas when natural gas prices go up. And vehicles can opt for ethanol or butanol when fossil diesel rates increase.

1.4.10 Less toxic

As a result of combustion, all forms of fuels, fossil fuels, and biofuels form carbon compounds. In the atmosphere, fossil fuels emit toxic carbon dioxide, especially in the presence of water vapor and methane gas. On the other side, biofuels' carbon occurs in nature and is used for photosynthesis by plants, serving as an energy source for plants.

1.4.11 Employment source for locals

Most of the bio plants are geographically set up, and human capital is required in the process, such as construction engineers, farmers, project managers, fuel distributors, and logisticians. This helps to generate new work opportunities for locals.

1.4.12 Lower levels of pollution

Using fossil fuels such as coal, sulfur, and lead can be produced along with acid rain. Unlike biofuels, sulfur is not contained in biofuels. Biofuels are renewable resources that emit fewer emissions into the environment. Nevertheless, this is only one reason why biofuels are promoted.

They emit lower levels of various contaminants, such as carbon dioxide than conventional diesel. It plays a part in lowering air pollution. Furthermore, biofuels are biodegradable, reducing the risk of soil degradation during transport, storage, or use.

Social and environmental studies reveal that biofuels minimize greenhouse gas emissions by up to 65%. When fossil fuels are burnt, they release huge amounts of greenhouse gases into the atmosphere, affecting the environment. The greenhouse gases absorb the sunlight, which causes the earth to be hot. Besides, burning coal and oil is a source of climate change. Various countries are opting to use biofuels as a way to reduce greenhouse gases.

1.4.13 Agricultural promotion

Increased demand for the production of biofuels will lead to further farming of the appropriate crops. Crops with high carbon and cellulose composition can be planted on a massive scale, and after harvesting the edibles, the rest of the plant components can be used for the production of biodiesel.

1.5 Disadvantages of biofuels

1.5.1 High production cost

Biofuels are very costly to manufacture in the current market, even with all the advantages associated with biofuels. The interest and capital investment put into the biofuels production are relatively low as of now, but it can balance demand.

If demand rises, then it will be a long-term process to raise the supply, which will be very costly. Such a downside also prevents the use of biofuels from becoming popular globally.

1.5.2 Monoculture

Monoculture is the method of growing the same crops year after year in a single field, rather than generating different crops in multiple fields. Although this may be lucrative for farmers, growing the same crop every year would deprive the soil of nutrients returned by cover crops and farming overused areas. The reasons for planting a single crop over large tracts of land are discussed. First of all, the environment changes when only one crop is grown, and pests can ruin the entire crop.

Besides, complete pest control can be accomplished with pesticides. Even certain pest insects would inevitably develop resistance to the chemicals we use to fight them, and they would be able to live in a single crop area.

As we intend to encourage insect resistance to our pest, the next obstacle comes with genetically modified species. The change is not likely to impact any species, and the related problem remains.

Biodiversity, which requires various varieties of plants and animals, is thus the key to healthy agricultural fields.

1.5.3 Application of fertilizers

Biofuels are derived from crops, and to grow better, these crops need fertilizers. The drawback to using fertilizers is that they can cause water contamination and have adverse effects on the surrounding environment. Nitrogen and phosphorus are found in fertilizers. It is possible to wash them away from the soil into surrounding lakes, rivers, or ponds.

1.5.4 Food scarcity

Biofuels are obtained from plants and crops which have high sugar levels in them. Many of these crops are also used as food crops. Even though plants' waste material may be used as raw resources, there will still be a need for such food crops. Other crops can take up farm space, which can cause several problems.

The use of existing biofuel land may not lead to acute food shortages, but it will undoubtedly pressure current plant growth. One big problem that people face is that the rising use of biofuels could also increase food prices.

Algae, which grows in rather inhospitable regions and has a small impact on land use, is favored by some people. The issue with algae, however, is water use.

1.5.5 Pollution in the industry

When burned, the carbon footprint of biofuels is smaller than the conventional sources of fuel. The method in which they are made makes up for that. Production depends to a large degree on lots of water and oil.

It is understood that large-scale industries intended for biofuel production produce large quantities of emissions and also cause small-scale water pollution. The total carbon pollution would not have a very significant dent in it unless more effective production means are placed.

1.5.6 Extensive use of water

In order to irrigate biofuel crops, large quantities of water are needed and can, if not handled wisely, place a strain on local and regional water supplies. Vast amounts of water that could place unnecessary pressure on local water supplies have been used in order to manufacture maize-based ethanol to satisfy consumer demands for biofuels.

1.5.7 Future price hike

The existing technology used for biofuel production is not as effective as it should be. Scientists are interested in the creation of better measurements that enable us to extract this fuel. However, testing and potential installation expense means that a significant increase will be seen in biofuels' price.

As of now, gasoline prices are equivalent and are still practicable. The use of biofuels can be as tough on the economy as the rising gas prices are doing right now.

1.5.8 Land use changes

Land must be cleared of natural vegetation if it is used to produce a biofuel feedstock, contributing to ecological harm done in three ways.

First, the harm is caused by community habitat loss, animal dwellings, microecosystems, and the general wellbeing of the region's resources will be diminished.

In extracting CO_2 from the atmosphere, the native forest is almost always better than a biofuel feedstock, partially because the CO_2 stays trapped and is never extracted by burning as with the fuel stock.

Secondly, the damage of the generated carbon debt is significant. This contributes to the production of greenhouse gases as it is necessary to deforest an area and prepare it for agriculture as well as to grow a crop, and puts the region at a net positive

DOI: http://dx.doi.org/10.5772/intechopen.104984

development of GHG even before the production of a specific biofuel. Estimates have shown that a carbon debt that can take up to 500 years to repay can actually be created by deforestation of native land.

Finally, almost always converting land to an agricultural status means that fertilizers can be used to get the most yields per area. Runoff and other agricultural emissions are a problem.

1.5.9 Global warming

The biofuels, which mainly burn hydrogen and carbon, create carbon dioxide that causes global warming. Biofuels generate less GHG emissions than fossil fuels, but that can only help slow down global warming and not avoid or reverse it.

Biofuels could therefore be able to help alleviate our energy requirements, but they will not solve all of our issues. In the short term, it can only act as a replacement as we invest in other technologies.

1.5.10 Weather issues

For use at low temperatures, biofuel is less satisfactory. It is more likely than fossil diesel to draw moisture, which in winter conditions causes problems. The engine that coats the engine filters also enhances microbial growth (Various Advantages and Disadvantages of Biofuels, 2020

2. Biorefinery technologies

2.1 What is biorefinery?

A biorefinery is a specialized facility that uses tools and materials for processing biomass into fuels, electricity, and useful chemicals. The biorefinery is situated next to the petrochemical industry, making various petroleum products, including gasoline and plastic. A biorefinery benefits from using multiple biomasses and intermediate items, thereby maximizing the value of biomass feedstocks. One potential example for biorefinery product is low-volume yet high-value chemical products and high volume but low-value fluid or liquid transportation fuel such as biodiesel or bioenergy. Because of its high efficiencies, energy efficient technology that generates electricity and captures the heat (CHP) technology can generate electricity for its own use and sell surplus electricity to the community. High-value products boost profitability, high-volume fuels help meet energy needs, and power generation helps lower energy costs and mitigate greenhouse gas emissions from traditional power plant installations.

Nevertheless, the production of fuel and chemicals in biorefinery is limited, which may become harder to do. Interdependent societies and trading firms demand vast quantities of fossil fuels to supply the bulk of their energy and chemical supplies. The manufacture and use of fossil fuels cause environmental degradation, resulting in toxins, greenhouse gases, and dangerous materials. The increasing demand for energy and chemicals makes the environment more dependent. The amount of waste produced is also continuously rising alongside the growth in our global population. Our country generates roughly 250 million tons of municipal solid waste per year (MSW). Overall, 35% of MSW is recycled and composted, 13% of this waste is used to generate power, and 53% of MSW is buried in landfills.

It is important to find alternative resources to make energy and chemicals due to limited fossil fuel resources and the increasing demand for energy and chemicals. In this process, biomass has been recognized as a possible future source of chemicals and energy and addresses the environmental risks of burning fossil fuels. Various biomass sources are available but can also be collected from various waste sources such as agricultural waste, municipal solid waste, and industrial contaminants such as paper factories and pulp factories. In a holistic waste management plan, biomass waste valorization also plays a prominent role in waste recycling. A biorefinery that uses renewable biomass as a feedstock for the production of chemicals is more sustainable as opposed to using fossil fuels. A biorefinery may contribute to economic expansion while also reducing air pollution in the environment [21].

2.2 Biorefinery technology, product, and application

To create an integrated biorefinery that aims to build on the biomass conversion process in such a way that the maximum added value can be derived from the sustainable biomass feedstock, it integrates a range of different technologies. New methods of bio-refining are being explored in hopes of saving the world. Biorefineries combine/integrate different technologies for the conversion of biomass into a variety of products (i.e., food, feed, chemicals, materials, petroleum, coal, heat, and/or electricity) and are defined as 'sustainable processing of biomass into a marketable commodity and energy spectrum' by IEA Bioenergy Task 42. The concept focuses on the method of how various petroleum products are refined and made into usable fuels.

As it is subject to unpredictable circumstances, such as when farmers use various farming methods, and climate changes, a biorefinery's precise technical specification can vary from case to case. Environmental and social factors decide which feedstock is available for processing. Numerous varieties of switchgrass, sugarcane, wheat, corn, wood, crop waste, sugar cane, surplus food, straw, freshwater biomass, and the biomass component of municipal and other sources of waste (MSW) can be used in a bio-refinery. Chemicals, biofuels, energy and heat, materials, food and feed, minerals and CO_2, are the main product groups in a biorefinery (Biorefinery, ctc).

Biorefineries can be classified based on the number of main characteristics they have. Various feedstocks for biofuel production include perennial grasses, starch crops (e.g., wheat and maize), sugar crops (e.g., beet and cane), lignocellulose crops (e.g., controlled forest, short growing coppice, switchgrass), lignocellulose residues (e.g., stover and straw), oil crops (e.g., palm and oilseed rape), and inorganic matter (e.g., stover and straw) (e.g., industrial, commercial and post-consumer waste).

Feedstocks can be treated on a number of platforms used by biorefineries. These platforms include biogas, consisting of single carbon molecules such as methane and carbon dioxide, starch, sucrose, or cellulose carbon carbohydrates; a mixed stream of 5 and 6 hemicellulose-derived carbon carbohydrates, lignin, oils (plant-based or algal), organic grass solutions, pyrolytic liquids. By integrating biological, thermal, and chemical processes, these main platforms can be changed to produce different products. Awareness of the feedstock, platform, and product that a biorefinery uses allows the business to be represented critically. Biorefinery practice creation helps compare biorefinery systems, improves understanding of global biorefinery growth, and allows technology differences to be developed.

The biorefinery classification examples include:

- C6 biorefinery sugar generating ethanol and animal feed from staple crops
- Syngas biorefinery from cellulosic biomass yielding FT-diesel and naptha
- Biorefinery of C6 and C6/C5 sugar and syngas containing ethanol, FT-diesel, and furfural from cellulose and hemicellulose crops

2.3 Biorefinery pathways

Biomass can be used as food, heating fuel, or converted into a liquid or gaseous form that can then be used for energy resources. There are different methods to transform biomass into biofuels. A distinction is made between biochemical conversion and thermochemical conversion. Anaerobic digestion, saccharification, and hydrolysis are common conversion technologies used throughout industries. 5 distinct sub-categories of thermochemical conversion include gasification, pyrolysis, liquefaction, gasification, and combustion.

As oxygen is completely removed, the aerobic decomposition of organic carbon diminishes organic nonwoody content. It sells easier, less volatile chemicals, including methane and carbon dioxide. However, the biochemical conversion process takes a lot of time and uses just a biomass portion. Thermochemical conversion technologies are more generally regarded as being superior for their flexibility and efficacy.

Most second-generation biofuels are generated by concentrating lignocellulosic biomass into different products. Containing three main constituents: cellulose, hemicellulose, and lignin. Under conditions between 200 and 380°C, hemicellulose is the easiest to break down, and cellulose decomposes between 320 and 400°C. The most stable substituent that breaks down when heated to 400°C is cellulose.

Temperature, heating rate, and residence time are three important moving parts in chemical-reaction thermochemistry. Combustion is currently the leading source of energy (approximately 80%) in the worldwide supply. Many alternative methods for pollution control, such as gasification and pyrolysis, are still in the research and development stage due to their high cost and low performance [22].

Biomass can be converted into an extended range of chemicals in two main ways:

a. Thermochemical pathway

b. Biochemical pathway

2.3.1 Thermochemical pathway

Thermochemical processing is used for two major routes:

One method that produces biomass is heating biomass with regulated oxygen quantities at high temperatures and pressure (a mixture of carbon monoxide and hydrogen). The process is called gasification. The gasification of solid biomass produces many industrial compounds. The effect of gasification on various liquid fuels will be addressed in the biofuels unit.

The second technical approach involves high-temperature heating of the biomass, but it works without relying on the atmosphere. As a method, pyrolysis is well-known. Glue must be used easily, so the reaction time must be short. If not, the top sector will be the carbon industry (char). This process is known as rapid pyrolysis, and the main product produced is organic oil [23].

The thermochemical conversion aims to minimize the entire biomass used in the chemical-making phase to steam. The Fischer–Tropsch process is a thermochemical conversion, which is why it is an example. Thermochemical biomass conversion is not a key application of chemical transformation. The main driver of this five-way conversion route is the output of thermal energy:

1. The Combustion

2. Torrefaction/Carbonization

3. Pyrolysis

4. Gasification

5. Liquefaction

The biomass is first transformed into syngas in the thermochemical pathway, converted by synthesis or some other method into ethanol.

2.3.1.1 Combustion

Considering that humans started with fire discovery, combustion was the first method of using living materials for energy production. Wood-burning forestry has taught people how to survive, heat and cook. Chemically, biomass combustion is an oxygen-based exothermic reaction. Here the biomass is oxidized by two major stable compounds, hydrogen and carbon dioxide. The heat generated by the reaction currently accounts for over 90% of energy consumption.

Bio-mass derived energy is mainly generated from heat and electricity. Biomass also provides renewable cooking fuel and heat in rural communities. Combustion of biomass also includes industrial heating and regional heating. Pellet stoves and wood-burning fireplaces are widely used in areas with cold climates. The use of biomass for electricity is important for modern-day environmental practices. Combustion of biomass in boilers and the power-producing steam turbine are the most common activities. Biomass is used as an alternative to fossil fuel in a boiler, typically for heating. The latter approach is more effective in lowering carbon dioxide emissions from a high-emission fossil-fuel plant than the prior solution.

2.3.1.2 Carbonization

Like torrefaction, carbonization is recommended for the productivity of biomass as a safe and efficient solid fuel. The biomass is continuously heated up to 200–300°C with little or no oxygen contact in torrefaction. This process changes the biomass hydrocarbon's chemical composition to increase its carbon content while reducing its oxygen content. Torrefaction also raises the density of biomass and increases hygroscopic biomass. These qualities thus increase the commercial value of the timber for electricity manufacturing and transport. The general goal for other carbonization processes is to form carbon-rich reliable products under different conditions.

2.3.1.3 Pyrolysis

Pyrolysis occurs in an oxygen-free environment, unlike combustion, even when using partial combustion to heat the reaction. It can be used to quickly and effectively transform biomass into gases, liquids, and solids.

Parts of biomass are broken down in the pyrolysis process. Slow pyrolysis end products include reliable charcoal and gas, while rapid pyrolysis produces only bio-oil. In order to turn biomass into liquid fuels, pyrolysis is suitable. It is not an endothermic reaction, while combustion releases heat.

2.3.1.4 Gasification

Solid, liquid, and gaseous fossil fuels are converted into usable gases. One needs a medium for gasification reactions, water, or steam. An air, oxygen, or both make up the gaseous environment.

Natural gas production through fossil fuel emissions is more common than bio-gasification to produce biogasoline. Gasification moves the fuel from one type of fuel to another. There are numerous reasons for the change from one form of language to another.

- The fuel's heating value is enhanced by eliminating non-combustible organic content such as nitrogen and water.
- To extract sulfur in the fuel gas so that it does not contribute to global warming.
- To lower the overall carbon to hydrogen (C/H) ratio in the gasoline.
- To optimize the fuel's hydrofluorocarbon (HFC) and perfluorocarbon (PFC) emissions

The higher the hydrogen content in fuel, the more likely the fuel will be in its gaseous state. The relative hydrogen content in the substance is produced by adding air to the material by using gasification or pyrolysis.

1. At high temperatures, exposure to hydrogen.
2. Indirect exposure to high temperature and pressure resulted in a hydrogen-rich product. The method also includes steam reforming.

Compared to the amount of oxygen in cellulosic biomass, the oxygen concentration in natural gas is decreased. Gasification decreases the overall carbon footprint and creates a more usable commodity.

Some natural gas is gasified in order to use as a source of energy and as a means of ammonia production. Nature gas reforming helps in steam output (a mixture of H_2 and CO).CO, which is present in the biogas, is indirectly hydrogenated through the smog to produce methanol. However, these systems use natural gas, causing more carbon dioxide to be produced than other systems. Biomass can be used as a replacement for fossil hydrocarbons in various manufacturing processes.

Changing liquid transportation fuels generates biomass gasification, providing a strong basis for carbon dioxide and hydrogen. The process may also generate methane, useful as a source of energy.

2.3.1.5 Liquefaction

There are several methods of bringing solid biomass into liquid fuels, including pyrolysis, gasification, and hydrothermal processes. The conversion of biomass into an oily liquid can be achieved by heating the biomass at higher temperatures (300–350°C) and under high pressure [24].

2.3.2 Biochemical pathway

Biochemical conversion requires the breakdown of biomass to make the carbohydrates usable for refining into sugars, which can then be transformed using microbes and catalysts into biofuels and bioproducts. The following are possible stocks of fuel mixtures and other bioproducts:

- Renewable gasoline
- Ethanol and other alcohols
- Renewable chemical products
- Renewable diesel

The significant challenges of breaking down the complex structures of cellulosic biomass include key challenges for biomass's biochemical conversion. To have access to these beneficial sugars, the Bioenergy Technologies Office explores more effective and affordable means of processing the sugars.

The critical challenge is to turn sugars into biofuels more efficiently and effectively. To achieve our target, the Bureau has developed new directions and technologies.

2.3.2.1 Step by step chemical conversion

In addition to heat and other chemicals, the biochemical conversion uses biocatalysts to convert the hemicellulose and cellulose into an intermediate stream of sugar. Such sugars are an intermediate stage in the manufacture of advanced biofuels and other biochemicals or are catalyzed chemically to generate useful substances. The whole method is made up of the following necessary steps.

A. Feedstocks are chosen because they contain the properties required for biochemical reactions. For fast, efficient plant operations, efficient feedstock handling systems are required.

B. Pretreatment: The biomass is heated to break down the fibrous cell walls and make it easier to hydrolyze cellulose and hemicellulose (see next step).

C. After hydrolysis, the sugars are separated from cellulose and hemicellulose in the pretreated content over the course of days.

D1. Bacteria are added to produce new chemicals, including fuels or building blocks for other chemicals, from sugars.

D2. Rather than chemical conversion, sugars can be converted using chemical catalysts to produce fuels and other useful items.

E. Recycling: When oils, solids, and residual impurities are separated from products.

F. Distribution: Fuels and other goods are shipped to refineries. Other products and intermediary products may be delivered to manufacturing plants for use in a wide range of consumer products.

G. The organic matter left is mostly comprised of lignin that can be combusted as natural gas and fuel [25].

2.4 Integration of biorefining in the processing industry

Due to the wide variety of biorefinery systems and their component selection, there are major energy properties variations. The process will be affected by the form of feedstock, crude oil metabolism, and end product. The freedom to select and change these parameters is a complex task that takes a great deal of thinking. The biorefinery portion should be considered to optimize the biorefinery's energy characteristics to better conform the overall process integration.

The model is dynamic and implemented several degrees of freedom. The optimization is difficult to do due to the substantial uncertainty in potential energy markets, investment costs, and, especially, the cost of CO_2. Optimization research on the different levels of these key parameters should also be carried out in order to find "robust" solutions, i.e., perfect ones for technical, environmental, and economic performance at multiple levels of these variables. The incorporation of a biorefinery model into a process industry is very comparable, in theory, to the foreground/background approach or, often, the Complete Site o ne for a given biorefinery design.

- It will combine two broad process components. It is possible to do this on many levels:
- Independent integration of processes within a biorefinery
- Independent process incorporation of the host process
- The complete study of the site or foreground/background of the two components
- The study of process integration considers all streams as components of one large process.

One experience from integrated biorefinery process integration studies is that a very non-integrated host process may be more appropriate for integration with the biorefinery than an integrated biorefinery process. Therefore, if a biorefinery in an organization is considered shortly, any planned energy-saving measures should be postponed or carefully evaluated. This will jeopardize the possibility of effective overall integration. It also proves that the above fourth alternative that combines all the streams must always be done as a first step.

The standard theoretical integration technique is used in the article. The third alternative will then be checked by deciding whether the flows could be integrated into structures of various kinds. The distance between the gas measurements indicates the ability to save electricity. It is the complicated approach that suits the problem and the straightforward solution that is easier to manage. Functional limitations can often make the most powerful solutions impractical, but an efficient targeting method is essential.

In certain cases, the integration possibilities often depend on the availability of energy or excess heat in the refinery and part of the biorefinery. Since heat is not extracted by convection in the original form, it is only cooled by the cheapest possible means of a cooling system. Therefore, it is important that the possible amounts of excess heat from the process can be analyzed when reasonable heat exchange is applied to increase these temperatures and provide a targeted protocol.

2.4.1 Biorefinery concepts in different types of process industry

In the process industry, there is a vast variety of proposed and researched ideas for biorefineries. Only some important examples of concepts are presented below. While the integration of processes would support all forms of biorefineries, refineries processing bulk goods will be of greater importance than chemical products. The examples are, therefore, all bulk goods.

Examples of Process Integration Results Studies

2.4.2 The industry for pulp and paper

- Biomass gasification and electricity processing, methanol or diesel Fischer-Tropsch
- Black liquor gasification with DME generation or green electricity generation
- Processing of ethanol using partially existing pulping machinery or docked to a pulp mill
- Precipitation and upgrading of lignin
- Hemicellulose precipitation and upgrading with water extraction Refineries for Oil
- Biomass gasification and hydrogen production, or diesel Fischer-Tropsch Industry of Petroleum
- Biomass gasification and methanol or SNG processing

The overall outcomes of these studies are:

- In almost all situations, energy-saving steps can be found by process integration between the biorefinery definition and the process industry.

- Energy-saving opportunities are usually strong for the overall system, up to 25%.

- The reduction of CO_2 emissions by integrating biorefineries with process industries depends on society's marginal energy production technology, particularly electricity. As long as coal convection plants are the marginal power generation technology, the global reduction in CO_2 emissions largely depends on the possibilities for biomass-based power generation as a by-product in integrated and non-integrated biorefinery networks.

- In almost all of the cases examined, the economic benefits of integrating biorefineries are fair or strong but depend entirely on future conditions about policy instruments, i.e., future levels of CO_2 charges (Berntsson).

3. Production of solid fuel biochar from waste biomass

Due to the possibility of exhausting fossil energy and the increase in climate change resulting from the excessive use of fossil fuels, there is a growing need to use renewable energy sources in future in place of fossil energy. Biomass energy is becoming an increasingly popular type of renewable energy, primarily due to its worldwide availability [1]. At present, biomass combustion alone (or co-combustion with coal) to produce heat and power in current coal-fired systems is widely accepted as a low-risk process and one of the least expensive methods of reducing CO_2 levels in the atmosphere [2]. However, raw biomass is not an effective energy carrier and there are significant barriers hindering the direct use of biomass due to its innate properties, including high moisture content, poor grindability and low energy density [3]. For instance, biomass has a fibrous structure that generates poor grindability, and this causes substantial increases in energy consumption and impairs the processes of fuel preparation and feeding. Moreover, in terms of high moisture content, it can reduce the maximum combustion temperature, which in turn reduces thermal efficiency and increases toxic emissions [3]. To address these issues, a pre-treatment process can be carried out to enhance the fuel quality of raw biomass before combustion.

To convert biomass feedstock into biofuels with a high-energy density, pyrolysis has been carried out. This process generates three primary product streams, namely liquid (bio-oil), solid (biochar) and gas products. The pyrolysis conditions largely determine the distributions and properties of these product streams. At present, a majority of research attention is focused on the liquid and gaseous products, and a number of different processes have been developed to generate increased yields and enhance the quality of the two target products [4–7]. Nonetheless, given the high instability and complexity of its composition, it is not possible to use bio-oil. Rather, more upgrading is necessary [8]. With regard to gaseous products, there is a low yield, whilst the separation and purification processes are largely complex. This ultimately limits the large-scale application of such products in practice [7]. In such cases, it can be beneficial to optimise the use of biochar, and this can ultimately enhance biomass utilization efficiency.

In comparison to liquid and gaseous products, very few studies have investigated the production of solid fuel biochars from waste biomass. In fact, a majority of these studies have focused on examining the improved physicochemical properties of woody biomass [9–16]. On the other hand, a few studies have examined the biochars produced from abundant agricultural wastes, although studies comparing the fuel quality of biochars obtained from woody biomass and agricultural residue are lacking. Moreover, biomass combustion has low thermal efficiency and produces extremely pollutant emissions. It also generates serious ash-related problems (i.e., fouling and slagging). Nonetheless, even though they are key issues when applying biomass as a solid fuel, very little research has examined the combustion qualities and ash issues relating to pyrolytic biochars in relevant literature [17, 18].

3.1 Biochar production

Several different biomass materials can be used to make biochar, including agriculture waste, animal waste, sewage sludge waste and algal waste. To produce biochar from biomass materials, a number of methods have been developed, such as pyrolysis, hydrothermal carbonisation and gasification.

3.1.1 Pyrolysis

Pyrolysis is a process carried out to thermally decompose biomass without oxygen. There are two stages involved in this decomposition process, namely the primary and secondary stages. Dehydration, dehydrogenation, and decarboxylation take place during the primary stage [19], after which a secondary reaction starts to take place, in which larger molecules are cracked and the solids are converted into gases and biochar. Moreover, there are two key types of pyrolysis, namely slow and fast pyrolysis and this is determined by the operating conditions. During slow pyrolysis, heat is set at a lower rate (0.1–1°C s^{-1}) and the reaction takes place over a long period of time (hours to days) at a temperature (300–900°C). This provides a favourable environment that facilitates the secondary processes and increases the production of biochar. By contrast, the biomass in fast pyrolysis is heated at a higher temperature (300–1000°C) and with a high heating rate (10–1000°C s^{-1}) for a short period of (0.5–2 s) [22], and this process results in the production of a solid (biochar), liquid (bio-oil) and gas (syngas) [24]. The solid carbonaceous substance biochar can be employed as either a catalyst, adsorbent or fuel. On the other hand, syngas (which is made up of CH_4, CO_2, H_2, CO, and other low molecular gases) is typically used in gas engines. Bio-oil is made up of water, phenolic compounds, alcohol, nitrogenous compounds (pyrazine, pyridine, and amines) and aliphatic and aromatic hydrocarbons. Thus, bio-oils are often used in boilers to produce heat [19]. The key properties of biochar (i.e., porosity, surface area and functional groups) are determined by the temperatures used during the pyrolysis process. At higher temperatures, the biochar's surface area and porosity increase. This is because the aliphatic alkyl and esters groups in the organic compounds break and this facilitates the removal of pore-blocking substances [25]. When lower temperatures are used during pyrolysis, the resultant biochar is hydrophilic and has a graphene structure with fewer functional groups on the surface. On the other hand, biochar created at higher temperatures are hydrophobic and with functional groups being reshuffled and new groups (carboxyl, lactone, phenol, pyridine) being introduced that serve as electron donors and acceptors [26]. In an experiment, Akinfalabi et al. [27] carried out pyrolysis at 400 °C for two hours to produce biochar from sugarcane bagasse biomass. During the process, the latter was sulphonated with $ClSO_3H$, which increased the surface area from 98 to 298 m^2 g^{-1}. When used as a catalyst to produce biodiesel production, a yield of 98.6% fatty acids methyl esters (FAMEs) was generated [28]. It is important to note that the yield and quality of the resultant biochar are determined by the types of biomass feedstock used. Conducting pyrolysis sing forestry plants generates a 30% biochar yield, whilst using lignin produces a slightly higher yield of 45.69%. Thus, this suggests that the biochar yield is largely determined by the lignin content [29]. Moreover, Zhang et al. [30] pyrolyzed a lotus stem at 800°C and found that it produced biochar with 55% greater surface area (1610 $m^2 g^{-1}$) than porous carbon made from leaves. This is because there is a greater number of metal ions in the stem.

3.1.2 Gasification

Gasification refers to the processes of converting biomass into gaseous fuel through decomposition (H_2, CO, CH_4, etc.). To do this, higher temperatures (500–1400 °C) and oxygen-deficient conditions are required. To enhance the production of the gaseous product, different gasification agents (e.g., steam, CO_2 and some gas mixtures) can be used. Approximately >50% of the biomass converted into gaseous fuel and biochar during this process was smaller in size, and resistant to chemical oxidation [28]. Temperature plays a critical role in the gasification process and results in the increased production of hydrogen and carbon monoxide. At higher temperatures, however, the levels of carbon dioxide, methane, and hydrocarbon are reduced [31]. In general, the surface area of biochar created during the gasification process is smaller and possesses fewer functional groups (i.e., hydroxyl, carbonyl and carboxyl groups) than that yielded during the pyrolysis process [32]. It is important to note that the equivalence ratio (ER) also impacts the yield and quality of biochar. A higher ER indicates that a high quantity of oxygen has been added to the gasifier and it can positively or negatively impact the properties of the resultant biochar. Yao et al. [33] conducted a study and found that there was a decrease in biochar yield from 0.22 to 0.14 kg g^{-1} and a reduction in carbon content from 88.17% to 71.6%. Additionally, the ER increased from 0.1 to 0.6 [34]. In general, if oxygen molecules are present in the compound, more ash content will be produced, whilst the yield and mechanical strength of the biochar will be reduced. A further study carried out by James et al. [35] investigated the impacts that airflow has on the properties of Pine woodchip biochar [22]. Their findings indicated that airflow of 8 to 20 L min^{-1} produced basic biochar (pH > 7.0) and that there were no acidic functional groups at a high airflow rate. The content of alkalis and alkaline earth metals is higher when biochars are created through gasification, although the exact content varies based on the type of biomass used [36]. There are several different gasifiers that can be used in such processes, including fixed bed, fluidized bed and circulating fluidized reactors. These will be discussed in more detail at a later stage [20].

3.1.3 Hydrothermal technology

During this process, wet biomass is thermochemical converted into hydrochar. Hydrochar can be produced in the same ways as biochar (i.e., using the methods discussed above). Moreover, the process is very much like the process of forming natural coal. Moderate temperatures (150–350°C) and conditions under 10–15 bar are established to perform the hydrothermal treatment process [25]. Interestingly, the properties of water change significantly at higher temperatures and pressure to become more of an organic solvent. In such cases, reactions involving acid-based catalysts are favourable in promoting biomass decomposition [37]. At present, the exact mechanism facilitating the hydrothermal treatment process is unknown.

However, it likely involves dehydration, hydrolysis, decarboxylation, aromatization and recondensation. Biomass breaks down to form saccharides and lignin throughout the hydrolysis process. Moreover, during the dehydration process, the hydroxyl group is eliminated, which ultimately removes water from the biomass. Meanwhile, during decarboxylation, all CO_2 is removed and this facilitates subsequent aromatization. A number of different compounds are produced during these processes, including phenols, aldehydes and acids and all such compounds are subjected to recondensation with aromatic polymers, which results in the production of hydrochar [38]. The key benefit of using hydrothermal technology is it can transform wet biomass into carbonaceous solids with no extensive drying required. Additionally, this produces a high yield. Other benefits include adaptable surface functionalities, conductive behaviour, the production of natural binders and high calorific value [39]. However, the hydrochar yield is reduced at higher temperatures (~350°C) (29%), as is the yield of bio-oil (31%). However, there is a substantial increase in gas fractions (67%) [40]. The O/C and H/C rations also decline with increased temperatures. Wang et al. [41] used a combination of thermal carbonization and activation to convert sunflower stalks into hydrochar [42]. The resultant hydrochar had a large surface area (1505 $m^2 g^{-1}$), as well as a 35.7 Wh kg^{-1} energy density. Several different chemical and physical methods can be used to produce hydrochar. For instance, KOH is a chemical activating agent that facilitates the production of hydrocar with a higher surface area than other chemicals ($ZnCl_2$, HCl, NaCl, and $MgCO_3$). This is because KOH can easily reach the outer surface layer of carbon material [21, 23, 25, 43–49].

4. Conclusion

Solar and wind energy have the potential to supplement the existing energy resources in order to meet the growing global demand for electricity. It is necessary to make renewable energy more economically and efficiently efficient in order to make it more accessible. Biochar has been investigated for use in the fabrication of electrode materials and catalysts for use in the catalysis of processes involved in the generation of biodiesel and biohydrogen. It is necessary to conduct research into the synthesis of biochar with the needed qualities in order to increase the efficiency of the biochar-mediated process. Recent advancements in biochar-based material research in the field of renewable energy indicate that it has the potential to be a source of future energy in the future. The scarcity of fossil fuel, environmental pollution and the rise of energy demand have triggered many researchers to find alternatives fossil fuels feedstock. Various sources of biomass have been studied and this present study has provided the most recent promising feedstock for the alternative liquid fuel production. Criteria of excellent biomass-derived biofuel were discussed. Price, availability, the total content of oil in seed and quality are critical factors. Edible palm oil and soybean oil were proven highly promising feedstock for biofuel production; nevertheless, due to the "*food versus fuel*" issue make the use of these edible oil as biofuel feedstock not viable. Non-edible oil derived from jatropha oil and WCO presented promising feedstock for biofuel production. Based on the chemical composition and physicochemical properties of raw vegetable feedstocks, all feedstocks have failed to meet diesel standards (ASTM D6751 and EN 14214) specification, which in turn strong affirmed that raw vegetable is unacceptable to be used in the diesel engines.

DOI: http://dx.doi.org/10.5772/intechopen.104984

Author details

Abdulkareem Ghassan Alsultan[1,2,3]*, Nurul Asikin-Mijan[2,4], Laith Kareem Obeas[5], Aminul Isalam[6], Nasar Mansir[7], Maadh Fawzi Nassar[8], Siti Zulaika Razali[9], Robiah Yunus[10] and Yun Hin Taufiq-Yap[1,2,3]

1 Faculty of Science and Natural Resources, Universiti Malaysia Sabah, Kota Kinabalu, Sabah, Malaysia

2 Faculty of Science, Catalysis Science and Technology Research Centre (PutraCat), Universiti Putra Malaysia, Serdang, Selangor, Malaysia

3 Universiti Malaysia Sabah, Kota Kinabalu, Sabah, Malaysia

4 Faculty of Science and Technology, Department of Chemical Sciences, Universiti Kebangsaan Malaysia (UKM), Bangi, Selangor Darul Ehsan, Malaysia

5 Technical Institute of Babylon, Al-Furat Al-Awsat Technical University (ATU), Iraq

6 Department of Petroleum and Mining Engineering (PME), Jashore University of Science and Technology, Jasho, Bangladesh

7 Faculty of Science, Department of Chemistry, Federal University Dutse, Dutse, Jigawa State, Nigeria

8 Faculty of Science, Department of Chemistry, Universiti Putra Malaysia, Serdang, Selangor, Malaysia

9 Institute of Nanoscience and Nanotechnology, Universiti Putra Malaysia – UPM, Serdang, Selangor, Malaysia

10 Institut Kajian Perladangan (IKP), Pejabat Pentadbiran, Universiti Putra Malaysia (UPM), Serdang, Malaysia

*Address all correspondence to: kreem.alsultan@yahoo.com

References

[1] Viêgas CV, Hachemi I, Freitas SP, Mäki-Arvela P, Aho A, Hemming J, et al. A route to produce renewable diesel from algae: Synthesis and characterization of biodiesel via in situ transesterification of Chlorella alga and its catalytic deoxygenation to renewable diesel. Fuel. 2015;**155**:144-154. DOI: 10.1016/j.fuel.2015.03.064

[2] Ayodele OB, Abbas HF, Daud WMAW. Catalytic upgrading of oleic acid into biofuel using Mo modified zeolite supported Ni oxalate catalyst functionalized with fluoride ion. Energy Conversion and Management. 2014;**88**:1111-1119. DOI: 10.1016/j.enconman.2014.02.014

[3] Aliana-Nasharuddin N, Asikin-Mijan N, Abdulkareem-Alsultan G, Saiman MI, Alharthi FA, Alghamdi AA, et al. Production of green diesel from catalytic deoxygenation of chicken fat oil over a series binary metal oxide-supported MWCNTs. RSC Advances. 2019;**10**(2):626-642. DOI: 10.1039/c9ra08409f

[4] Sidibe S, Blin J, Daho T, Vaitilingom G, Koulidiati J. Comparative study of three ways of using Jatropha curcas vegetable oil in a direct injection diesel engine. Scientific African. 2020;**7**:e00290. DOI: 10.1016/j.sciaf.2020.e00290

[5] Lam WY, Kulak M, Sim S, King H, Huijbregts MA, Chaplin-Kramer R. Greenhouse gas footprints of palm oil production in Indonesia over space and time. Science of the Total Environment. 2019;**688**:827-837. DOI: 10.1016/j.scitotenv.2019.06.377

[6] Mohana S, Shrivastava S, Divecha J, Madamwar D. Response surface methodology for optimization of medium for decolorization of textile dye Direct Black 22 by a novel bacterial consortium. Bioresource Technology. 2008;**99**(3):562-569. DOI: 10.1016/j.biortech.2006.12.033

[7] Míguez JL, López-González LM, Porteiro J, Paz C, Granada E, Morán JC. Contribution of renewable energy sources to electricity production in galicia (Spain). Energy Sources, Part A: Recovery, Utilization and Environmental Effects. 2006;**28**(11):995-1012. DOI: 10.1080/00908310600718882

[8] Demirbas A. Combustion systems for biomass fuel. Energy Sources, Part A Recovery, Utilization, and Environmental Effects. 2007;**29**:303-312. DOI: 10.1080/009083190948667

[9] Mollahoseini A, Tabatabaei M, Najafpour GD. Biofuel Production. Biochemical Engineering and Biotechnology, 2nd edn. Amsterdam: Elsevier; 2015. pp. 597-630

[10] Dave N, Selvaraj R, Varadavenkatesan T, Vinayagam R. A critical review on production of bioethanol from macroalgal biomass. Algal Research. 2019;**42**:101606. DOI: 10.1016/j.algal.2019.101606

[11] Marchetti JM. A summary of the available technologies for biodiesel production based on a comparison of different feedstock's properties. Process Safety and Environment Protection. 2012;**90**(3):157-163. DOI: 10.1016/j.psep.2011.06.010

[12] McNeil J, Day P, Sirovski F. Glycerine from biodiesel: The perfect diesel fuel. Process Safety and Environment Protection. 2012;**90**(3):180-188. DOI: 10.1016/j.psep.2011.09.006

[13] Rahman WU et al. Biodiesel synthesis from eucalyptus oil by utilizing waste egg shell derived calcium based metal oxide catalyst. Process Safety and Environment Protection. 2019;**122**:313-319. DOI: 10.1016/j.psep.2018.12.015

[14] Orozco LM, Echeverri DA, Sánchez L, Rios LA. Second-generation green diesel from castor oil: Development of a new and efficient continuous-production process. Chemical Engineering Journal. 2017;**322**:149-156. DOI: 10.1016/j.cej.2017.04.027

[15] Fadhil AB, Aziz AM, Al-Tamer MH. Biodiesel production from Silybum marianum L. seed oil with high FFA content using sulfonated carbon catalyst for esterification and base catalyst for transesterification. Energy Conversion and Management. 2016;**108**:255-265. DOI: 10.1016/j.enconman.2015.11.013

[16] Fauzan NA, Tan ES, Pua FL, Muthaiyah G. Physiochemical properties evaluation of Calophyllum inophyllum biodiesel for gas turbine application. South African Journal of Chemical Engineering. 2020;**32**(December):56-61. DOI: 10.1016/j.sajce.2020.02.001

[17] N. E. staff Writer. Top biofuel production countries. 2019

[18] Trindade M. Increased Biodiesel Efficiency. 2018

[19] Lee HV, Juan JC, Binti Abdullah NF, Nizah Mf R, Taufiq-Yap YH. Heterogeneous base catalysts for edible palm and non- edible Jatropha- based biodiesel production. Chemistry Central Journal. 2014;**8**(1):1-9. DOI: 10.1186/1752-153X-8-30

[20] Feedstocks for Biofuel Production. 2019. https://farm-energy.extension.org/feedstocks-for-biofuel-production/

[21] Takkellapati S, Li T, Gonzalez MA. An overview of biorefinery-derived platform chemicals from a cellulose and hemicellulose biorefinery. Clean Technologies and Environmental Policy. 2018;**20**(7):1615-1630

[22] Bhaskar T, Bhavya B, Singh R, Naik DV, Kumar A, Goyal HB. Thermochemical conversion of biomass to biofuels. Biofuels. 2011:51-77

[23] Riaz S, Rhee KY, Park SJ. Polyhydroxyalkanoates (PHAs): Biopolymers for biofuel and biorefineries. Polymers. 2021;**13**:253. DOI: 10.3390/POLYM13020253

[24] Basu P. "Biomass gasification, pyrolysis and torrefaction: Practical design and theory," Biomass Gasification, Pyrolysis Torrefaction Pract. Des. Theory. 2018. pp. 1-564

[25] Sweeney MD, F. Xu. "Biomass converting enzymes as industrial biocatalysts for fuels and chemicals: Recent developments." Catalysts. 2012;**2**(2):244-263. DOI:10.3390/catal2020244

[26] Chi NTL, Anto S, Ahamed TS, Kumar SS, Shanmugam S, Samuel MS, et al. A review on biochar production techniques and biochar based catalyst for biofuel production from algae. Fuel. 2021;**287**:119411

[27] Akinfalabi SI, Rashid U, Ngamcharussrivichai C, Nehdi IA. Synthesis of reusable biobased nano-catalyst from waste sugarcane bagasse for biodiesel production. Environmental Technology & Innovation. 2020;**18**: 100788

[28] Berntsson T. Integration of Biorefinery concepts in Process Industries What is a Biorefinery?

[29] Basu P. (2013) Biomass gasification, pyrolysis and torrefaction: Practical design and theory. Biomass Gasification, Pyrolysis and Torrefaction: Practical Design and Theory. 2018. p. 1564. DOI: 10.1016/C2016-0-04056-1

[30] Zhang L, He W, Yang J, Sun J, Li H, Han B, et al. Bread-derived 3D macroporous carbon foams as high performance free-standing anode in microbial fuel cells. Biosensors and Bioelectronics. 2018;**122**:217-223

[31] Demshemino I et al. Comparative analysis of biodiesel and petroleum diesel, 1, 2013

[32] The Concept of Biorefinery. 2021. https://www.bioenergyconsult.com/biorefinery/

[33] Yao Z, You S, Ge T, Wang CH. Biomass gasification for syngas and biochar co-production: Energy application and economic evaluation. Applied Energy. 2018;**209**:43-55

[34] First Generation Biofuels. http://biofuel.org.uk/first-generation-biofuel.html

[35] James RAM, Yuan W, Wang D, Wang D, Kumar A. The effect of gasification conditions on the surface properties of biochar produced in a top-lit updraft gasifier. Applied Sciences. 2020;**10**(2):688

[36] Facts on Health and the Environment. https://www.greenfacts.org/en/index.htm

[37] Hochman G, Rajagopal D, Timilsina GR, Zilberman D. Impacts of biofuels on food prices. In: The Impacts of Biofuels on the Economy, Environment, and Poverty: A Global Perspective. 2014. p. 4764. ISBN: 9781493905188

[38] Liquid Biofuels for Transport Prospects, risks and opportunities. https://www.greenfacts.org/en/biofuels/l-2/5-food-security-poverty.htm

[39] Nunez C. Biofuels, explained. 2019. https://www.nationalgeographic.com/environment/global-warming/biofuel/

[40] Ahmed I, Zia MA, Afzal H, Ahmed S, Ahmad M, Akram Z, et al. Socio-economic and environmental impacts of biomass valorisation: A strategic drive for sustainable bioeconomy. Sustainability. 2021;**13**(8):4200

[41] Wang X, Yun S, Fang W, Zhang C, Liang X, Lei Z, et al. Layer-stacking activated carbon derived from sunflower stalk as electrode materials for high-performance supercapacitors. ACS Sustainable Chemistry & Engineering. 2018;**6**(9):11397-11407

[42] Seetho A, Ton Kan A, Teo Cher KG, Swa Liang H, Koh Qi SF, Olaganathen R. Potential and technological advancement of Biofuels. International Journal of Advanced Scientific and Technical Research. 2014;**4**:1229

[43] Pankin K et al. Comparison of the physicochemical characteristics of biofuels and petroleum fuels. Chemistry and Technology of Fuels and Oils. 2011;**47**:7-11. DOI: 10.1007/s10553-011-0247-6

[44] Picazo-Espinosa R, Gonzalez-Lopez J, Manzaner M. Bioresources for Third-Generation Biofuels. Biofuel's Engineering Process Technology. 2011. DOI: 10.5772/17134

[45] Islam A, Roy S, Khan MA, Mondal P, Teo SH, Taufiq-Yap YH, et al. Improving valuable metal ions capturing from spent Li-ion batteries with novel materials

and approaches. Journal of Molecular Liquids. 2021;**338**:116703

[46] Second Generation Biofuels. http://www.biofuel.org.uk/second-generation-biofuels.html

[47] Various Advantages and Disadvantages of Biofuels. 2020. https://www.conserve-energy-future.com/advantages-and-disadvantages-of-biofuels.php

[48] Various Types and Benefits of Biofuels. 2019. https://www.conserve-energy-future.com/types-benefits-biofuels.php

[49] Kamil FH, Salmiaton A, Shahruzzaman RMHR, Omar R, Alsultsan AG. Characterization and application of aluminum dross as catalyst in pyrolysis of waste cooking oil. Bulletin of Chemical Reaction Engineering & Catalysis. 2017;**12**(1):81-88

Chapter 13

Biochar Development as a Catalyst and Its Application

Stephen Okiemute Akpasi, Ifeanyi Michael Smarte Anekwe, Jeremiah Adedeji and Sammy Lewis Kiambi

Abstract

Biochar is a carbon-rich pyrogenic material that is made from carbon-neutral sources (i.e., biomass). It offers key strategies for carbon capture and storage (CCS) as well as being an environmentally friendly means of soil amendment. The recent recognition of biochar as a versatile media for catalytic applications has prompted preliminary research into biochar's catalytic capacity and mechanistic practices via various routes. This chapter provides a review of biochar production technologies, biochar's catalyst development, and its application in various catalytic processes as well as descriptions of the benefits and drawbacks of the various applications currently available. The characteristics of biochar-based catalysts, challenges of effective application of this catalyst system, emerging application, prospects, and future work consideration for effective utilization of biochar-based catalysts were presented.

Keywords: biochar, biodiesel, biomass, catalyst, pyrolysis, tar reforming, wastewater treatment

1. Introduction

With energy shortages and pollution escalating worldwide, renewable feedstocks are crucial for human long-term development. There are many natural sources of animal fats, including lignocellulosic biomass, crops, aquatic culture, biowaste generated by waste management, and domestic and urban waste recycling [1]. Utilizing thermochemical decomposition processes like gasification or pyrolysis, biofuels (bio-oil and syngas) can be produced from biomass and a carbon-based solid residue called biochar [2].

As a porous solid with high carbon content, biochar is formed during the thermal decomposition of biomass at moderate temperatures (e.g., 350–700°C) and under oxygen-limited conditions [3–7]. Despite its chemical and physical properties, biochar's thermochemical process and the intrinsic properties of biomass feedstock are two of the factors that influence its properties [4]. Due to its porosity and large surface area, biochar is classified as activated carbon (AC), yet it also contains numerous surface functional groups (carbon monoxide, hydroxyl, carbonyl, carboxylic acid, among others) that can be easily tuned and used to make various functionalized carbon materials. As well as being used for AC production and soil amendment, biochar serves as an adsorbent for pollutants in water and air [5].

Recent research has revealed that biochar is widely utilized as support for metals in catalysis, due to its feedstock availability, large surface area (for good metal phase dispersion and stability), low cost, and stability in basic and acidic media [6]. In addition to catalysis, biochar's excellent performance in supporting and catalyzing a wide range of reactions has been demonstrated: electrochemical reactions, hydrolysis, gasification/pyrolysis, catalytic reforming/cracking, esterification/transesterification, peroxide/peroxynmonosulfate oxidation, and many more.

Biochar-based catalysts have been utilized for a variety of applications, including water and soil remediation. On the other hand, current perspectives tend to concentrate on applications designed to remediate soils, revegetate, and restore them, convert energy, and remove contaminants from water and wastewater. Despite this, there is still a lack of understanding regarding the synthesis, development, and novel applications of biochar-based catalysts. This chapter provides a comprehensive overview of recent developments in the production, application, and limitations of biochar-based catalysts. Various emerging catalytic applications of biochar-based catalysts are also addressed in this chapter. Further, the benefits of using biochar as catalysts and catalyst supports, as well as the correlations between structural and physical properties of biochar, which provide insights into the development of effective and promising biochar-based catalysts will be highlighted. The challenges and future advancements of using biochar-based catalyst materials are further discussed.

1.1 Properties of biochar

Biochar is a form of organic material that is mostly rich in carbon and other elements such as nitrogen (N), oxygen (O), and hydrogen (H). Biochar has a carbon (C) content ranging from 380 to 800g kg^{-1} and has both alkyl and aromatic structures [7]. Biochar is also composed of inorganic elements including phosphorus (P), calcium (Ca), aluminum (Al), potassium (K), and silicon (Si), whose quantities vary according to the feedstock used [8]. It has been reported that acidic pH can occur during pyrolysis, depending on conditions of production and the raw materials [9]. Other factors can affect the biochar pH, ranging from neutral to alkaline [10]. In general, biochar has a pH between 5 and 12, and its pH tends to increase in response to increased pyrolysis temperature as bionic acid decomposes, and mineral alkali elements increase [11]. Also, the high pH of biochar can be attributed to the functional organic groups present in it, namely hydroxyl-, aldehyde, and ketone- [12]. As a buffer between acid and bases, these functional organic groups influence biochar's hydrophobicity and hydrophilicity as well as its adsorption properties [8]. The functional organic groups have the effect of lowering the negative charge on biochar, and therefore, enhancing its cation exchange capacity (CEC) [13].

Due to its high carbon content, biochar has a complex microstructure with numerous pores, which maximizes its surface area [14]. Biochar's surface area and total pore volume typically range from 8 to 132 m^2/g and 0.016–0.083 cm^3/g, respectively. Using the right precursor and pyrolysis parameters, biochar can have surface areas and pore volumes as high as 490.8 m^2/g [15] and 0.25 cm^3/g [16]. Following effective post-treatments, such as potassium hydroxide (KOH) activation, the surface area and total pore volume of biochar can be enhanced to 3263 m^2/g and 1.772 cm^3/g, respectively [17], which is comparable to or even greater than commercial activated carbon.

Biochar's surface area and porosity are greatly affected by the pyrolysis temperature [16]. Biochar with a higher pyrolysis temperature within a certain temperature range has a greater surface area [12]. As temperature rises in biochar pyrolysis, volatile substances are forced out of the char, causing pores to form a larger surface area [17]. Due to its high porosity/high amount of residual pores and large surface area, biochar can retain a large quantity of water [14, 18–20]. In contrast, a high pyrolysis temperature diminishes the polar functional groups found in biochar, thereby increasing its hydrophobicity [18]. According to the above characteristics, biochar can influence the, pH, soil water-holding capacity, as well as base saturation, and CEC [14]. It is generally possible to modify the properties of biochar by modifying its conditions of preparation [19], as outlined in the next section.

1.2 Biochar production

To produce biochar from different feedstocks, several approaches have been developed. Torrefaction, pyrolysis, gasification, hydrothermal carbonization (HTC), and flash carbonization are the most prominent thermochemical conversion technologies (**Figure 1**).

1.2.1 Torrefaction

Torrefaction is a mild pre-treatment consisting of slow heating at 200–300°C, followed by a short retention time before gasification or pyrolysis [20]. Often, the resultant solid product is porous, low density, and carbon-enriched, with low moisture content and O/C ratio, an increase in energy density, and improved grindability, making it easier to store and deliver [21]. Its carbon yield can be affected by temperature, retention times, raw material types, and furnace atmosphere [22]. At 200°C, for example, beech lignin began to degrade, the majority of biomass developed at 230°C and cellulose only degraded over 270°C [22]. Using a pilot process, hardwood and

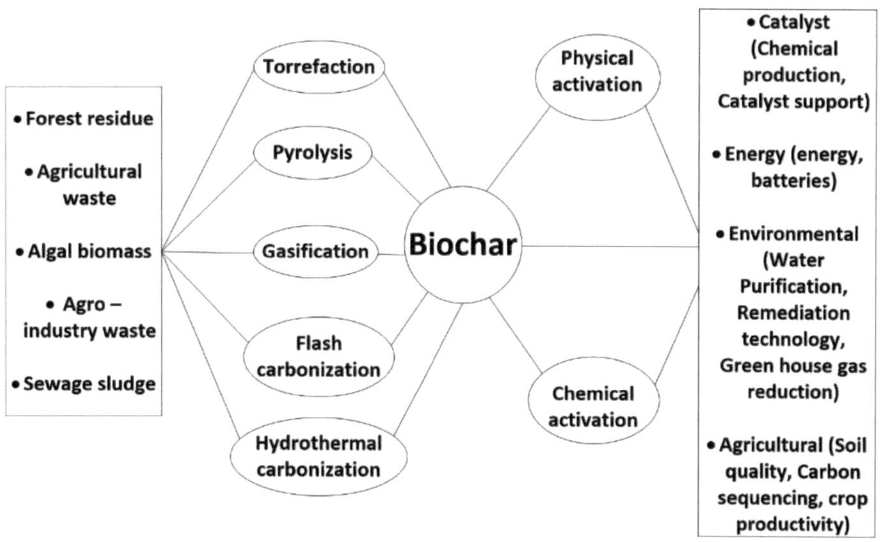

Figure 1.
Overview of biochar-based system production and applications.

switchgrass pellets produced solid yields above 77 wt% [23]. Oil palm fiber pellets were torrefied in an inert atmosphere for 30 min and in an oxidizing atmosphere for 30 min at 275–350°C to yield 43 and 65 wt% biochar, respectively [24].

1.2.2 Pyrolysis

Pyrogenic carbons are produced by the decomposition of biomass at 300–1200°C without oxygen (or with limited oxygen). During pyrolysis, biochar is produced at temperatures ranging from 300 to 700°C. A pyrolysis process can be classified into slow, fast, intermediate, flash, and vacuum modes [25].

1.2.2.1 Slow pyrolysis

In slow pyrolysis, the process temperature is lower (400–600°C), the heating rate is slower (~10°C min^{-1}), the vapor residence time is much longer (5–30 minutes), and the holding time is long (hours to days) [25]. Biochar typically yields 20–40 wt%, with yields decreasing with increasing pyrolysis temperature and heating rate [26], however, biochar characteristics are also affected by the procedure and feedstock used [27]. Comparing biochars derived from the wood stem and bagasse with palm kernel shell, paddy straw, and cocopeat, biochar derived from the wood stem and bagasse exhibited a wide range of pores and a high surface area. Biochar develops a significant surface area structure and pore structure at around 500°C [28] with a wide range of mineral compositions and high thermal stability [29].

1.2.2.2 Fast pyrolysis

Fast pyrolysis refers to the treatment of biomass at high temperatures without oxygen [30]. It is usually necessary to dry and grind the feedstock to facilitate effective heat exchange and conversion. This technique produces high liquid yields (bio-oil) rather than solid char (15–25 wt%) [31]. In contrast to the slow pyrolysis of wheat straw, fast pyrolysis generated biochar with a labile un-pyrolyzed carbohydrate fraction (8.8%) rather than carbonized completely [32]. There was a significant difference in the pH, particle size, and specific surface area for biochars produced using these two methods at 400°C, as well as a significant increase in surface area at 500°C (175.4 $m^2\ g^{-1}$), in comparison to 300°C (2.9 $m^2\ g^{-1}$) and 400°C (4.8 $m^2\ g^{-1}$) [21, 33].

1.2.2.3 Intermediate pyrolysis

Intermediate pyrolysis produces 15–35 wt% dry and brittle biochar at temperatures between slow and fast pyrolysis, i.e., solid residence durations of 0.5–25 min, vapor residence times of 2–4 s, and moderate temperatures up to 500°C [34]. Utilizing barley straw and wood pellets, a pilot-scale production yielded 30 wt% char with a carbon content of 75 wt% [25]. The process produces 51.7 wt% char from the organic fraction of municipal solid waste as a result of inert fractions in the biomass [35]. **Table 1** illustrates the product yield of pyrolysis processes.

1.2.3 Gasification

Carbonaceous materials are turned into char, tars, and syngas through gasification at high temperatures (~800°C) in the presence of a gaseous active medium (e.g., carbon

Process	Bio-oil (liquid)	Biochar [37]	Syngas (gas)
Fast pyrolysis Short hot residence time (<2 s) Moderate temperature (~500°C)	75% (25% water)	15–25%	10–25%
Intermediate pyrolysis Moderate hot vapor residence time Low-moderate temperature (300–400°C)	50% (50% water)	25%	20–30%
Slow pyrolysis Long residence time Low-moderate temperature (200–300°C)	50% (50% water)	35%	35%
Gasification Long vapor residence time High temperature (>700°C)	5% tar (5% water)	10%	85%

Table 1.
Summary of product yield of pyrolysis processes [36].

dioxide, air, nitrogen, oxygen, steam, or gas mixtures) [37, 38]. During this process, the material is dried, pyrolyzed, partially oxidized, and reduced. Generally, char only makes up 5–10 wt% of the mass of the feedstock [39]. As a by-product of large-scale processes, biochar is produced in large quantities every day. Biochar produced through gasification usually has smaller particles than biochar produced by pyrolysis, lower surface area, and a lower total pore volume [40]. Since the aromatic rings are condensed, gasification chars contain little carbon (20–60 wt%) but are highly stable, preventing microbial mineralization and chemical oxidation; however, their surface chemistry is constrained by their absence of functional groups [34]. Biodiesel generation, catalytic tar decomposition, soil amendment, anode materials for direct carbon fuel cells, and anaerobic digestion additives are just a few of the uses for gasification char [41].

1.2.4 Hydrothermal carbonization

Biomass can be processed using a thermochemical process called hydrothermal carbonization (HTC). In closed vessels with liquid water and autogenous pressure of 2–10 MPa, the feedstock is heated from 200 to 300°C and hydrochar is produced [42]. The thermal stability of hydrochar is improved by high temperatures (300°C). Wet torrefaction or wet pyrolysis are other terms for HTC [43, 44]. In comparison with biochar, hydrochar contains less carbon, ash, surface area, and a smaller pore volume [39].

1.2.5 Flash carbonization

Through flash carbonization, biomass can be transformed into biocarbon (i.e., charcoal) rapidly and efficiently, typically by starting and controlling a flash fire at a high temperature within a packed bed (~1 MPa) [45]. The biomass is transformed into gas and charcoal in less than 30 min when the combustion flame flows in the opposite direction of the airflow. Charcoal yields are typically approximately 40 wt% [45].

1.3 Biochar as a promising catalyst

Biochar can serve as catalyst support. Besides stabilizing and dispersing nanoparticles, biochar can also provide more active sites for catalytic degradation reactions [46].

Biochar's mesoporous structure enhances the proper dispersion of immobilized metal particles while also preventing particle aggregation owing to intra-particle interaction [47]. The incorporation or fixing of metal elements, for example, magnesium (Mn), copper (Cu), cobalt (Co), and iron (Fe) into biochar pores result in no or minimal metal escape into the aqueous phase [48].

As a heterogeneous catalyst or support, biochar offers many advantages including large surface area, lower cost, functional group tailoring, etc., which makes it highly beneficial for many catalytic applications. There are several intrinsic properties of biochar that contribute to its effectiveness as a catalyst [49]. It has a good thermal, stable structure, mechanical stability, and a chemically hierarchical structure that originates from biomass. Biochar-based catalysts have the following distinctive characteristics: (i) heterogeneity, i.e., the reaction mixture can be easily isolated from other reactants; (ii) bifunctionality, i.e., transesterification and esterification are involved; (iii) recyclable; (iv) porous; (v) non-graphiţable, i.e., it does not form crystal at high temperatures [50]. Comparing biochar-based catalysts with other solid-based catalysts, biochar has the advantages of being cost-effective, eco-friendly, easy to produce, reusable, and biodegradable.

Furthermore, biochar as a catalyst can be used in many different fields, including agriculture, environment, and energy, for biodiesel production, tar removal, waste management, production of syngas, production of chemicals, and removal of contaminants, etc. [45, 51]. Biochar is an excellent catalyst with several beneficial properties. Biochar, for instance, is catalytically active in cracking tar because of its presence of inorganic elements including Fe and K [47]. A biochar-supported metal catalyst can be synthesized by adsorbing metal precursors on its surface functional groups [52]. Despite this, biochar has some properties that preclude it from functioning as a catalyst, such as poor porosity and low surface area. Considering that biochar contains more functional groups, it must have a large surface area for catalysis. A functional group, such as OH, adsorbs norfloxacin. Adsorption of ammonium is possible through C=O and −OH groups. To endow biochar with specific properties, it is necessary to develop a variety of modification strategies. Furthermore, several processes can be used to activate feedstocks, control synthesis conditions, functionalize materials on the surfaces, form composites with other materials [53], etc.

1.4 Characteristics of biochar based catalyst

In addition to its properties, biochar's potential for specific applications is dependent on both the biomass source and the conditions of preparation. Biochar, for instance, is suitable as an electrode material because it is electrically conductive and porous [54]. It has been proven that structurally bound nitrogen groups and high porosity biochar make superior supercapacitor electrode materials [55]. However, the intrinsic inorganics, matrix nature, and surface functionality of biochar have a significant influence on its catalytic performance.

1.4.1 Bulk element and inorganics

The carbon content of activated carbon from coal is approximately 80–95%; however, that content is lower for biochar (45–60 wt%) than carbon black (98%) [5]. Biochar also contains substantial amounts of hydrogen and oxygen. Another characteristic of biochar is that it contains small amounts of inorganic elements like potassium, sodium, calcium, magnesium, sodium, iron, and calcium. The nature of raw biomass greatly affects the amount and composition of inorganics. Woody biomass, as well as herbaceous and hydrophyte biomass, usually have a much lower inorganic content than biochar made from these sources [56, 57].

The inorganic components of biochar are crucial to many of the biochar's catalytic applications [47], including tar cracking [58], methane decomposition, and bio-oil upgrading [59].

1.4.2 Chemistry of biochar matrix

Amorphous crystalline sheets of high-conjugated aromatics make up most of the biochar matrix. As shown in **Figure 2**, these aromatic sheets are crosslinked randomly. In response to rising processing temperatures, biochar crystallites increase in size, and order is created throughout the entire structure [62]. The aromatic structure of biochar may also contain heteroatoms, including N, P, and S. These heteroatoms have a different electronegativity from the aromatic C, which results in biochar's chemical heterogeneity. This plays a key role in catalytic applications [58].

1.4.3 Surface functional groups

Comparing biochar to other carbon materials including (activated carbon and carbon black), **Figure 3** shows that it typically contains large numbers of surface functional groups. Biochar can be functionalized using its surface functional groups. Moreover, biochar has been shown to facilitate the loading of metal precursors onto metal catalysts as part of the synthesis of a metal catalyst supported by biochar [52]. Biochar-based catalysts can also work better for certain reactions if they contain some surface functional groups. Biochar-based solid-acid catalysts are typical examples. Kitano, Yamaguchi [63] demonstrated that sulfonated carbon is more effective at hydrolyzing cellohexaose, than sulfonic acid (SO_3H)—bearing resins. Adsorption sites, in this case, were found in the carboxylic acid (COOH) and hydroxyl (OH) groups of phenolic groups in the carbon material. Researchers found that the combination of functional groups on biochar-based solid acids was efficient for hydrolyzing cellulose and 1,4-glucan.

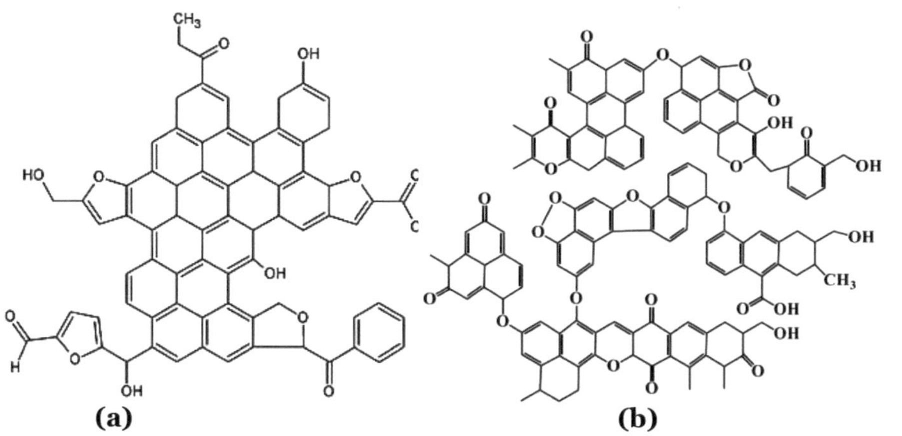

Figure 2.
Chemical structures of (a) pyrochar and (b) hydrochar (adapted from Shi & Lee) [60, 61].

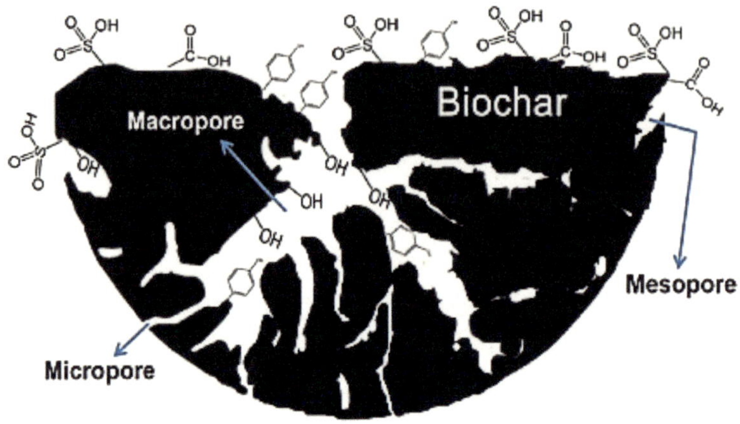

Figure 3.
A porous biochar model with multiple functional groups (adapted from Yang et al [61]).

2. Preparation of biochar-based catalyst

Biochar has been activated and functionalized in various ways to adjust its physicochemical properties, leading to enhanced reactivity in a range of processes and applications [48]. Impregnation and physical or chemical activation are the most popular methods. *In-situ* or post-synthesis methods are employed in such modifications. Biochar-based catalysts have the potential to be a feasible alternative to metal-based catalysts and carbon catalysts driven by fossil fuels. **Table 2** lists the types of biomass used to make biochar-based catalysts.

2.1 Impregnation

This technique involves mixing feedstock and metallic precursors (in-situ) into biochar structures to incorporate active metallic species into them [64]. With the use of biochar, lignin magnetite pellets were synthesized into zero-valent iron at 900°C [64]. It was possible to remove trichloroethylene by both adsorptive and degradative mechanisms due to the macro-porosity developed. Rice straw biochar was impregnated with cobalt nitrate ($Co(NO_3)_2$), then hydrothermally treated and calcined to produce the composite [65]. In comparison to pure biochar (43.0 $m^2 g^{-1}$, 0.081 $cm^3 g^{-1}$) and cobalt (II, III) oxide (CO_3O_4 (37.0 $m^2 g^{-1}$, 0.184 $cm^3 g^{-1}$), the composite showed greater SBET (62.7 $m^2 g^{-1}$) and total pore volume (0.207 $cm^3 g^{-1}$). The catalyst was shown to be effective for oxidatively degrading ofloxacin (over 90% removal in 10 min) using peroxymonosulfate (PMS). An X-ray photoelectron spectrometer (XPS) study revealed that the rich mesoporous support contains many CO−OH groups, which are important for activation. The obtained pristine biochar may also contain metal species varying in amounts and characteristics, depending on the biomass source. Despite this, impregnation typically produces composites rather than carbonaceous biochar, so one could compare biochar with impregnated composites and exhausted catalysts.

2.2 Physical activation

A physical activation process involves exposing the pyrolyzed biochar materials to a streamflow control or carbon dioxide or a mixture of both when temperatures

Feedstock	Production method (biochar)	Production method (catalyst)	Type of catalyst	Ref.
Banana	Carbonization	Wet impregnation	Solid alkali	[49]
Vegetable oil asphalt, coconut shell, oat hull, glucose, peanut shell	Carbonization	Sulfonation	Solid acid	[50]
Rice husk	Carbonization	Chemical activation	Solid acid	[51]
Peat	Carbonization	Wet impregnation	Solid alkali	[52]
Irul wood saw dust	Slow pyrolysis	Sulfonation	Solid acid	[53]
Pamelo peel, shelled palm kernel	Carbonization	Wet impregnation Calcination	Solid alkali	[54]

Table 2.
Production methods and feedstocks for biochars and biochar-based catalysts.

exceed 700°C. Gaseous activation agents, depending on the degree of $C-H_2O$ and/or $C-CO_2$ gasification that occurs at such high temperatures, are capable of partially eroding carbon atoms in the as-prepared biochar matrix [66]. By physically activating the carbonized material, most of the reactive carbon parts can be eliminated and the enclosed pores in the biochar matrix can be opened and interconnected [67]. Consequently, the surface area of biochar increases significantly, resulting in an improved micropore structure and a lower mesopore content [68]. **Figure 4** illustrates the process for producing biochar-based catalysts.

Activated biochars differ significantly from one another in terms of a specific area, pore size distribution, and porosity based on the type of biomass, reaction parameters, and activating gas [66]. Lima et al. [62] for example, evaluated the effects of steam activation on the surface areas and porosities of different biochars, as well as their metal ion adsorptive capabilities. They found that steam-activating biochars at 800°C for 45 minutes dramatically increased the surface area and micropore volume from less than (5 m^2 g^{-1}) to (136–793) m^2 g^{-1}. In addition, due to the increased porosity and surface area, these biochars were able to improve their metal ion adsorption performance to varying degrees after activation [69]. In addition, Kołtowski et al. [60] utilized steam and CO_2 to activate biochar produced from the slow pyrolysis of willow. Their findings revealed that both steam and carbon dioxide activation considerably increased the porosity and surface area of biochar. Additionally, steam-activated biochar (840.6 m^2 g^{-1}) and CO_2-activated biochar (512.0 m^2 g^{-1}) showed significantly larger surface areas than those of unactivated biochar (11.4 m^2 g^{-1}). In contrast with the CO_2-activated biochar, steam-activated biochar was found to have higher specific surface areas and pores [60].

2.3 Chemical treatment

Chemical activation involves mixing freshly prepared biochar with activation agents (e.g., KOH, $ZnCl_2$, K_2CO_3, H_2SO_4, H_3PO_4, etc.). The biochar is subsequently heated at high temperatures in an inert gas flow [70]. While the mechanism for chemical activation is still unclear, chemical activation is more corrosive than physical activation [71]. However, high temperatures can significantly enhance the corrosion properties of chemical activation substances. Aside from removing some carbon atoms from the biochar matrix, these chemicals might suppress tar formation and/or

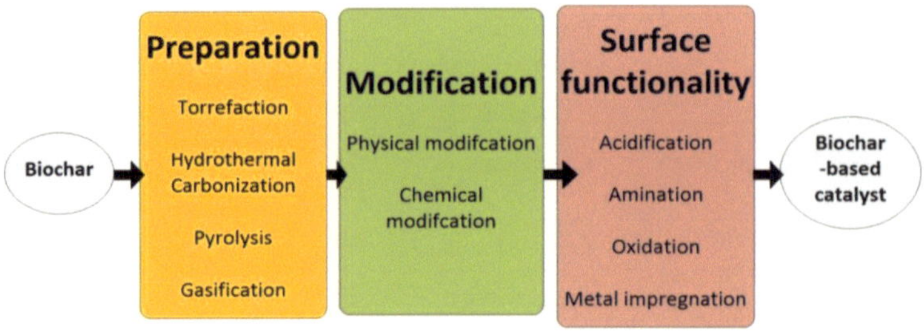

Figure 4.
Method involved in producing biochar-based catalyst.

facilitate the formation of volatile compounds [67]. It was reported by Liu et al. [59], that chemical erosion and physical activation lead to large surfaces and high porosities in KOH-activated biochar, and metallic K intercalation. Chemical activation generally results in a higher activation efficiency than physical activation, and chemical activation may be performed at a relatively lower temperature, resulting in a more porous and higher surface area biochar [43]. Although chemical activation leaves biochar with improved surface area and porosity, it is usually necessary to wash it to remove impregnating agents and salts [50]. The use of chemical activation is, therefore, affected to some extent by several factors, including corrosion of equipment, chemical recycling, secondary pollution, etc. [43].

Several factors affecting the chemically activated biochar, including the temperature of activation, feedstock type, the type, and concentration of the activating agent, etc., are significant [66]. Biochar impregnated with KOH solution has been investigated by Dehkhoda et al. [65] to determine how activation temperature (685–700°C) influenced the electrosorption performance, porosity, and surface area. In their study, there was an increase in the surface area of the biochar from (1.66 $m^2 g^{-1}$) to (614–990 $m^2 g^{-1}$), as well as its porosity, which increased from negligible to 0.6–0.9 $m^3 g^{-1}$. Additionally, as the temperature rises, a decrease in biochar surface area is observed, by collapsing and burning off the micropore walls or causing the formation of graphite-like structures in the matrix. Since biochar activated at 675°C contains more micropores and oxygen-containing functional groups, its overall electrosorption capacitance was more than twice as high as that of activated biochar at 1000°C [68].

3. Application of biochar catalyst

The growing discovery of biochar as a diverse material for catalytic activities has prompted preliminary study into the catalytic potential of biochar as well as applications in different processes.

3.1 Biodiesel production on biochar catalysts

It has been demonstrated that biodiesel can be used as a renewable alternative to traditional petrochemical-derived diesel [67, 72]. The application of traditional catalysts in the synthesis of biodiesel from biomass (vegetable oils) has been extensively explored. However, the manufacture of such catalysts necessitates the use of costly metal precursors. Because of their low cost and versatility, sulfonated biochars have been utilized to produce biodiesel. It has been demonstrated that sulfonated biochar can produce the maximum productivity (88%) of biodiesel products from vegetable oil in the esterification of FFAs (free fatty acids) and transesterification of TGs (triglycerides) carried out simultaneously at 100°C for 15 h [72–74]. It was observed that after five recycles of the catalyst, the output of methyl esters reduced from 88% to 80%, due to the leaching of $-SO_3H$ functional groups [74]. Using a biochar catalyst made from palm kernel shells to transesterify sunflower oil, Kostić et al. [69] investigated the catalytic activity. With the deposition of 3 wt% catalysts into a reaction, the production of methyl esters was 99% at 65°C [75]. The solid acid/base biochar catalysts mentioned above resulted in a significant synthesis of biodiesel from a variety of edible oils. In contrast, both catalysts exhibited signs of deactivation after many re-uses in the laboratory. While transesterification was taking place, the base catalyst was contaminated by undesired secondary products formed by CaO and the feed oil interactions [75]. The ester output (from TGs and FFAs) is comparable to that obtained from non-biochar catalysts. However, to make biochar catalysts for biodiesel generation more realistic, the stability of biochar catalysts must be increased to prevent the need for post-treatment processes to remove S or Ca from the catalyst [75]. The biodiesel production efficiency of different biochar and non-biochar-based catalysts is shown in **Table 3**.

Biochar-based catalyst	Feedstock	Temp.	Biodiesel yield	Ref.
Wood biochar-mixture	Canola oil	423 K	44%	[76]
Peanut hull-biochar	Palmitic + stearic acid + soybean oil	333 K	70%	[77]
Husk of rice	Oleic acid + canola oil	423 K	48%	[78]
Rice husk-biochar	Cooking oil waste	383 K	88%	[79]
Biochar-palm kernel shell	Sunflower oil	333 K	99%	[80]
Non-biochar catalysts				
$Al(HSO_4)_3$	Vegetable oil waste	493 K	81%	[81]
Zeolite beta	Cooking oil waste	353 K	25%	[82]
SO_4^{2-}/ZrO_2	Cooking oil waste	353 k	44%	[83]

Table 3.
A comparison of biochar and non-biochar-based catalysts for biodiesel production.

3.2 Biomass hydrolysis on biochar catalysts

Biochar catalysis has been applied in biomass hydrolysis. The fact that most biochar-based catalysts are more effective than commercially available and traditional catalysts has long been recognized. According to Ormsby et al. [75], pinewood chips and peanut hulls that were sulfonated with H_2SO_4 were used as the raw materials for biochar. When used to hydrolyze xylan, the sulfonated pine chip-biochar catalyst demonstrated an 85% transformation rate in 2 h at 393 K. On the other hand, while having a greater surface area (1391 m^2g^{-1}) than the biochar catalyst (365 m^2g^{-1}), industrial activated carbon only achieved a 57% transformation in 24 hours [84]. Furthermore, biochar catalysts showed greater starting process rates for the hydrolysis of cellobiose and xylan when compared to other catalysts (activated carbon and Amberlyst-15) [84], indicating that they were more efficient than the other two catalysts. Moreover, the hydrolysis of maize stover, switchgrass, and prairie cordgrass biomass was accomplished using a corn stover-biochar mixture [85]. Compared to a traditional homogeneous H_2SO_4 catalyst, the catalyst exhibited a stronger preference for glucose and xylose, confirming its superior efficiency in biomass hydrolysis. The existence of sulfonated corn stover-based biochar increased the production of glucose and xylose from lignocellulosic biomass [48]. The glucose output was 8–10% and the xylose yield was 23–41% when compared to the equivalent polysaccharide [85]. The findings were equivalent to those obtained from the hydrolysis of model substances using a similar catalyst: cellulose yielded 3% glucose and xylan yielded 40% xylose. This indicated that the biochar was able to sustain good efficiency even when exposed to contaminants and a complex matrix of biomass materials. The performance of different biochar based catalysts for hydrolysis is shown in **Table 4**.

Feedstock	Condition of catalyst preparation	Feedstocks	Condition of reaction	Catalyst performance	Ref.
Forestry wood waste	Slow pyrolysis at 700°C for 15 h; sulfonated with 30 w/v% H_2SO_4.	Fructose or maltose	C_{cata} = 25 w/v%, T = 1 h (maltose: 160°C and fructose: 180°C), C_F = 5 w/v%	Selectivity = 60.4%; HMF yield = 42.3%; selectivity = 88.2%. Glucose yield = 85.4%	[86]
Palm kernel shells	Carbonized at 550°C for 4 h; sulfonated at 120°C for 6 h with concentrated sulfuric acid	Cellobiose	C_{cata} = 50 g/L, T = 24 h, Temp. = 110°C C_F = 100 g/L	Glucose yield ≈ 100%	[87]
Bamboo	Carbonates of alkali molten molten at 450°C; pyrolyzed at 450°C; sulfonated at 150°C for 6 h	Cellulose	$C_{cata} \approx$ 16.67 g/L, T = 12 h, Temp. = 150°C $C_F \approx$ 16.67 g/L	RSs yield = 52.8%, Glucose yield = 43.5%	[88]
Bamboo	80% sulfuric acid treatment; immersion in oleum and heating with N2; ultrasonic vibration treatment with NaCl saturated aqueous solution; treatment with an excess of IL-Cu in anhydrous MeCN	Bamboo and cellulose	Bamboo: $C_F \approx$ 133.33 g/L, $C_{cata} \approx$ 66.67 g/L, T = 2 h, Temp. = 110°C, Microwave radiation = 750 W. Cellulose: $C_F \approx$ 133.33 g/L, $C_{cata} \approx$ 66.67 g/L, T = 2 h, Temp. = 90°C, Microwave radiation = 350 W	TON: 3.54 (cellulose), 2.42 (bamboo), RSs yield = 35.6% (cellulose), 22.5% (bamboo)	[89]

Hydroxymethyl furfural = HMF, reducing sugars = RSs, TON = turnover number, temperature = Temp, time = T, catalyst amount = C_{cata}, feedstocks concentration = C_F, imidazolium chloride = IL-Cu; anhydrous MeCN = anhydrous acetonitrile.
Source: Adapted from Shan et al. [90].

Table 4.
Biochar catalyst for hydrolysis.

3.3 Production of biogas

3.3.1 Tar reforming (syngas synthesis)

Tar reforming is the process of converting the hydrocarbon combination that is inevitably generated following the gasification and pyrolysis of biomass into useful syngas (combination of CO and H_2). Syngas is a multipurpose intermediate and/or beginning raw material for the synthesis of fuels and chemicals. As a result of this fact, several studies have investigated the potential involvement of biochar catalysts in the generation of syngas in recent years [58]. Biochar comprises catalytic centers that are similar to those found in traditional catalysts, such as dolomites ($MgCO_3{\cdot}CaCO_3$), olivine ($(Mg^{2+}, Fe^{2+})_2SiO_4$), and Ni- and alkali metal-based catalysts, could be efficient for tar reforming [78]. The switchgrass biochar that had been activated by KOH demonstrated the highest efficacy, with around 90% elimination of toluene. This was likely owing to the increased surface area of the switchgrass biochar. Iron calcined biochar [79] and nickel nanoparticle-embedded biochar [80] have also been shown to be efficient. Ren et al. [58] noted that the application of a biochar catalyst improved the quantity of syngas produced during biomass pyrolysis. At 480°C, it was discovered that the syngas output increased from 15 wt% to 46 wt% in the absence and presence of biochar catalyst respectively. According to Ren et al. [58, 81], the hydrogen content in syngas rose significantly with the addition of the biochar catalyst (27 vol%), in contrast to when the catalyst was not employed. A current investigation shows that biochar can be applied in the dry reforming process [82]. The dry reforming of CH_4 was carried out on a tungsten carbide [83] Based on a biochar (WC-biochar) catalyst. As the CH_4/CO_2 ratio rose, the CH_4 transformation reduced, while the CO_2 transformation improved. Increases in the CH_4/CO_2 ratio and temperature resulted in greater H_2 production, and the WC-biochar catalyst remained stable for 500 hours after being introduced into the system [82].

3.3.2 Tar elimination

The gasification of biomass is a viable sustainable energy pathway since it has the potential to enhance the generation of large quantities of syngas. A consequence of its synthesis, however, is the formation of condensable hydrocarbons (tar). Tars can accumulate in pipelines throughout a system, causing them to become clogged and potentially inhibiting downstream operations [84]. To commercialize biomass gasification for syngas generation, the elimination and/or mitigation of tar is a vital first stage in the procedure [85, 90]. In reality, catalytic tar cracking was carried out at 823–1173 K, with dolomite, olivine, and base metals including nickel [78], serving as catalysts. These conventional tar cracking catalysts, on the other hand, were susceptible to deactivation as a result of coking and contamination [91]. It has been attempted numerous times to degrade tars using a secondary reactor containing noble metal catalysts (e.g., platinum, palladium, and rhodium) [92], but the restoration of the catalyst has remained a difficult process. The introduction of an affordable catalyst for tar breakdown is therefore preferable in this situation. In this regard, biochar was found to be superior to traditional catalysts when used as a catalyst to remove tar [93]. The tar removal efficiency of biochar catalysts is summarized in **Figure 5**. The majority of investigations have relied on model processes of tar disintegration with toluene, naphthalene, and phenol. Moreover, the biochar-based metal catalysts (e.g., Nickel and Iron) outperformed the typical mineral catalysts in terms of tar removal

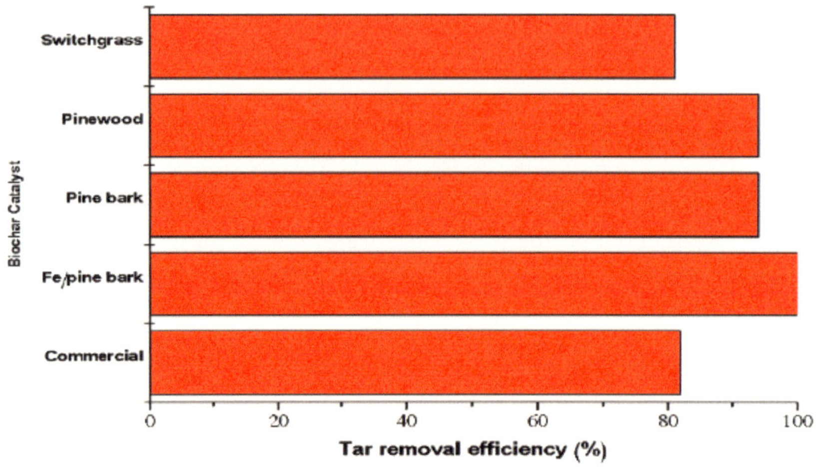

Figure 5.
Evaluation of tar elimination using biochar-based catalysts at 973–1173 K (adapted from Lee et al. [84] with modifications).

efficiency. For example, a catalyst constituted of a combination of NiO and wood-biochar eliminated 97% of the genuine tars formed during sawdust gasification, resulting in an improvement in syngas synthesis attributed to the catalytic reformation of the tars [94]. According to Shen et al. [79], bimetallic catalysts based on rice husk-biochar generated seven times fewer tars in the biomass combustion process than monometallic catalysts and raw biochars during the pyrolysis of biomass. The NiO-biochar catalyst combination remained stable for an 8-h time in the stream (TOS). One of the limitations linked to biochar and metal-biochar catalysts for tar reduction is the process temperature, as tar elimination occurs at >973 K. At reduced temperatures (i.e., 843 K) with the typical nickel catalyst, tar removal can be commenced [92], however, biochar is not yet efficient at these lower temperatures [95]. To overcome these restrictions and broaden the scope of biochar's application as a catalyst, future work must concentrate on overcoming these constraints.

3.4 Wastewater treatment

Due to its ability to remedy environmental pollutants, biochars are becoming highly significant for enhancing environmental quality in the world today [96]. Wastewater, which is a result of household, commercial, and agricultural operations, has long been a global concern since it affects everyone. Biochars offer a significant deal of promise for use in wastewater remediation applications. Biochar's applications in the cleanup of different wastewaters are the primary focus of this section.

3.4.1 Industrial wastewater remediation

Industrial wastewater originates from a variety of sources. In addition, heavy metals and organic contaminants are the most prevalent contaminants in industrial wastewater. It has been demonstrated that biochars can be used in the treatment of industrial effluent. It is possible to cast membranes, beads, and solutions from a biochar-chitosan combination that has been cross-linked. It has the potential to be used efficiently as an adsorbent for the adsorption of heavy metals in industrial wastewater. The amount of chitosan and biochar used in the adsorption of Cu, Pb, As, Cd and other heavy metals in industrial wastewater would depend on the ratio of the two materials [97]. Gliricidia biochar has shown promise in the elimination of crystal violet (CV) from aquatic environments in dye-based industries. A biochar's pH value, surface area, and pore volume are all important factors to consider throughout the CV sorption process [98]. Biochar made from bagasse was employed to absorb lead from the effluent of the battery production sector. The maximal adsorption ability can attain 13 mg/g, and the adsorptive activity is dependent on the moderate pH value, contact time, and concentration [99]. So far, the majority of the trials on the utilization of biochar in the clean-up of contaminants from industrial wastewater have been carried out in a laboratory environment; however, additional study and deployment in the actual situation are required.

3.4.2 Treatment of municipal wastewater

Biochar can be employed alone or in combination with other techniques for municipal wastewater treatment, resulting in the retrieval of labile nitrogen and phosphorus [100]. Engineered biochar containing aluminum oxyhydroxides (AlOOH) was used to recover and restore phosphorus from tertiary remediated wastewater [101]. The adsorption strategy of phosphorus is mostly based on electrostatic interaction. Phosphorus adsorbed on manufactured biochar has the potential to be used as a slow-release fertilizer for agricultural activities. Biochar generated from digested sludge was employed as an adsorbent for the elimination of NH_4 from municipal wastewater. Biochar produced at 723 K has the maximum NH_4 reduction capability due to its increased functional group density and surface area, and the procedure is governed by chemisorption [102]. This shows that biochar derived from waste sludge can be utilized to ozonate refinery effluent and achieve a significant reduction rate of total organic carbon (TOC) [103].

3.4.3 Wastewater treatment in the agricultural sector

Because of the rapid development of the agriculture sector, agricultural pollution is getting extremely serious. As a result, pesticides and toxic heavy metals are released into croplands in large quantities, the situation is becoming increasingly worrisome [104, 105]. The use of biochar and its modified forms in the remediation of agricultural wastewater pollution has been investigated. Pesticides such as atrazine and pentachlorophenol are two of the most often used in agriculture. Adsorption of atrazine and imidacloprid from agricultural wastewater by rice straw biochar and phosphoric acid-modified rice straw biochars is much higher than that of adjusted rice straw biochar [106]. Corn straw and soybean biochars both exhibit strong atrazine reduction potentials, with the adsorption efficiency owing mostly to the pH value and pore volume of the biochars [107]. Steam-activated biochar is efficient at eliminating sulfamethazine, and the rate at which it absorbs the substance is reliant on the pH value [108]. The presence of hazardous heavy metals in agricultural wastewater is yet another widespread issue.

4. Emerging advances in the applications of biochar catalyst

Recent advancements in the use of biochar for processes other than agriculture have been linked to biochar's various properties. Among other characteristics that are suitable for electrode materials, biochar has high porosity and high electrical conductivity [54]. It is preferred to use biochar with structurally bound nitrogen groups and high porosity as electrode materials for supercapacitors [55]. During catalysis, surface functionality, matrix nature, and intrinsic inorganic components are all important factors [49]. Unlike activated carbon derived from coal, biochar has a considerable amount of other organics present in it based on the biomass feedstock. These organics aid its compatibility, utilization, and effectiveness for varying applications than activated carbon.

There are several advantages to using biochar as a catalyst or catalyst support. Firstly, since biomass resources are sustainable and synthesis techniques have been developed, the process for producing biochar is simple and inexpensive. Secondly, the physicochemical properties of biochar can be easily tuned through a variety of methods. As a third consideration, biochar may be of interest in catalytic applications because of its surface functional groups, a hierarchical structure derived from the biomass matrix, and the presence of inorganic species [48]. Additionally, active metals and biochar support may, in some cases, have synergistic effects on catalysis [49].

4.1 Energy storage and conversion

Due to excess energy generation, energy storage is becoming more popular in some developed countries, and stored energy can also be used as a backup in the event of an emergency. The increased use of electric vehicles necessitates the continuous development of batteries with greater energy storage capacity. Despite continuous battery development, there are times when an unplanned situation may occur in electric vehicles. To alleviate such a situation, supercapacitors, which are energy storage devices primarily made of carbon materials, have been applied as continuous power sources in digital communications systems and electric vehicles. Because of its wide availability and low environmental impact, carbon materials with a high surface area and a rich porous structure are the primary raw materials for making super-capacitors [98]. It is crucial to the development of the supercapacitor industry to produce attractive, high-quality carbon materials at a reasonable price [99].

The utilization of biochar as material for supercapacitors has been tested by researchers with incredible results obtained. Biochar is made from paper cardboard and woody biomass. Based on the pyrolysis of woody biomass, the biochar supercapacitor electrodes exhibited a potential window of about 1.3 V, and fast charging-discharging behaviors with about 14 F/g gravimetric capacitance [100]. The authors also enhanced the performance of woody biochar by activating it with nitric acid. According to the researchers, the nitric acid treatment helped increased the capacitance from 14 to 115 F/g with 5000 usage cycles [100]. Likewise, Liu et al. [98] also created a high-performance supercapacitor out of biochar-derived carbon monolith, which was created by pyrolyzing poplar wood at 900°C for 6 h and then surface-modifying with nitric acid. The supercapacitor was discovered to have a highly consistent structure as well as a high porosity. The maximum specific capacitance was high (234 F/g) and cyclic stability was excellent [98, 99].

With the recent development of direct carbon fuel cell (DCFC) which converts carbonaceous material directly into electricity. The DCFC directly oxidizes solid carbon to produce electricity by using the chemical energy contained therein. Fuel utilization can reach nearly 100% if fuel feed and product gases are separated easily. The use of biochar as an energy source for this fuel cell has shown tremendous results. In a study by Kacprzak et al. [101], nine different carbonaceous fuels were tested, including commercial graphite, a carbon black, two commercial types of hard coal, and four biochars made by the authors, and one commercial biochar. At 0.5 V, commercial biochar had the second-highest current density (64.22 mA/cm^2) and the third-highest power density (32.8 mW/cm^2). Biochar produced in the laboratory had a high current density (36–44.6 mA/cm^2) and power density (18–22.4 mW/cm^2) [102].

4.2 Challenges and prospects of biochar-based catalyst applications

The use of biochar just as any other material has some limitations in its application for energy storage, conversion, and electrocatalyst. In terms of energy storage, the performance efficiency of tested biochar is still low when compared to its counterparts, though the biochar is easy to access and economical. Likewise, in the use of biochar in DCFC, it has been reported that upon consumption of the carbon content, the ash content present in biochar blocks the active surface area thereby impeding the effectiveness of the whole process [102]. In terms of reusability as a catalyst, further work still needs to be done as biochar from some feedstocks is reusable after the second attempt. For electrochemical oxidation of fuel, an ideal anode should have a large surface area, high porosity, and a continuous frame to ensure mechanical strength. Boosting the DCFC's power output and durability is therefore possible by improving its anode material [99].

Along with biochar's widespread use in wastewater remediation, scientists should consider its possible adverse impact on the ecosystem. To effectively employ biochar, one of the most significant features that must be considered is its capacity to maintain its stability throughout time. The aromaticity and extent of aromatic condensation of biochar are two factors that influence the stability of biochar [103]. When biochar is employed for wastewater detoxification, the possible emission of carbon from the biochar can cause the carbon concentration of the solution to be treated to rise. Moreover, the discharge of heavy metals from biochar formed from sludge is a possibility, particularly for biochar generated from sludge. Huang et al. [105] demonstrated that the dissolution of organic materials from biochar into an aqueous solution is caused by the biochar's instabilities. In addition, it was discovered that the stability of the biochar deteriorated after multiple cycles when it was employed as a support for a catalyst. This can be attributed to variations in the carbon framework of the biochar. It is usually acknowledged that the stability of biochar relies on the type of the starting feedstock as well as the experimental settings utilized during its thermal transformation. As a result, it is required to establish a relationship between these two factors and the stability of the biochar. Another significant element to consider is the renewal and restoration of biochar after it has been utilized. The adsorption procedure is characterized by the transition of pollution from the liquid stage to the solid material/adsorbent phase in most cases. As a result, it is critical to transforming the hazardous pollutants that are bonded to biochar into non-toxic conditions to control them effectively [101].

5. Challenges, prospects, and future perspectives

5.1 Challenges and prospects of effective application of biochar-based catalyst

The use of biochar-based catalysts can be beneficial in several catalytic processes, including biodiesel production, bio-oil up-gradation, reforming, and various organic reactions involving specialty or functional chemicals. These are currently in their infancy and must be scaled up. Biochar production systems must be set up on an industrial scale to enable the scaling up of these processes. The biggest barriers to scaling up biochar production are multiple competing end-users, as well as the collection and transportation of raw materials to the facilities that manufacture biochar. Homagain [107] studied the sensitivity of transportation distance and distinct carbon offset values and found that the system is financially viable at 200 km with good biomass availability. Furthermore, the seasonal biomass production cycle makes it difficult to maintain a steady supply of sustainable and reliable fuel.

The moisture content and particle size are other critical parameters in the synthesis of biochar. The biochar production method requires a lot of energy to process feedstocks with a high moisture content or large particle size. During biochar production, it is necessary to pre-process feedstock by drying and reducing its size. The heat resistance of feedstocks, on the other hand, limits heat transfer during biochar formation. Due to temperature differences, this phenomenon causes unconverted feedstock to accumulate on the inner walls of reactors, posing a significant barrier to the widespread production of uniform biochar [103].

Biochar's properties can also be difficult to fine-tune once it has been produced to achieve the required transformation. Following the proper design of biochar-based catalysts, the resulting materials will have real-world applications and will be able to replace catalysts that are expensive, non-renewable, and harmful to the environment. These conditions can be met by conducting mechanistic investigations during the char activation/synthesis/loading of necessary metals and catalytic processes. It is critical to comprehend two key factors in the catalytic process. The first is the interaction between biochar's physicochemical properties and its catalytic activity. The second step is to tune physicochemical parameters during the char production and activation process based on catalytic activity. Regarding this, the investigation of high surface area, active sites, and optimal pores is critical to managing the combined impacts of important production process variables (e.g., reagent gas, duration, heating rate, and temperature) and activation process variables (e.g., chemical, and physical). Just a few experiments have been conducted to control the physicochemical parameters of biochar for catalytic applications. However, the biorefinery of the future will require a single-step method for producing biochar with effective porous structure and functionality that is closely related to the production of biochemicals, biogas, and biofuels.

5.2 Future perspectives

Although biochar has many applications, biochar-based catalysts are still in the very early stages of development. Therefore, it is imperative to develop a method that can maximize catalytic activity. Researchers are currently exploring the modifications that can be carried out on biochar-based catalysts to apply them in future fields such as catalysis, environmental pollution, energy storage and conservation, and even chromatography.

Laboratory research is still underway for biochar-based catalysts. A purpose-driven synthesis and modification will be necessary for the future of an industrial application. Mechanistic studies may help to achieve this. A first step would be to investigate how biochar's catalytic properties relate to its physicochemical properties. To accomplish this, advanced characterization techniques of catalytic materials can be combined with theoretical modeling of the mechanisms involved. Second, it is critical to determine how biochar's properties are affected by synthesis conditions and feedstock. It is extremely difficult to work with biomass because of its complex composition and complex formation mechanism. The application of advanced characterization techniques, such as pyrolysis/gas chromatography/mass spectrometry (Py/GC/MS), and thermogravimetric analysis/Fourier-transform infrared spectroscopy/mass spectrometry (TGA/FTIR/MS), is potentially vital for the future.

In terms of process optimization, the role of catalysts in biochar synthesis must be given much more thought. The presence of some inorganic species in biomass feedstock can catalyze pyrolysis. However, their autocatalysis is not enough to ignite the process. A catalyst must achieve at least one of the following goals: (1) to reduce reaction temperature or residence time so that biochar can be produced more efficiently; (2) to make biochar with desirable properties in a single step instead of having modification and synthesis done separately. In the future, we may be able to produce biochar-supported catalysts directly from biomass using catalysts that can produce effective functional groups and porous structures in a single step. A biomass refinery would also be able to produce biofuels and biochemicals in close coordination with manufacturing biochar-based catalysts, allowing for a more integrated and environmentally sustainable process for using biomass.

Biochars intended for use as catalysts require a functionalization and/or activation process because of their limited porosity, surface area, and surface functional groups. According to the activation technique, biochar can have varying physicochemical properties, such as surface area or porosity. Activated biochar can be endowed with specialized properties via the addition of functional groups or substances, such as selectivity, catalysis, and selective adsorption. Although biochars vary significantly according to the type of biomass they are produced from, as well as their production conditions and functionalization or activation. Future research should focus on the production of biochar with stable properties on an industrial scale.

6. Conclusion

The use of biochar-based catalysts in environmental applications has excellent catalytic properties. Recent achievements of biochar catalyst preparation procedures, as well as their performance, were examined from a range of applications. Additionally, the catalytic properties of biochar were examined further by its production and activation methods. Through various chemical and/or physical treatments, biochar can be modified in terms of morphology and surface functionality. Therefore, biochar has a strong potential for replacing costly and non-renewable conventional catalysts.

It has been demonstrated that biochar-derived catalysts are effective in a variety of reactions, including the production of biodiesel from biomass, removal of tars from bio-oil and syngas, and production of syngas. However, biochar catalyst properties (including surface functionality, surface area, porosity, and acidity) vary widely with biomass origin, biochar synthesis conditions, and pre/post-treatment. Yet, there is limited information about how biochar's properties can be controlled to enable its catalytic applications. Therefore, further research is needed to develop the catalytic properties of biochar to design active, stable, and selective biochar catalysts. Also, if biochar is to be considered as an industrial heterogeneous catalyst, the development of a method that allows for the manufacture of biochar on an industrial scale is extremely desirable. For large-scale production, it is also challenging to secure stable sources of raw biochar materials. To meet these challenges, biochar catalysts must be stimulated and facilitated to be used in real-world applications to replace costly, non-environmentally benign catalysts, which have been used for a wide range of applications until now.

Author details

Stephen Okiemute Akpasi[1*], Ifeanyi Michael Smarte Anekwe[2], Jeremiah Adedeji[3] and Sammy Lewis Kiambi[1]

1 Department of Chemical Engineering, Durban University of Technology, Durban, South Africa

2 School of Chemical and Metallurgical Engineering, University of the Witwatersrand, Johannesburg, South Africa

3 Department of Chemical Engineering, School of Engineering, University of KwaZulu-Natal, Durban, South Africa

*Address all correspondence to: stephenakpasi48@gmail.com

References

[1] Lam E, Luong JH. Carbon materials as catalyst supports and catalysts in the transformation of biomass to fuels and chemicals. ACS Catalysis. 2014;**4**(10):3393-3410

[2] Czernik S, Bridgwater A. Overview of applications of biomass fast pyrolysis oil. Energy & Fuels. 2004;**18**(2):590-598

[3] Laird DA, Brown RC, Amonette JE, Lehmann J. Review of the pyrolysis platform for coproducing bio-oil and biochar. Biofuels, Bioprod Biorefining. 2009;**3**:547-562

[4] Manyà JJ, Manya JJ. Pyrolysis for biochar purposes: A review to establish current knowledge gaps and research needs. Environmental Science & Technology. 2012;**46**:7939-7954

[5] Liu WJ, Jiang H, Yu H-Q. Development of biochar based functional materials: Toward a sustainable platform carbon material

[6] Ok YS, Chang SK, Gao B, Chung HJ. Smart biochar technology. A shifting paradigm towards advanced materials and health care research. Environmental Technology and Innovation. 2015

[7] Xu Y, Fang Z. Advances on remediation of heavy metal in the soil by biochar. Environmental Engineering. 2015;**33**(2):156-159

[8] Yuan P et al. Review of biochar for the management of contaminated soil: Preparation, application and prospect. Science of the Total Environment. 2019;**659**:473-490

[9] Qi F et al. Effects of acidic and neutral biochars on properties and cadmium retention of soils. Chemosphere. 2017;**180**:564-573

[10] Li J et al. A comparison of biochars from lignin, cellulose and wood as the sorbent to an aromatic pollutant. Journal of Hazardous Materials. 2014;**280**:450-457

[11] Windeatt JH et al. Characteristics of biochars from crop residues: Potential for carbon sequestration and soil amendment. Journal of Environmental Management. 2014;**146**:189-197

[12] Lee JW et al. Characterization of biochars produced from cornstovers for soil amendment. Environmental Science & Technology. 2010;**44**(20):7970-7974

[13] Wang H, Feng L, Chen Y. Advances in biochar production from wastes and its applications. Chemical Industry and Engineering Progress. 2012;**31**(4):907-914

[14] Glaser B, Lehmann J, Zech W. Ameliorating physical and chemical properties of highly weathered soils in the tropics with charcoal—A review. Biology and Fertility of Soils. 2002;**35**(4):219-230

[15] Chen B, Zhou D, Zhu L. Transitional adsorption and partition of nonpolar and polar aromatic contaminants by biochars of pine needles with different pyrolytic temperatures. Environmental Science & Technology. 2008;**42**(14):5137-5143

[16] Yang H et al. Biomass-based pyrolytic polygeneration system for bamboo industry waste: Evolution of the char structure and the pyrolysis mechanism. Energy & Fuels. 2016;**30**(8):6430-6439

[17] Liu D et al. A green technology for the preparation of high capacitance rice husk-based activated carbon. Journal of Cleaner Production. 2016;**112**:1190-1198

[18] Pandian K et al. Effect of biochar amendment on soil physical, chemical and biological properties and groundnut yield in rainfed Alfisol of semi-arid tropics. Archives of Agronomy and Soil Science. 2016;**62**(9):1293-1310

[19] Huang H et al. Effects of pyrolysis temperature, feedstock type and compaction on water retention of biochar amended soil. Scientific Reports. 2021;**11**(1):1-19

[20] Razzaghi F, Obour PB, Arthur E. Does biochar improve soil water retention? A systematic review and meta-analysis. Geoderma. 2020;**361**:114055

[21] Kim KH et al. Influence of pyrolysis temperature on physicochemical properties of biochar obtained from the fast pyrolysis of pitch pine (*Pinus rigida*). Bioresource Technology. 2012;**118**:158-162

[22] Chen W-H et al. Thermal pretreatment of wood (Lauan) block by torrefaction and its influence on the properties of the biomass. Energy. 2011;**36**(5):3012-3021

[23] Boateng A, Mullen C. Fast pyrolysis of biomass thermally pretreated by torrefaction. Journal of Analytical and Applied Pyrolysis. 2013;**100**:95-102

[24] Chen W-H et al. Product characteristics from the torrefaction of oil palm fiber pellets in inert and oxidative atmospheres. Bioresource Technology. 2016;**199**:367-374

[25] Yang Y et al. Intermediate pyrolysis of biomass energy pellets for producing sustainable liquid, gaseous and solid fuels. Bioresource Technology. 2014;**169**:794-799

[26] Carrier M et al. Production of char from vacuum pyrolysis of South-African sugar cane bagasse and its characterization as activated carbon and biochar. Journal of Analytical and Applied Pyrolysis. 2012;**96**:24-32

[27] Chellappan S et al. Experimental validation of biochar based green Bronsted acid catalysts for simultaneous esterification and transesterification in biodiesel production. Bioresource Technology Reports. 2018;**2**:38-44

[28] Lee Y et al. Comparison of biochar properties from biomass residues produced by slow pyrolysis at 500 C. Bioresource Technology. 2013;**148**:196-201

[29] Waqas M et al. Development of biochar as fuel and catalyst in energy recovery technologies. Journal of Cleaner Production. 2018;**188**:477-488

[30] Mohan D, Pittman CU Jr, Steele PH. Pyrolysis of wood/biomass for bio-oil: A critical review. Energy & Fuels. 2006;**20**(3):848-889

[31] Bridgwater A. The production of biofuels and renewable chemicals by fast pyrolysis of biomass. International Journal of Global Energy Issues. 2007;**27**(2):160-203

[32] Bruun EW et al. Effects of slow and fast pyrolysis biochar on soil C and N turnover dynamics. Soil Biology and Biochemistry. 2012;**46**:73-79

[33] Bruun EW et al. Influence of fast pyrolysis temperature on biochar labile fraction and short-term carbon loss in a loamy soil. Biomass and Bioenergy. 2011;**35**(3):1182-1189

[34] Hansen V et al. Gasification biochar as a valuable by-product for carbon sequestration and soil amendment. Biomass and Bioenergy. 2015;**72**:300-308

[35] Yang Y et al. Intermediate pyrolysis of organic fraction of municipal solid waste and rheological study of the pyrolysis oil for potential use as bio-bitumen. Journal of Cleaner Production. 2018;**187**:390-399

[36] Perez M, Perez R. Update 2022—A fundamental look at supply side energy reserves for the planet. Solar Energy Advances. 2022:100014

[37] Lee M-S, Park S-J. Silica-coated multi-walled carbon nanotubes impregnated with polyethyleneimine for carbon dioxide capture under the flue gas condition. Journal of Solid State Chemistry. 2015;**226**:17-23

[38] Tripathi M, Sahu JN, Ganesan P. Effect of process parameters on production of biochar from biomass waste through pyrolysis: A review. Renewable and Sustainable Energy Reviews. 2016;**55**:467-481

[39] Sun Y et al. Effects of feedstock type, production method, and pyrolysis temperature on biochar and hydrochar properties. Chemical Engineering Journal. 2014;**240**:574-578

[40] Peterson SC, Jackson MA. Simplifying pyrolysis: Using gasification to produce corn Stover and wheat straw biochar for sorptive and horticultural media. Industrial Crops and Products. 2014;**53**:228-235

[41] You S et al. A critical review on sustainable biochar system through gasification: Energy and environmental applications. Bioresource Technology. 2017;**246**:242-253

[42] Mumme J et al. Hydrothermal carbonization of anaerobically digested maize silage. Bioresource Technology. 2011;**102**(19):9255-9260

[43] Kambo HS, Dutta A. A comparative review of biochar and hydrochar in terms of production, physico-chemical properties and applications. Renewable and Sustainable Energy Reviews. 2015;**45**:359-378

[44] Libra JA et al. Hydrothermal carbonization of biomass residuals: A comparative review of the chemistry, processes and applications of wet and dry pyrolysis. Biofuels. 2011;**2**(1):71-106

[45] Gomez-Eyles JL et al. Effects of biochar and the earthworm Eisenia fetida on the bioavailability of polycyclic aromatic hydrocarbons and potentially toxic elements. Environmental Pollution. 2011;**159**(2):616-622

[46] Gan L et al. Cellulose derived carbon nanofiber: A promising biochar support to enhance the catalytic performance of $CoFe_2O_4$ in activating peroxymonosulfate for recycled dimethyl phthalate degradation. Science of the Total Environment. 2019;**694**:133705

[47] Hervy M et al. Multi-scale characterisation of chars mineral species for tar cracking. Fuel. 2017;**189**:88-97

[48] Xiong X et al. A review of biochar-based catalysts for chemical synthesis, biofuel production, and pollution control. Bioresource Technology. 2017;**246**:254-270

[49] Jitjamnong J et al. Response surface optimization of biodiesel synthesis over a novel biochar-based heterogeneous catalyst from cultivated (*Musa sapientum*) banana peels. Biomass Conversion and Biorefinery. 2021;**11**(6):2795-2811

[50] Dehkhoda AM, Ellis N. Biochar-based catalyst for simultaneous reactions of esterification and transesterification. Catalysis Today. 2013;**207**:86-92

[51] Guo F et al. Evaluation of the catalytic performance of different activated biochar catalysts for removal of tar from biomass pyrolysis. Fuel. 2019;**258**:116204

[52] Wang S et al. Reactivity and deactivation mechanisms of toluene reforming over waste peat char-supported Fe/Ni/Ca catalyst. Fuel. 2020;**271**:117517

[53] Chellappan S et al. Synthesis, optimization and characterization of biochar-based catalyst from sawdust for simultaneous esterification and transesterification. Chinese Journal of Chemical Engineering. 2018;**26**(12):2654-2663

[54] Zhao C et al. Biodiesel synthesis over biochar-based catalyst from biomass waste pomelo peel. Energy Conversion and Management. 2018;**160**:477-485

[55] Lawrinenko M et al. Macroporous carbon supported zerovalent iron for remediation of trichloroethylene. ACS Sustainable Chemistry & Engineering. 2017;**5**(2):1586-1593

[56] Chen L et al. Biochar modification significantly promotes the activity of Co_3O_4 towards heterogeneous activation of peroxymonosulfate. Chemical Engineering Journal. 2018;**354**:856-865

[57] Cha JS et al. Production and utilization of biochar: A review. Journal of Industrial and Engineering Chemistry. 2016;**40**:1-15

[58] Ren S et al. Hydrocarbon and hydrogen-rich syngas production by biomass catalytic pyrolysis and bio-oil upgrading over biochar catalysts. RSC Advances. 2014;**4**(21):10731-10737

[59] Liu S, Wang M, Sun X, Xu N, Liu J, Wang Y, et al. Facilitated oxygen chemisorption in heteroatom-doped carbon for improved oxygen reaction activity in all-solid-state zinc-air batteries. Advanced Materials. 2018;**30**(4):1704898

[60] Kołtowski M et al. Effect of biochar activation by different methods on toxicity of soil contaminated by industrial activity. Ecotoxicology and Environmental Safety. 2017;**136**:119-125

[61] Yang X et al. Characterization and ecotoxicological investigation of biochar produced via slow pyrolysis: Effect of feedstock composition and pyrolysis conditions. Journal of Hazardous Materials. 2019;**365**:178-185

[62] Lima IM, Boateng AA, Klasson KT. Physicochemical and adsorptive properties of fast-pyrolysis bio-chars and their steam activated counterparts. Journal of Chemical Technology & Biotechnology. 2010;**85**(11):1515-1521

[63] Kitano M et al. Adsorption-enhanced hydrolysis of β-1, 4-glucan on graphene-based amorphous carbon bearing SO3H, COOH, and OH groups. Langmuir. 2009;**25**(9):5068-5075

[64] Angın D, Altintig E, Köse TE. Influence of process parameters on the surface and chemical properties of activated carbon obtained from biochar by chemical activation. Bioresource Technology. 2013;**148**:542-549

[65] Dehkhoda AM, Ellis N, Gyenge E. Electrosorption on activated biochar: Effect of thermo-chemical activation treatment on the electric double layer capacitance. Journal of Applied Electrochemistry. 2014;**44**(1):141-157

[66] Um B-H, Kim Y-S. A chance for Korea to advance algal-biodiesel technology. Journal of Industrial and Engineering Chemistry. 2009;**15**(1):1-7

[67] Bazargan A et al. A calcium oxide-based catalyst derived from palm kernel shell gasification residues for biodiesel production. Fuel. 2015;**150**:519-525

[68] Li M et al. Biodiesel production from waste cooking oil using a heterogeneous catalyst from pyrolyzed rice husk. Bioresource Technology. 2014;**154**:345-348

[69] Kostić MD et al. Optimization and kinetics of sunflower oil methanolysis catalyzed by calcium oxide-based catalyst derived from palm kernel shell biochar. Fuel. 2016;**163**:304-313

[70] Yu JT, Dehkhoda AM, Ellis N. Development of biochar-based catalyst for transesterification of canola oil. Energy & Fuels. 2011;**25**(1):337-344

[71] Kastner JR, Miller J, Geller DP, Locklin J, Keith LH, Johnson T. Catalytic esterification of fatty acids using solid acid catalysts generated from biochar and activated carbon. Catalysis Today. 2012;**190**(1):122-132

[72] Lou W-Y, Zong M-H, Duan Z-Q. Efficient production of biodiesel from high free fatty acid-containing waste oils using various carbohydrate-derived solid acid catalysts. Bioresource Technology. 2008;**99**(18):8752-8758

[73] Ramachandran K, Sivakumar P, Suganya T, Renganathan S. Production of biodiesel from mixed waste vegetable oil using an aluminium hydrogen sulphate as a heterogeneous acid catalyst. Bioresource Technology. 2011;**102**(15):7289-7293

[74] Luque R, Clark JH. Biodiesel-like biofuels from simultaneous transesterification/esterification of waste oils with a biomass-derived solid acid catalyst. ChemCatChem. 2011;**3**(3):594-597

[75] Ormsby R, Kastner JR, Miller J. Hemicellulose hydrolysis using solid acid catalysts generated from biochar. Catalysis Today. 2012;**190**(1):89-97

[76] Li S et al. Biochar based solid acid catalyst hydrolyze biomass. Journal of Environmental Chemical Engineering. 2013;**1**(4):1174-1181

[77] Anekwe IMS, Khotseng L, Isa YM. The place of biofuel in sustainable living; prospects and challenges. Journal: Comprehensive Renewable Energy. 2022:226-258

[78] Asadullah M et al. Biomass gasification to hydrogen and syngas at low temperature: Novel catalytic system using fluidized-bed reactor. Journal of Catalysis. 2002;**208**(2):255-259

[79] Shen Y et al. In situ catalytic conversion of tar using rice husk char/ash supported nickel–iron catalysts for biomass pyrolytic gasification combined with the mixing-simulation in fluidized-bed gasifier. Applied Energy. 2015;**160**:808-819

[80] Wang D, Yuan W, Ji W. Char and char-supported nickel catalysts for secondary syngas cleanup and conditioning. Applied Energy. 2011;**88**(5):1656-1663

[81] Shen Y et al. In-situ catalytic conversion of tar using rice husk char-supported nickel-iron catalysts for biomass pyrolysis/gasification. Applied Catalysis B: Environmental. 2014;**152**:140-151

[82] Mani S, Kastner JR, Juneja A. Catalytic decomposition of toluene

using a biomass derived catalyst. Fuel Processing Technology. 2013;**114**:118-125

[83] Wang B, Gao B, Fang J. Recent advances in engineered biochar productions and applications. Critical Reviews in Environmental Science and Technology. 2017;**47**(22):2158-2207

[84] Lee J, Kim K-H, Kwon EE. Biochar as a catalyst. Renewable and Sustainable Energy Reviews. 2017;**77**:70-79

[85] Wathukarage A et al. Mechanistic understanding of crystal violet dye sorption by woody biochar: Implications for wastewater treatment. Environmental Geochemistry and Health. 2019;**41**(4):1647-1661

[86] Xiong X et al. Sulfonated biochar as acid catalyst for sugar hydrolysis and dehydration. Catalysis Today. 2018;**314**:52-61

[87] do Couto Fraga A et al. Biomass derived solid acids as effective hydrolysis catalysts. Journal of Molecular Catalysis A: Chemical. 2016;**422**:248-257

[88] Wei Y et al. Study on reaction mechanism of superior bamboo biochar catalyst production by molten alkali carbonates pyrolysis and its application for cellulose hydrolysis. Science of the Total Environment. 2020;**712**:136435

[89] Zhang C et al. Chlorocuprate ionic liquid functionalized biochar sulfonic acid as an efficiently biomimetic catalyst for direct hydrolysis of bamboo under microwave irradiation. Industrial & Engineering Chemistry Research. 2013;**52**(33):11537-11543

[90] Shan R, Han J, Gu J, Yuan H, Luo B, Chen Y. A review of recent developments in catalytic applications of biochar-based materials. Resources, Conservation and Recycling. 2020;**162**:105036

[91] Cole AJ et al. Good for sewage treatment and good for agriculture: Algal based compost and biochar. Journal of Environmental Management. 2017;**200**:105-113

[92] Zheng Y et al. Reclaiming phosphorus from secondary treated municipal wastewater with engineered biochar. Chemical Engineering Journal. 2019;**362**:460-468

[93] Tang Y et al. Influence of pyrolysis temperature on production of digested sludge biochar and its application for ammonium removal from municipal wastewater. Journal of Cleaner Production. 2019;**209**:927-936

[94] Chen C et al. Activated petroleum waste sludge biochar for efficient catalytic ozonation of refinery wastewater. Science of the Total Environment. 2019;**651**:2631-2640

[95] Mandal A, Singh N. Optimization of atrazine and imidacloprid removal from water using biochars: Designing single or multi-staged batch adsorption systems. International Journal of Hygiene and Environmental Health. 2017;**220**(3):637-645

[96] Zhao X et al. Properties comparison of biochars from corn straw with different pretreatment and sorption behaviour of atrazine. Bioresource Technology. 2013;**147**:338-344

[97] Rajapaksha AU et al. Enhanced sulfamethazine removal by steam-activated invasive plant-derived biochar. Journal of Hazardous Materials. 2015;**290**:43-50

[98] Liu M-C et al. Porous wood carbon monolith for high-performance supercapacitors. Electrochimica Acta. 2012;**60**:443-448

[99] Qian K et al. Recent advances in utilization of biochar. Renewable and Sustainable Energy Reviews. 2015;**42**:1055-1064

[100] Jiang J et al. Highly ordered macroporous woody biochar with ultra-high carbon content as supercapacitor electrodes. Electrochimica Acta. 2013;**113**:481-489

[101] Kacprzak A et al. The effect of fuel type on the performance of a direct carbon fuel cell with molten alkaline electrolyte. Journal of Power Sources. 2014;**255**:179-186

[102] Munnings C et al. Biomass to power conversion in a direct carbon fuel cell. International Journal of Hydrogen Energy. 2014;**39**(23):12377-12385

[103] Wiedemeier DB et al. Aromaticity and degree of aromatic condensation of char. Organic Geochemistry. 2015;**78**:135-143

[104] Wei D et al. Biochar-based functional materials in the purification of agricultural wastewater: Fabrication, application and future research needs. Chemosphere. 2018;**197**:165-180

[105] Huang M et al. Application potential of biochar in environment: Insight from degradation of biochar-derived DOM and complexation of DOM with heavy metals. Science of the Total Environment. 2019;**646**:220-228

[106] Li X et al. Effect of cassava waste biochar on sorption and release behavior of atrazine in soil. Science of the Total Environment. 2018;**644**:1617-1624

[107] Homagain K et al. Life cycle cost and economic assessment of biochar-based bioenergy production and biochar land application in Northwestern Ontario, Canada. Forest Ecosystems. 2016;**3**(1):1-10

[108] Mašek O et al. Consistency of biochar properties over time and production scales: A characterization of standard materials. Journal of Analytical and Applied Pyrolysis. 2018;**132**:200-210

Section 4

Biochar Unveiled: Advanced Investigation

Chapter 14

Biochar and Application of Machine Learning: A Review

Kingsley Ukoba and Tien-Chien Jen

Abstract

This study discusses biochar and machine learning application. Concept of biochar, machine learning and different machine learning algorithms used for predicting adsorption onto biochar were examined. Pyrolysis is used to produce biochar from organic materials. Agricultural wastes are burnt in regulated conditions to produce charcoal-like biochar using pyrolysis. Biochar plays a major role in removing heavy metals. Biochar is eco-friendly, inexpensive and effective. Increasing interest in biochar is due to stable carbon skeleton because of ease of sourcing the precursor feedstock and peculiar physicochemical. However, artificial intelligence is a process of training computers to mimic and perform duties human. Artificial intelligence aims to enable computers to solve human challenges and task like humans. A branch of artificial intelligence that teaches machine to perform and predict task using previous data is known as machine learning. It uses parameters called algorithms that convert previous data (input) to forecast new solution. Algorithms that have been used in biochar applications are examined. It was discovered that neural networks, eXtreme Gradient Boosting algorithm and random forest for constructing and evaluating the predictive models of adsorption onto biochar have all been used for biochar application. Machine learning prevents waste, reduces time and reduces cost. It also permits an interdisciplinary means of removing heavy metals.

Keywords: review, machine learning, biochar, AI, adsorption

1. Introduction

The world is embracing the fourth industrial revolution and adapting technology in every sphere of human endeavours. 4IR is adjusting ways humans engage, work and live [1]. Its ushers humanity into a new phase caused by incredible technological advancements comparable to the first, second and third industrial revolutions. Machine learning has been deployed simply in different aspects of human lives to living and cost [2, 3]. It is gaining interest in biochar. Biochar is a produced using pyrolysis. Forestry and agricultural wastes are burnt in regulated conditions to produce biochar [2, 3]. This study examines the various algorithms used in machine learning to predict adsorption in biochar.

Fourth Industrial Revolution will alter patterns of key sectors. This includes technological shift, deviation in societal patterns and processes caused by increased interconnection among other features [4]. It hopes to transform the ways things are done. Things will communicate via networks, data sharing and the likes. It is an era that will see machines perform tasks more than before. The machines will learn using previously generated data and transform those learning to solve human challenges. This is all-encompassing, including in biochar.

Biomass conversion without oxygen produces a solid product (biochar) [5–7]. Stability of biochar is responsible for carbon sequestration [8]. It could be a way to combat climate change [9, 10]. Biochar improves soil fertility. It increases agricultural yield in acidic soils [11, 12]. Biochar is made from various organic waste feedstocks, including agricultural waste and sewage sludge [13, 14]. Biochar has many applications, including heat and power generation and a soil amendment. Process parameters and feedstock influence the characteristics of carbonised biomass. Selection of acceptable conditions to manufacture a char with the necessary qualities thus necessitates quantitative and qualitative knowledge of interdependence and affecting factors [15].

In machine learning, input is a set of instructions (algorithms) used to generate result. It learns from previous data to perform and optimise operations. Attempts have been made to adapt machine learning in biochar [16, 17].

There have been attempts to implement machine learning in various aspects of biochar [18], review machine learning [19, 20] and review biochar [21]. However, there is limited literature focusing on the review of machine learning in biochar. This forms the basis of this study. The concept of biochar is examined and, after that, machine learning. This is closely followed by examining biochar and machine learning.

2. Biochar: history, properties and applications

2.1 History of biochar

The term 'biochar' is a late-twentieth-century English neologism. It is from a Greek words 'o, bios' or 'life' and 'char' or 'clarification' (charcoal produced by carbonisation of biomass) [22]. It is charcoal, prevalent in soil, aquatic ecosystems and animal digestive systems and participates in biological processes. Biochar usage for soil nutrient retention and improvement started in the Brazilian Amazon about 2000 years ago [23]. John Miedema, a commercial fisherman, organic farmer and inventor, first learned about biochar 5 years ago while looking for a better solution to clean up effluent from a dairy manure digester [24]. Biochar was made by pre-Columbian Amazonians by covering burning biomass with soil in ditches [25]. Terra preta de Indio was the name given to it by European settlers [26].

2.2 Production of biochar

Biochar is made by heating biomass without oxygen, either completely or partially [27, 28]. The most common process for making biochar is pyrolysis, which can also be found in the early stages of gasification ad combustion [29]. Biochar is made from different biomass sources, including solid wastes, plant materials, biomass from wood, agricultural residues and so on [30, 31]. Pyrolysis is a typical technique to produce biochar. The process is performed between 400 and 1000°C [32, 33]. Pyrolysis, hydrothermal carbonisation, gasification, flascarbonisation and torrefaction are some of the most prevalent thermochemical processes used to make biochar [34–36]. Pyrolysis is the most common biochar production method of all of these [37]. The process is depicted in Eq. (1).

$$C_6H_{8.7}O_4 \rightarrow C_3H_{1.4}O_{0.4} \tag{1}$$

$$\text{Biomass} \rightarrow \text{Biochar}$$

Biochar is created the same way as charcoal, but it is meant to be used as an adsorbent and a soil amendment [38]. The end use of the material is, in essence, the key. If it is meant to be used as a fuel, it is called charcoal, and it is made with the best fuel qualities possible.

2.3 Properties of biochar

Biochar's efficacy as a soil amendment is influenced by its chemical and physical qualities. As biochar interacts with bacteria, mineral substances and soil organic and plant roots, its characteristics alter. The biochar qualities affect its performance as a soil amendment.

Biochar comes in various forms, each with its own set of characteristics. Biochar's qualities impact how well it works as a soil amendment [39]. It can be altered by conditioning, which includes adding minerals, nutrients and/or microorganisms to the biochar after it has been made [40]. Biochar from clean biomass differs from biochar produced with field residue in terms of the qualities. This is because the field residue biochar has been mixed with fertilisers, soil and manure. The characteristics of biochar are altered when it is mixed with soil organic, mineral substances and bacteria. Biochar improves with age.

Biochar properties are influenced by the type of biomass used [41]. As long as the biomass is not polluted with hazardous compounds, it can be used to make biochar (e.g. heavy metals, PCBs). Biochar feedstocks include plant residues, grasses, industrial wastes, woods, seaweed, manures, MSW, food waste [42]. **Figure 1a** shows the pyrolysis of seaweed to produce a biochar. **Figure 1b** shows the evolution of biochar from biomass.

The properties of biochar are grouped under chemical and physical [45] in **Table 1**.

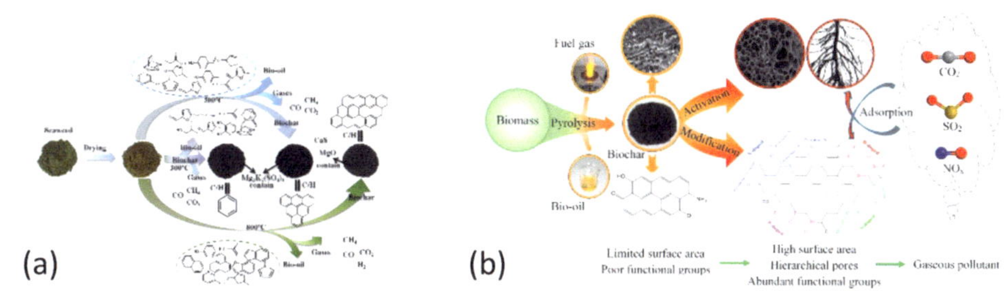

Figure 1.
(a) Process of seaweed pyrolysis to biochar [43] and (b) biomass to biochar [44].

Properties	Parameters
Physical	Particle size, bulk density, hydrophobicity, water holding capacity, macro and micro-porosity, particle density and grindability
Chemical	Electrical conductivity, micro and macro-nutrient content, toxic compounds, soluble organic compounds, cation and anion exchange capacity, heavy metals, proton activity and liming value

Table 1.
Summary of biochar properties.

2.3.1 Physical properties of biochar

Biochar's physical features influence its environmental mobility, interactions with minerals, soil water, nutrients and usefulness as an ecological niche for soil microorganisms and mycorrhizal fungus by soil microorganisms mycorrhizal fungus providing surfaces, growing space and predator protection [46]. Physical parameters such as particle density and size, porosity, bulk density and surface area are numerical and action connected. Porosity affects particle density and surface area [47]. Biochar with high porosity and low density may hold more water. However, wind and water easily remove such biochar. The quality of biochar is affected by heating rate, biomass type [48] as enumerated in **Figure 2**.

Grass biochar has a particle density of 0.25–0.3 g/cm^3, while wood biochar has 0.47–0.6 g/cm^3 [49]. Particle density of biochar affects the loss and movement in water or wind [50]. Biochar with a low bulk density can be used to remediate wall gardens and compacted soils. Pore sizes can vary by six orders of magnitude and are classed as macro-, meso- and micro-pores, with varied implications for biochar interactions with the environment [51, 52]. Most woody biochar has low bulk densities, medium-to-high surface area and porosity [53, 54]. The process utilised to make biochar has an impact on porosity.

Hydrophobicity impacts biochar's water uptake, its water holding capacity and microbial interactions. Tars (aliphatic chemicals) condensing on the charcoal surface during pyrolysis induce hydrophobicity. Biochar has high hydrophobic at low temperatures. However, longer pyrolysis times can lessen hydrophobicity. Hydrophobicity may diminish as biochar mixes with soil.

A low Hardgrove Grindability Index (HGI) indicates that the material is difficult to grind, whereas a high HGI value suggests that the material is easy to grind [55, 56]. HGI of 80–120 can be achieved for woody biochar having volatile matter content of about 20%, which is commonly achieved at temperatures around 600°C, defining charcoal as easily grindable.

2.3.2 Chemical properties of biochar

Persistent carbon is composed of carbon ring structures, with some nitrogen and oxygen thrown in. Structures' ring sizes are determined by temperature of biochar production. Biochars' water-soluble and mineralisable chemicals can nourish bacteria and can boost seeds and plant nutrient and yield. Water-extractable organics are substantially more abundant in low-temperature biochars. Total and bioavailable polycyclic aromatic hydrocarbons (PAH) have maximum acceptable limits. A common (90%) PAH in biochar is naphthalene. Many biochars at 350–500°C have included mineralisable organic molecules that benefit plants and soil [57, 58]. Low dosages of

Figure 2.
Factors affecting biochar quality.

phenols, butenolide (a component of tobacco), carboxylic and fatty acids and even PAH can encourage plant development. In contrast, high quantities can inhibit or kill it, a phenomenon known as hormesis.

2.4 Merit and demerit of biochar

Biochar continues to attract interest owing to its vast potential and benefit. However, there are some disadvantages associated with it. Discussed below are the merit and demerit of biochar.

2.4.1 Merit of biochar

Biochar is a carbon-rich substance, some scientists believe that it is the secret to soil renewal [59]. Biochar, which is relatively light and porous, can act as a sponge and provide a home for various beneficial soil microbes useful for soil and plant health. It increases agricultural production. Biochar can remove CO_2 from the atmosphere for long periods and provide other environmental benefits [60]. Plants transform carbon dioxide from the air into organic material, or biomass, through photosynthesis. It helps in climate change mitigation [10].

2.4.2 Demerit of biochar

It absorbs nutrients, resulting in a nutrient deficit in growing plants [47, 61]. Biochar application regularly creates soil compaction, which reduces crop yield. Land loss is also due to erosion, pollution risk, agricultural residue removal and worm life rate reduction.

2.5 Application of biochar

Biochar is useful in several applications [62]. It is used to enhance soil health via soil amendment. It also serves as microbial carrier immobilising agents for remediation of toxic metal and organic contaminant in water and soil. It is catalyst for industrial application, porous materials for mitigating greenhouse gas emission and odorous compound. It is used as feed supplements to improve nutrient intake efficiency, animal health and hence productivity [63]. **Figure 3** shows the influence of biochar properties on the agriculture and soil conditions.

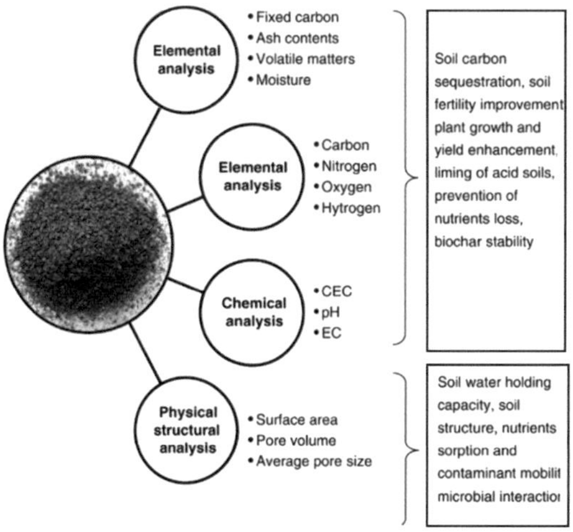

Figure 3.
Impact of biochar properties on soil conditions and agriculture [48].

2.5.1 Biochar for soil amendment

Biochar has a lot of potential as a long-term product for improving agricultural soil health and fertility. The manufacture of biochar and its impact on soils can help to reduce the need for commercial fertilisers. Diverse research has also reported that addition of biochar to agricultural soil can aid in reducing greenhouse gas emission [64–67].

Biochar is utilised as an agricultural soil amendment because it has a lot of fascinating properties, such as high carbon content, a high pH, high stability, a high porosity and a high surface area [68, 69]. Over the last few years, multiple research studies have been conducted to analyse the global impact of biochar on diverse agricultural soils [70, 71]. Biochar has improved soil's chemical, physical and biological qualities, enhancing crop productivity [72, 73]. Furthermore, biochars with a high surface can be utilised as soil remediation technique to adsorb both inorganic and organic contaminants, for instance, heavy metals, and pesticides, hence minimising leaching into waterway. Once applied to carbon in biochar, soils, that are highly stable, can be sequestered for more than 1000 years.

2.5.1.1 Application of biochar for soil amendment

When utilised as soil amendments, biochar is incorporated into the plant's root zone – the area of soil surrounding a plant's roots – ideally into 4–6 inches of soil depth. Increasing the time nutrients stay in the soil by mixing up to one part compost with one part biochar, most gardeners start with a ratio of 10 parts compost to one part biochar to ensure that plants tolerate it well.

Several materials such as green waste [74], rice straw [75], poultry litter [76] and other materials have been used for producing biochar using vacuum pyrolysed and other methods of biochar for soil amendments [77].

2.5.2 Carbon sink

A carbon sink is any natural or artificial reservoir that indefinitely gathers and stores carbon-containing chemical compounds [78]. Also, anything that absorbs more carbon from the atmosphere than it releases, such as plants, the ocean and soil, is a carbon sink. Oceans are the primary natural carbon sinks, absorbing over half of all carbon released [79]. Carbon dioxide is sucked from the atmosphere by plants for use in photosynthesis. On the other hand, a carbon source is anything that releases more carbon into the atmosphere than it absorbs, such as fossil fuel combustion or volcanic eruptions [80]. Carbon is deposited on our planet in four major sinks: (1) organic molecules in living and dead organisms in the biosphere; (2) carbon dioxide in the atmosphere; (3) organic matter in soils; and (4) fossil fuels and sedimentary rock deposits such as limestone and dolomite in the lithosphere. Because the process takes a supposedly carbon-neutral phase of naturally decaying, biochar reduces CO_2 in the environment.

Growing plants or collecting waste biomass, converting it to biochar and adding it to soils remove carbon dioxide (CO_2) from the environment: plants growth eliminates CO_2 from the atmosphere and produces additional biomass; the carbon in that biomass is transformed into a stable form [81, 82]. Biochar production can offset about 12% of world's greenhouse gas emissions. At $30–120 per ton of CO_2, biochar might sequester 0.5–2 $GtCO_2$ per year by 2050 [83, 84]. According to the scholarly literature, sequestration rates range from 1 to 35 $GtCO_2$ each year, with a potential of 78–477 $GtCO_2$ in this century [85, 86].

2.5.3 Biochar for water retention

Water retention refers to how much water a soil can keep for its crops, allowing plants to have more water available. Biochar can improve the soil's water retention and holding ability due to its porous structure. An agriculturally applicable biochar amendment of 5% biochar (approximately 100 metric tons/ha) leads to a 24% increase in water retention capacity over unamended soil or a 50% increase [87]. Researchers have understudied the impact of biochar on water retention [88], on sandy soil [89], clay [90], the application in different agricultural soil [91] and the relationship between plant and water [92]. There has also been the study of southeastern coastal soil [93] and midwestern agricultural soil [94].

2.5.4 Biochar for stock fodder

Stock fodder, also known as provender, is an agricultural feed used to feed domesticated animals such as cattle, rabbits, sheep and horses [95]. Fodder crops are divided into two categories: temporary and permanent. Fodder is used to describe the crops gathered and utilised for stall feeding. Forage is a vegetative matter used as animal feed, whether fresh or stored. Grasses, legumes, crucifers and other forage crops are farmed and utilised as hay, grazing, fodder and silage.

Xie et al. [96] provided a thorough investigation of biochar's technical features and possible applications as an engineered material for environmental remediation. Mandal et al. [97] presented quantitative data and discussed the benefits of biochar composites over pure biochar. The synthesis of nano-metal-aided biochar and its features and applications in soil improvement and heavy metal removal are discussed. Shakoor et al. [98] discuss how to boost biochar's heavy metal sorption capability by activating it with steam or acids/bases and impregnating biochar-based composite with mineral, organic compound and carbon-rich material. Biochars' chemical/physical activation of biochar can improve their surface area, resulting in better functionality, while pretreatment/modification techniques aid in developing new sorbent with efficient surface attribute for heavy metal removal from aqueous solution using biochar as a supporting media. This is essential because heavy metal sorption is driven by type of biochar, heavy metal species and various processes, including physical binding, complexation, ion exchange, surface precipitation and electrostatic interactions. Efforts were also made to review the application of biochar to remove heavy metals and toxic elements in water and wastewater [99, 100].

2.6 Future outlook of biochar

Wood-based biochar is the most popular product, accounting for approximately 64% of the market. Soil conditioner is the most popular application, accounting for almost 82% of the market.

The global biochar market is expected to be worth USD 314.6 million in 2022, with a readjusted size of USD 524.7 million by 2028, representing an 8.9% CAGR (compound annual growth rate) over the research period. From 2021 to 2030, the global biochar markets are expected to increase at a CAGR of 13.2%, from $170.9 million in 2020 to $587.7 million in 2030. Carbon Gold, The Biochar Company (TBC), Biochar Supreme, Cool Planet, Black Carbon and Swiss Biochar GmbH, among others, are global biochar significant players. The top three firms account for roughly 20% of the market [101].

3. Machine learning: history, algorithm and application

Machine learning (ML) is a process of predicting values using a previous learning. It is a subset of AI. It uses set of instructions called algorithm. ML uses algorithm to emulate variable or humanity. AI is used to solve complex tasks like how humans solve problems. There are four types of algorithms. They are reinforcement, unsupervised, semi-supervised and supervised. Python, Java, C++, R and JavaScript are among the top five programming languages and libraries for machine learning. Python is the language of choice for machine learning engineers, with more than 60% of them adopting and prioritising it for development since it is simple to learn. A little coding knowledge is required for the effective deployment of machine learning.

3.1 History of machine learning

An American IBMer (Arthur Samuel) was first to use machine learning in 1959 [102, 103]. Another term used is 'self-teaching computer' [104, 105]. A book on machine learning for pattern categorisation by Nilsson dominated the1960s [106]. Pattern recognition continued till the 1970s [107]. An approach for teaching neural network using 40 character recognition by computer terminal was documented in 1981 [108, 109]. This terminal included 4 special symbols, 26 letters and 10 digits. Tom Mitchell opined 'A computer program is said to learn from experience E for some class of tasks T and performance measure P if its performance at tasks in T, as measured by P, improves with experience E'. This became accepted machine learning definition [110, 111]. However, the definition provided operational description of the ML tasks instead of cognitive. It aligns with Alan Turing's method 'Computing Machinery and Intelligence', replacing 'Can machine think' with 'Can machines do what we (as thinking creatures) can achieve' [112].

The goal of modern ML is to classify data using standard models and generate predictions about future outcomes using these models. A stock trading machine learning system may provide the trader with future prospective predictions [113, 114].

3.2 Theory of machine learning

Most beginners' main goal is to generalise what they have learned [115]. Generalisation is ML ability to execute precisely, previously unseen data using algorithm. Data (training) originate from new probability distribution. It represents space of occurrences. Optimisation prediction requires general model development. Computational learning theory is analysis of performance of algorithms. Training sets are limited because of future uncertainty. Learning theory rarely provides guarantees about algorithm performance. Probabilistic performance bounds are tremendously widespread. Bias-variance decomposition is used for generalisation error.

For the best generalisation outcomes, the hypothesis' complexity needs reflect the intricacy of the functions behind the data. If the assumption is fewer intricate than the functions, the system will under-fit the data. Increment in the complexity of the model reduces training error. Poor generalisation due to overfitting is caused by complicated hypothesis of model [116]. Learning theorists look at the temporal intricacy and feasibility of learning in addition to performance bounds. A computation is deemed viable in computational learning theory if it can be completed in polynomial time [117].

3.3 Classification of machine learning approach

ML is classified as reinforcement, unsupervised and supervised based on feedback or signal as depicted in **Figure 4** [118, 119].

Optimisation problem is solved using reinforced and unsupervised learning [118–120]. Although, supervised learning uses trained labelled data to produce result [121, 122]. Unsupervised learning uses unguided structure to solve problem [123]. Unsupervised learning is either intended or a means to an end (finding hidden patterns in data) (feature learning). It is used to obtain hidden pattern or future learning. Reinforcement learning is the third type. It is interaction in a dynamic circumstance. An example is driving on the road on the computer. Another example is engaging an opponent in competitive game [124]. Incentives (data) are fed to the software to help solve problem.

Unsupervised learning exposes latent patterns and structures from unlabelled data. Supervised learning solves problem using guided learning [125]. **Figure 5** depicts the most often used supervised algorithms.

Deep learning is used to clean heavy metal by constructing improved adsorption models. Machine learning or deep learning can develop models depending on data complexity, dimensionality and end use [127]. However, challenges of complexity and dimensionality are improved by deep learning with encoder.

3.4 Models of machine learning

Machine learning entails building a model that has been guided by training data. It can subsequently process more data to produce prediction. For machine learning systems, different models have been utilised and investigated. These are shown in **Figure 6**. The models include artificial neural networks, decision trees,

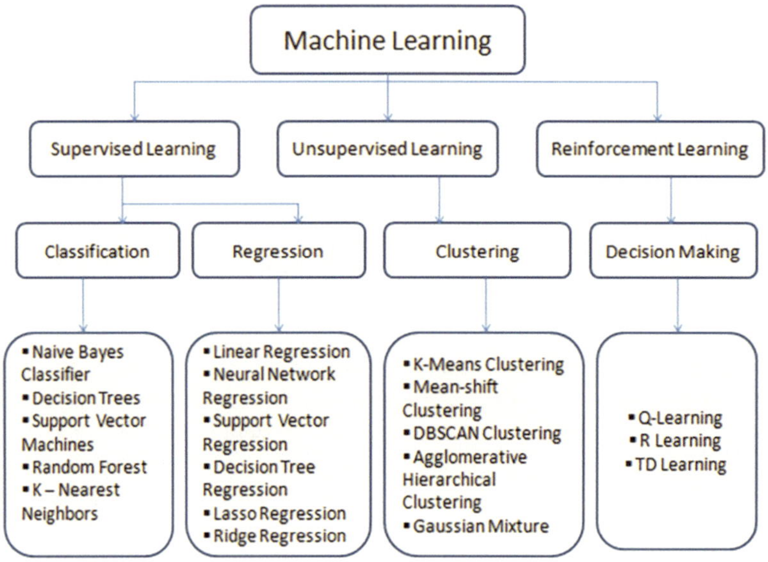

Figure 4.
Classification of machine learning.

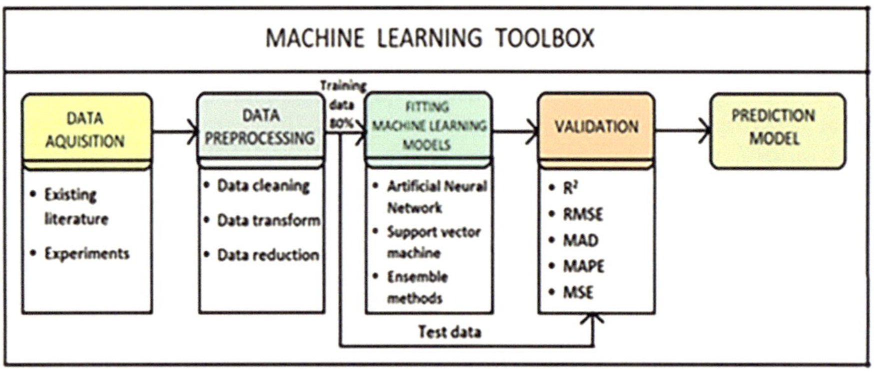

Figure 5.
Flowchart of supervised machine learning procedure [126].

support-vector machines, regression analysis, genetic algorithms, Bayesian networks, training models and federated learning [129–131].

The following models have been used in biochar applications. An overview is given for understanding the models.

i. Artificial Neural Networks (ANN) have become increasingly popular [132, 133]. ANN mimics the human brain with parallel processing to develop complex relationship between independent and dependent variables by developing structures for the model training via experimental data and the tool forming pattern between output and input data. It is a great tool because of its benefits in non-linear system adaptations and approximation without knowing the variables' relationship and ease of use [134].

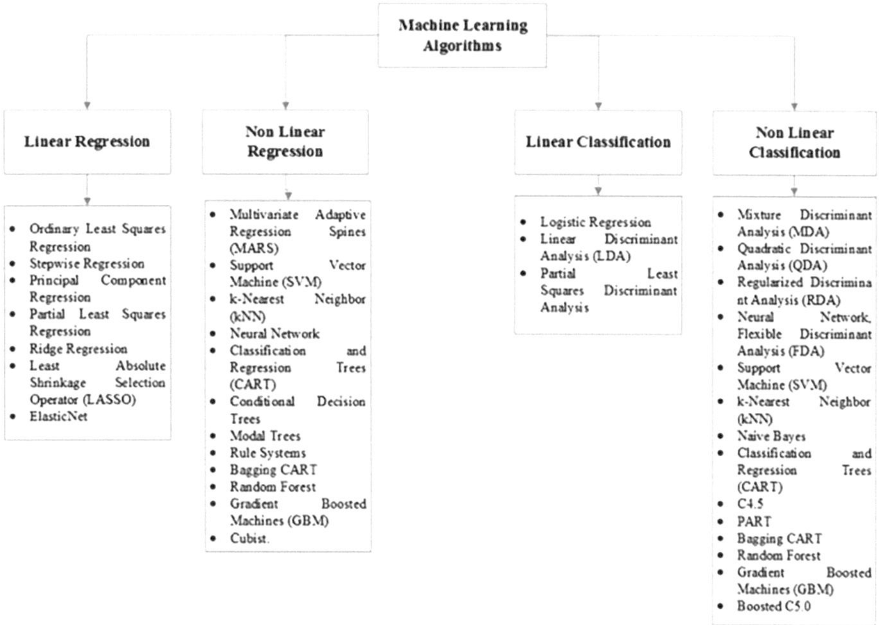

Figure 6.
Algorithms of machine learning [128].

ii. Random forest (RF) models are machine learning models that use the results of a series of regression decision trees to predict the output. Each tree is built independently and is based on a random vector sampled from the input data, with the same distribution across the forest. Using bootstrap aggregation and random feature selection, the predictions from the forests are averaged [135]. RF models are reliable predictors for small sample numbers and high-dimensional data. The RF classifier is an ensemble approach for training several decision trees parallel with bootstrapping and aggregation, often known as bagging [136].

iii. Support-vector machine

A support-vector machine (SVM) is a supervised machine learning model that uses classification techniques [137]. SVM models can categorise new text after being given sets of labelled training data for each category. Though we might also argue regression difficulties, categorisation is the best fit. The SVM algorithm aims to find the optimum line or decision boundary for categorising n-dimensional space into classes so that additional data points can be readily placed in the correct category in the future [138, 139]. A hyperplane is a name for the optimal choice boundary. The goal of the SVM algorithm is to find a hyperplane in an N-dimensional space that categorises data points. In SVM, a kernel is a function that aids in problem-solving. They give shortcuts to help avoid doing complicated mathematics. The amazing thing about kernel is that it allows us to go to higher dimensions and execute smooth calculations. Kernels allow us to go up to an infinite number of dimensions. SVM is used for regression and classification of problems. It is a linear model. It can solve both linear and nonlinear problems and is useful for a wide range of applications.

C is a hypermeter that is set before the training model to control error, and Gamma is another hypermeter that is placed before the training model to give the decision boundary curvature weight.

iv. eXtreme Gradient Boosting Model

Gradient boosting is a machine learning technique used for various applications, including regression and classification [140, 141]. Extreme Gradient Boosting (XGBoost) is an open-source package that implements the gradient boosting technique efficiently and effectively. Extreme Gradient Boosting is a tree-based method that belongs to Machine Learning's supervised branch. It's a machine-learning algorithm that can predict classification or regression. It returns a prediction model in the form of an ensemble of weak prediction models, most commonly decision trees [142].

3.5 Applications of machine learning

The following are some machine learning applications. Image and speech recognition, traffic prediction, self-driving cars, product recommendation, online fraud detection, stock market trading, medical diagnosis, automatic language translation, email spam and malware filtering, Alexa, Google assistant and Google Maps [119].

3.5.1 Image recognition

Image recognition is one of the most common machine learning applications [143]. It's utilised in identifying things such as people, places and digital photograph. Automatic buddy tag suggestion is a commonly used facial identification and picture recognition. Facebook has tools that suggest friends auto-tagging. When we submit photos with our friends Facebook, we obtain automatic tags recommended with their names powered by machine learning's face identification and algorithm recognition. It is based on the 'Deep Facia' Facebook projects that manage face recognition and individual identification in photos.

3.5.2 Speech recognition

The user of Google has the option to 'Search by voice', which falls under recognition of speech and is a prominent machine learning application. Recognition of speech, frequently referred to as 'Computer speech recognition' or 'Speech to text', is the turning process of voice instruction to text. Machine learning technique is now used widely in speech recognition application [144]. Technology of speech recognition is utilised by Alexa, Google Assistant, Siri, and Cortana to obey voice command.

3.5.3 Google Maps is used when visiting a new location or using an app hailing taxi

The map provides the best route with the shortest routes and forecasts traffic condition. It utilises two techniques in anticipating traffic condition, such as whether traffic is clear, extremely congested or sluggish moving: The vehicle's location is tracked in real time via the Google Map app and sensor. At the same time, the average time has been taken on previous days. Everyone making use of Google Maps contributes to the improvement of the apps. It collects data from the users and transmits it back to the database to improve its performance.

3.5.4 Product suggestions

Different entertainment and e-commerce organisations, for instance, Netflix, Amazon and others use machine learning to make products recommendation to user. We begin to receive advertisements for the same goods while browsing the internet on the same browser, because of machine learning, whenever we look for a product on Amazon [145]. Google deduces the user's interests and recommends products based on those interests using multiple machine learning techniques. Likewise, when we use Netflix, we receive recommendations for series of entertainment, movies and other contents, which is also based on machine learning.

3.5.5 Self-driving automobiles

Self-driving cars are one of the most intriguing machine learning applications [146]. In self-driving automobile, machine learning plays key roles. Tesla, the well-known automobile manufacturer, is developing self-driving vehicles. It trains automobile model to recognise people and object while driving using an unsupervised learning method.

3.5.6 Medical diagnosis

In medical science, machine learning is used to diagnose disorders [147, 148]. Therefore, medical technology is evolving rapidly, and 3D model that can predict the exact lesions location in the brain is now possible. It facilitates the brain cancers detection and other brain-related illness.

3.5.7 Automatic language translation

Machine learning aids in translation by transforming text into familiar language. This feature is provided by Google Neural Machine Translation (Google's GNMT), a Neural Machine Learning that translates text into native language automatically. Sequence-to-sequence learning methods are the technology behind automatic translation, coupled with translation of text from one language to another and picture recognition.

3.6 Limitations of machine learning

Machine learning has proved transformative in several domains, yet it frequently fails to produce the promised outcomes [149]. There are various reasons for this, including a lack of (appropriate) data, data access issues, data bias, privacy issues, poorly designed tasks and algorithms, incorrect tools and personnel, a lack of resources and evaluation issues [150]. In 2018, an Uber self-driving car failed to identify a person, and the pedestrian (Elaine Herzberg) was killed due to the incident [151, 152]. Even after years of effort and billions of dollars, IBM Watson's attempts to employ machine learning in healthcare failed to deliver [153]. Machine learning has been utilised in updating evidence concerning systematic reviews and increased reviewer concerns due to the biomedical literature development. When students 'learn the wrong lesson', they can be disappointed. An image classifier trained just on photographs of brown horses and black cats, for example, may conclude that all brown patches are most likely horses [106]. In the real world, unlike people, existing image classifiers frequently do not make decisions based on the spatial relationships between picture component and instead study associations between pixels that human is unaware of but correlates with specific sorts of image of real object. Modifying this pattern on lawful images can cause the algorithm to misclassify the image as 'adversarial' non-linear systems, or non-pattern disturbances can potentially lead to adversarial vulnerabilities. Several systems are so fragile that single change hostile pixel causes misclassification.

3.7 Ethics of machine learning

Machine ethics (also known as machine morality, computational morality or computational ethics) is a branch of artificial intelligence ethics concerned with enhancing or ensuring the moral behaviour of man-made machines that employ artificial intelligence, also known as artificial intelligent agents [154, 155]. Privacy and surveillance, bias and discrimination and perhaps the deepest, most difficult philosophical question of the era, the role of human judgement, are three major ethical concerns for society, according to Sandel, who teaches a course on the moral, social and political implications of new technologies [156, 157].

3.8 Hardware of machine learning

More effective techniques in training deep neural network (machine learning specific subdomain) that incorporate various non-linear hidden unit layers have been developed since the 2010s, thanks to developments in computer technology and machine learning algorithms [158]. By 2019, GPUs had supplanted CPUs as the most common way of training large-scale commercial cloud AI, frequently with AI-specific upgrades [159]. From AlexNet (2012) to AlphaZero (2017), OpenAI calculated the amount of hardware computing required in large deep learning project and discovered 300,000-fold increase in the required computing amount, with 3.4-month doubling-time trendline [160].

There are embedded machine learning and neuromorphic or physical neural networks.

3.8.1 A physical neural network

A physical neural network also known as a neuromorphic computer, is an artificial neural network in which an electrical changeable substance emulates the neural synapse function. The term 'physical' neural network refers to physical hardware to simulate neurons rather than software-based techniques. Other artificial neural networks that use memristor or other electrical adjustable resistance materials to imitate neural synapse are also known as memristor networks [161, 162].

3.8.2 Embedded machine learning

Embedded Machine Learning is a sub-field of machine learning that uses embedded system with low computing capabilities, for instance, microcontrollers, wearable computers and edge devices to run machine learning models. Running machine learning models in embedded device eliminates the necessity to transport and store data on cloud server for processing further, resulting in fewer data breach and privacy leak and less theft of intellectual property, personal data and company trading secrets.

Embedded Machine Learning can be implemented using various methods, including hardware acceleration, approximation computation and machine learning model optimisation [163].

3.9 Software of machine learning

Different software suites having various algorithms have been used for machine learning. Some are free and open-source, and others are proprietary. The open-source and free software includes Caffe, ELKI, Deeplearning4j, Microsoft Cognitive Toolkit and DeepSpeed. However, KNIME and RapidMiner are the most popular open-source proprietary software [164], alongside R tool and Weka [165]. R tool is free and used for environmental statistics. RapidMiner is a complete data science platform focusing on delivering business value [166]. It brings together data preparation, machine learning and model operations to boost users' productivity of all skill levels within an organisation. The Konstanz Information Miner (KNIME) is a free and open-source platform for data analyses, reporting and integration [167]. Through its modular data pipelining 'Building Blocks of Analytics' concept, KNIME integrates multiple components for machine learning and data mining. The paid proprietary includes Angoss Knowledge STUDIO, Ayasdi, Amazon Machine Learning, IBM Watson Studio, Azure Machine Learning, IBM SPSS Modeler, Google Prediction API, Mathematica, KXEN Modeler, STATISTICA Data Miner, LIONsolver, Oracle Data Mining, MATLAB, Oracle AI Platform Cloud Service, Neural Designer, NeuroSolutions, SAS Enterprise Miner, Splunk, SequenceL, PolyAnalyst and RCASE.

4. Machine learning and biochar: past and the future

4.1 Classification of machine learning algorithms

For new users, selecting 'which algorithm to study' can be tough. Machine learning algorithms have their own set of advantages and disadvantages. Some excel with textual data, others excel at visuals and others at other data types. Many characteristics, such as resemblance, behaviour, data kinds and others, can be used to classify machine learning algorithms [168, 169].

Linear Regression, Logistic Regression, Decision Tree, SVM (Support Vector Machine) Algorithm, Naive Bayes Algorithm, KNN (K-Nearest Neighbours) Algorithm, K-Means and Random Forest Algorithm are some of the most used machine learning algorithms [170–172] as shown in **Figure** 7.

4.2 Machine learning algorithms used in biochar

Some selected works have been done using machine learning in biochar optimisation, which is dependent on the design of experiments for identifying pyrolysis parameters and optimising processes, which are all influenced by interconnected elements. The literature optimisation is separated into two categories: production and use. The optimisation procedure maximises the biochar's adsorption capacity and effectiveness for environmental and water remediation by antibiotics, extracting heavy metals and other contaminants from industrial effluent [174]. The three most significant process parameters in biochar manufacture are the heating temperature, heating time and heating rate [175]. The gaseous environment and particle size

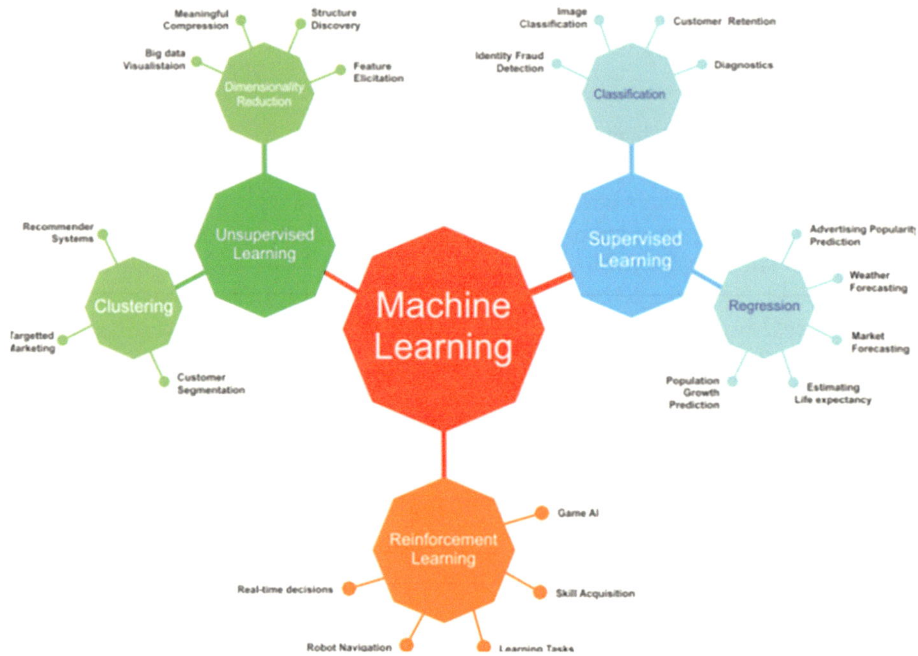

Figure 7.
Classification of machine learning algorithms [173].

employed in the biochar production variable such as the moisture contents, presence of inorganic/organic elements that catalyse certain reaction were included as feedstock factors for optimisation.

4.2.1 Yield prediction via machine learning

The algal biochar yield was predicted via extreme gradient algorithms. The XGB (eXtreme Gradient Boosting) machine-learning algorithm was used for prediction of algal biochar composition and yield in this study. In the XGB model, an intensive grid search strategy was designed to evaluate all of the available input parameter combination for forecasting biochar yield. Thirteen distinct pyrolytically significant input parameters combination were compared with the combination indicated by the model's techniques selection feature to predict biochar yield. The ash content, N/C, pyrolysis temperature, H/C and duration are essential parameters in determining the algal biochar output in this feature selection technique, where N, H and C are the nitrogen, hydrogen and carbon biomass content, respectively. Once the model was trained with the training data set, the highest R^2 of 0.84 was attained between model predictive and experimental biochar yield for the data set test. A Pearson correlation coefficients matrix showed the link between the biochar yield and input parameters. The Feature Temperature was the most significant element in plots. The interactive influence of other input parameter and temperature on algal charcoal output was represented using Shapley Additive exPlanations (SHAP) Dependence Plot. The plots' summary revealed the relevant features combined with SHAP and feature values.

The created XGB model adds to our understanding of the input parameter impact on algal biochar yield prediction.

Zhu et al.'s [176] machine learning was utilised in this study to construct prediction models for yield and carbon content of biochar (C-char) based on pyrolysis data of lignocellulosic biomass and investigate the inner information underlying the models. Based on biomass properties and pyrolysis circumstances, the results revealed that random forests could reliably forecast biochar output and C-char. Furthermore, for both yield (65%) and C-char, the proportional contribution of pyrolysis conditions was higher than that of biomass characteristics (53%). Structural information was more significant than element compositions for biomass characteristics for effectively estimating biochar yield, and the opposite was true for C-char. In the pyrolysis process, the partial dependence plot analysis revealed the impact of each important component on the target variable and the interactions between these elements. The study added the biomass pyrolysis process knowledge and improved biochar yield and C-char quality.

Sun et al. [177] studied the application of machine learning methods to predict metal immobilisation remediation by biochar amendment in soil. The work began by compiling and categorising data from published literature to develop a biochar soil remediation database, which now contains 930 data sets with 74 biochars and 43 soils. Then, based on biochar characteristics, soil physicochemical properties, incubation conditions (e.g. water holding capacity and remediation time) and the initial state of heavy metals, it modelled the remediation of five heavy metals and metalloids (lead, cadmium, arsenic, copper and zinc) by biochars using machine learning (ML) methods such as artificial neural network (ANN) and random forest (RF) to predict remediation efficiency. The ANN and RF models surpass the accuracy and predictive performance of the linear model ($R^2 > 0.84$). Meanwhile, the anticipated outputs of the models investigated model tolerance for missing data and interpolation reliability. Both the ANN and the RF models performed admirably, with the RF model having a higher tolerance for missing data. Finally, the contribution of factors employed in the model was assessed using ML models' interpretability. And the findings revealed that the type of heavy metals, the pH value of biochar and the dosage and remediation period were the most influential elements of remediation. The relative importance of variables could point researchers on the proper path for better heavy metal cleanup in soil.

Cao et al. [178] employed SVM (support-vector machine) approach for estimation of the biochar output from cattle dung pyrolysis in their study. The parameters employed for modelling were moisture content, pyrolysis temperature, biochar yield, biochar mass, sample mass and heating rate, and they were based on a data set of 33 experimental data. The following metrics were used to assess the performance: Magnitudes of root mean square error (RMSE), average percent relative error (APRE), average absolute percent relative error (AAPRE) and coefficient of discrimination (R^2). To compare the resilience and properties of SVM, an ANN model was created. Surprisingly, SVM outperformed ANN with an R^2 score of 0.9625, whilst ANN's R^2 value was 0.8040.

Li et al. [179] compiled information from prior studies to create a predictive model for biochar qualities depending on feedstock and pyrolysis settings. Though significant biochar properties such as pH, yield, specific surface area, cation exchange capacity, volatile matter content, ash content and elemental compositions are affected by different factors, there is strong link between biochar properties, feedstock type and pyrolysis temperature.

4.2.2 Distributing heavy metal via machine learning

Heavy metal testing using traditional spectral approaches is time-consuming and impossible to detect for huge amounts of effluent. Based on remote sensing imagery, geographical data and spatial distribution, machine learning algorithm may be utilised to forecast effluents metal distribution. RF, SVM and ANN have been used for this.

RF and ANN machine learning algorithm were utilised in predicting the heavy metals concentration present in soil using visible and infrared spectroscopy data [180]. Also, Zhang et al. [181] used geographical distribution data, and the concentrations of Cd, As, Cu, Zn, Pb, Cr, Hg and Ni in the soil were predicted via SVM, RF and ANN algorithms. Hu et al. [182] utilised RF to find the regulating factors in heavy metal bioaccumulation in soil-crop systems. ANN is a simple method for determining the link between the heavy metal pollutants removal and process parameter [133, 183].

4.2.3 Pyrolysis parameter

In recent literature, ANN has been primarily utilised to optimise pyrolysis parameters, but techniques such as the Taguchi approach have also been applied. This application creates orthogonal matrices using a basic statistical tool to conceptualise an integrated experimental design to discover crucial factors in an optimised operation [175]. For effective optimisation, ANN is employed in conjunction with other technologies. In Lakshmi et al. [126], a unique approach is described that combines several types of ANN in conjunction with techniques such as particles swarm optimisation to almost always guarantee global optimum without local minimum trapping. Particles swarm optimisation is novel, efficient, rapid, robust and simple when tackling non-linear, multi-variable problems. Razzaghi et al. [183] employ genetic algorithms to optimise the generated ANN, resulting in process parameter values.

4.2.4 Metal remediation and machine learning

Machine learning could be useful in developing predictive models for heavy metals cleanup utilising modified biochar. ML models are useful in the adsorption process because of their ability to analyse intricate correlations between factors [184]. ML models are an effective modelling tool in the adsorption process because of their capacity to improve analysed relationships among numerous parameters [185]. The performance of adsorption is affected by operational parameters such as heating rate, temperature, dosage, adsorbent surface area, particle size, starting concentration, pH and contact time value. Taking all of this into account, constructing adsorption models is time-consuming and takes a lot of experimentation. To avoid this tedium, ML can be used to create robust models in evaluating the heavy metals adsorption process [186–189].

Wong et al. [184] examined the operational parameters effect such as dosage, contact time, operating temperature and biochar initial concentration on the process of adsorption using rambutan peel biochar to remove Cu(II) from water body. They used AI models such as Multi-Layer Regression, ANN and ANFIS to study the impact of the above-mentioned operational parameters (MLR). Adaptive neuro-fuzzy inference system (ANFIS) is a Neuro-Fuzzy intelligent modelling and control technique for ill-defined and unpredictable systems. The system's input/output data pairs under examination form the basis of ANFIS. The ANFIS model was the most accurate, with 90.24% score, followed by 88.27% ANN and 59.14% MLR. For Pb(II) adsorption on ethylenediaminetetraacetic acid (EDTA) treated biochar, Li et al. [190] constructed an AI model utilising the SVM algorithm.

Nath and Sahu [155] employed iron oxides infused mesoporous rice-husk nano-biochar in removing arsenic. Using ANN and RSM methodologies, they obtained a removal efficiency of 96%. Six AI models was developed by Afridi [173] with different architectures network in ANN for prediction of heavy metal adsorptions on modified biochar. The six models were effective, with R^2 values greater than 0.99 between predicted and expected variables. Chakraborty and Das [191] developed an ANN model to estimate Cr (VI) absorption efficiency on sawdust biochar nano-composite. The ANN model assisted them in determining an appropriate adsorption mechanisms and the most excellent feasible Cr (VI) equations for absorption on biochar modified.

Zhao et al. [192] demonstrated a new method to establish sensitive parameter impacting the process of adsorption and develop strong predictive model using AI. For prediction of the efficiency of six metal ions adsorption, the authors used kernel extreme learning machine, with SVM and Kriging model subset. These models accurately identified sensitive parameters, such as T, pH water, ionic radius, total carbon ratio and pH solute, with R^2 above 0.9, and could provide the necessary framework for developing predictive models for various scenarios.

Zhu et al. [193] investigated the application of machine learning methods to predict metal sorption onto biochars. The study used 353 data sets of adsorption studies from works of literature, the adsorption of six heavy metals (lead, cadmium, nickel, arsenic, copper and zinc) on 44 biochars was predicted using artificial neural networks (ANNs) and random forests (RF). The regression models were trained and refined to estimate adsorption capacity based on biochar properties, metal sources, environmental factors (temperature and pH) and the initial metal-to-biochar concentration ratio. The study discovered that RF model was more accurate than the ANN model.

Machine learning may be used to forecast and automate the remediation process and optimise process variables and feedstock conditions for optimal heavy metal removal efficiency. Machine learning may be utilised to create kinetic models and hybrid isotherm, which will accurately model for multicomponent systems and reduce error making the removal of heavy metal more cost-effective and efficient time.

5. Conclusions

The study was able to draw a relationship between biochar and machine learning. A review of biochar from history to application and challenges was discussed. Remediation of heavy metal is critical to avoid bioaccumulation, soil degradation and environmental contamination. Biochar is a practical and inexpensive method for removing heavy metal from waste effluent. Various approaches can improve the removal heavy metals effectively from pristine biochar. The paper also gave an overview of machine learning. Various algorithms of machine learning were discussed. After that, selected algorithms used for biochar were reviewed, and areas of opportunities were discussed. Artificial neural networks, support-vector models and random forests have been deployed in the machine learning of biochar. The ANN and RF models surpass the accuracy and predictive performance of the linear model. It was seen that random forest models perform better than artificial neural network models for predicting and generalisation. Machine learning will lead to a greater understanding of biochar's effectiveness and applications in more sectors.

Acknowledgements

The authors acknowledge the funding from URC of the University of Johannesburg and the National Research Foundation of South Africa.

Conflict of interest

The authors declare that there is no conflict of interest.

Author details

Kingsley Ukoba* and Tien-Chien Jen
University of Johannesburg, Johannesburg, South Africa

*Address all correspondence to: ukobaking@yahoo.com

References

[1] Skilton M, Hovsepian F. The 4th Industrial Revolution. Springer; 2018

[2] Granatstein D et al. Use of Biochar from the Pyrolysis of Waste Organic Material as a Soil Amendment. Washington State University; 2009

[3] Kim Y et al. Modification of biochar properties using CO_2. Chemical Engineering Journal. 2019;**372**:383-389

[4] Morrar R, Arman H, Mousa S. The fourth industrial revolution (Industry 4.0): A social innovation perspective. Technology Innovation Management Review. 2017;**7**(11):12-20

[5] Ramola S, Belwal T, Srivastava RK. Thermochemical conversion of biomass waste-based biochar for environment remediation. In: Handbook of Nanomaterials and Nanocomposites for Energy and Environmental Applications. 2020. pp. 1-16

[6] Parvez AM et al. Utilization of CO2 in thermochemical conversion of biomass for enhanced product properties: A review. Journal of CO2 Utilization. 2020;**40**:101217

[7] Imoisili PE et al. Production and characterization of hybrid briquette from biomass. British Journal of Applied Science & Technology. 2014;**4**(10):1534

[8] Kofi A-K. Design, Construction and Testing of a Biocharring Unit. Kwame Nkrumah University of Science And Technology Kumasi; 2014

[9] Yousaf B et al. Investigating the biochar effects on C-mineralization and sequestration of carbon in soil compared with conventional amendments using the stable isotope (δ13C) approach. GCB Bioenergy. 2017;**9**(6):1085-1099

[10] Woolf D et al. Sustainable biochar to mitigate global climate change. Nature Communications. 2010;**1**(1):1-9

[11] Ding Y et al. Biochar to improve soil fertility. A review. Agronomy for sustainable development. 2016;**36**(2):1-18

[12] Gupta R et al. Rice straw biochar improves soil fertility, growth, and yield of rice–wheat system on a sandy loam soil. Experimental Agriculture. 2020;**56**(1):118-131

[13] Aup-Ngoen K, Noipitak M. Effect of carbon-rich biochar on mechanical properties of PLA-biochar composites. Sustainable Chemistry and Pharmacy. 2020;**15**:100204

[14] Man KY et al. Use of biochar as feed supplements for animal farming. Critical Reviews in Environmental Science and Technology. 2021;**51**(2):187-217

[15] Weber K, Quicker P. Properties of biochar. Fuel. 2018;**217**:240-261

[16] Palansooriya KN et al. Prediction of soil heavy metal immobilization by biochar using machine learning. Environmental Science & Technology. 2022

[17] Zhu X et al. Machine learning exploration of the direct and indirect roles of Fe impregnation on Cr (VI) removal by engineered biochar. Chemical Engineering Journal. 2022;**428**:131967

[18] Paula AJ et al. Machine learning and natural language processing enable a data-oriented experimental design

approach for producing biochar and hydrochar from biomass. Chemistry of Materials. 2022

[19] Singh A, Thakur N, Sharma A. A review of supervised machine learning algorithms. In: 2016 3rd International Conference on Computing for Sustainable Global Development (INDIACom). IEEE; 2016

[20] Crisci C, Ghattas B, Perera G. A review of supervised machine learning algorithms and their applications to ecological data. Ecological Modelling. 2012;**240**:113-122

[21] Wang J, Wang S. Preparation, modification and environmental application of biochar: A review. Journal of Cleaner Production. 2019;**227**:1002-1022

[22] Ronsse F, Nachenius RW, Prins W. Carbonization of biomass. In: Recent Advances in Thermo-Chemical Conversion of Biomass. Elsevier; 2015. pp. 293-324

[23] Hayes M. Biochar and biofuels for a brighter future. Nature. 2006;**443**(7108):144-144

[24] Fulton W et al. Biochars as a potting amendment in greenhouse nurseries: How to eliminate unwanted ethylene 1 emissions 2. Agronomy for Sustainable Development. 2013;**33**(3)

[25] Dooley J et al. Moderate-Scale Biochar Production Across Forested Landscapes

[26] Bezerra J et al. The promises of the Amazonian soil: Shifts in discourses of Terra Preta and biochar. Journal of Environmental Policy & Planning. 2019;**21**(5):623-635

[27] Guo M, He Z, Uchimiya SM. Introduction to biochar as an agricultural and environmental amendment. Agricultural and Environmental Applications of Biochar: Advances and Barriers. 2016;**63**:1-14

[28] Boateng AA et al. Biochar production technology. In: Biochar for Environmental Management. Routledge; 2015. pp. 95-120

[29] Kan T, Strezov V, Evans TJ. Lignocellulosic biomass pyrolysis: A review of product properties and effects of pyrolysis parameters. Renewable and Sustainable Energy Reviews. 2016;**57**:1126-1140

[30] Tripathi M, Sahu JN, Ganesan P. Effect of process parameters on production of biochar from biomass waste through pyrolysis: A review. Renewable and Sustainable Energy Reviews. 2016;**55**:467-481

[31] Palanivelu K, Ramachandran A, Raghavan V. Biochar from biomass waste as a renewable carbon material for climate change mitigation in reducing greenhouse gas emissions—A review. Biomass Conversion and Biorefinery. 2021;**11**(5):2247-2267

[32] Suman S, Gautam S. Pyrolysis of coconut husk biomass: Analysis of its biochar properties. Energy Sources, Part A: Recovery, Utilization, and Environmental Effects. 2017;**39**(8):761-767

[33] Wang Z et al. Biochar produced from oak sawdust by lanthanum (La)-involved pyrolysis for adsorption of ammonium (NH4+), nitrate (NO3−), and phosphate (PO43−). Chemosphere. 2015;**119**:646-653

[34] Cha JS et al. Production and utilization of biochar: A review. Journal of Industrial and Engineering Chemistry. 2016;**40**:1-15

[35] Velusamy K et al. A review on nano-catalysts and biochar-based catalysts for biofuel production. Fuel. 2021;**306**:121632

[36] Martinez CLM et al. Evaluation of thermochemical routes for the valorization of solid coffee residues to produce biofuels: A Brazilian case. Renewable and Sustainable Energy Reviews. 2021;**137**:110585

[37] Bridgwater AV. Review of fast pyrolysis of biomass and product upgrading. Biomass and Bioenergy. 2012;**38**:68-94

[38] Tenenbaum DJ. Biochar: Carbon Mitigation from the Ground Up. National Institute of Environmental Health Sciences; 2009

[39] Lin Q et al. Assessing the potential of biochar and aged biochar to alleviate aluminum toxicity in an acid soil for achieving cabbage productivity. Ecotoxicology and Environmental Safety. 2018;**161**:290-295

[40] Jaiswal AK et al. Activating biochar by manipulating the bacterial and fungal microbiome through pre-conditioning. New Phytologist. 2018;**219**(1):363-377

[41] Tomczyk A, Sokołowska Z, Boguta P. Biochar physicochemical properties: Pyrolysis temperature and feedstock kind effects. Reviews in Environmental Science and Bio/Technology. 2020;**19**(1):191-215

[42] Janu R et al. Biochar surface functional groups as affected by biomass feedstock, biochar composition and pyrolysis temperature. Carbon Resources Conversion. 2021;**4**:36-46

[43] Zhang C et al. Evolution of the functional groups/structures of biochar and heteroatoms during the pyrolysis of seaweed. Algal Research. 2020;**48**:101900

[44] Chen Y et al. The structure evolution of biochar from biomass pyrolysis and its correlation with gas pollutant adsorption performance. Bioresource Technology. 2017;**246**:101-109

[45] Fellet G et al. Application of biochar on mine tailings: Effects and perspectives for land reclamation. Chemosphere. 2011;**83**(9):1262-1267

[46] Brewer CE et al. New approaches to measuring biochar density and porosity. Biomass and Bioenergy. 2014;**66**:176-185

[47] Atkinson CJ, Fitzgerald JD, Hipps NA. Potential mechanisms for achieving agricultural benefits from biochar application to temperate soils: A review. Plant and Soil. 2010;**337**(1):1-18

[48] Al-Wabel MI et al. Impact of biochar properties on soil conditions and agricultural sustainability: A review. Land Degradation & Development. 2018;**29**(7):2124-2161

[49] Monnie F. Effect of Biochar on Soil Physical Properties, Water Use Efficiency, and Growth of Maize in a Sandy Loam Soil. University of Ghana; 2016

[50] Liao W, Drake J, Thomas SC. Biochar granulation enhances plant performance on a green roof substrate. Science of the Total Environment. 2022;**813**:152638

[51] Gupta S, Kua HW. Carbonaceous micro-filler for cement: Effect of particle size and dosage of biochar on fresh and hardened properties of cement mortar. Science of the Total Environment. 2019;**662**:952-962

[52] Yang C, Liu J, Lu S. Pyrolysis temperature affects pore characteristics

of rice straw and canola stalk biochars and biochar-amended soils. Geoderma. 2021;**397**:115097

[53] Esmaeelnejad L et al. Impacts of woody biochar particle size on porosity and hydraulic conductivity of biochar-soil mixtures: An incubation study. Communications in Soil Science and Plant Analysis. 2017;**48**(14):1710-1718

[54] Anderson N et al. A comparison of producer gas, biochar, and activated carbon from two distributed scale thermochemical conversion systems used to process forest biomass. Energies. 2013;**6**(1):164-183

[55] Shang L et al. Changes of chemical and mechanical behavior of torrefied wheat straw. Biomass and Bioenergy. 2012;**40**:63-70

[56] Singh M et al. Co-combustion properties of torrefied rice straw-sub-bituminous coal blend and its Hardgrove Grindability Index. In: Biomass Conversion and Biorefinery. 2021. pp. 1-15

[57] Zhao L et al. Heterogeneity of biochar properties as a function of feedstock sources and production temperatures. Journal of Hazardous Materials. 2013;**256**:1-9

[58] Rehrah D et al. Production and characterization of biochars from agricultural by-products for use in soil quality enhancement. Journal of Analytical and Applied Pyrolysis. 2014;**108**:301-309

[59] Rashid M et al. Prospects of biochar in alkaline soils to mitigate climate change. In: Environment, Climate, Plant and Vegetation Growth. Springer; 2020. pp. 133-149

[60] Lehmann J et al. Biochar in climate change mitigation. Nature Geoscience. 2021;**14**(12):883-892

[61] Oni BA, Oziegbe O, Olawole OO. Significance of biochar application to the environment and economy. Annals of Agricultural Sciences. 2019;**64**(2):222-236

[62] Verheijen F et al. Biochar application to soils. A critical scientific review of effects on soil properties, processes, and functions. EUR. 2010;**24099**:162

[63] Bolan N et al. Multifunctional applications of biochar beyond carbon storage. International Materials Reviews. 2021:1-51

[64] Brassard P et al. Biochar for soil amendment. In: Char and Carbon Materials Derived from Biomass. Elsevier; 2019. pp. 109-146

[65] Kamali M et al. Biochar for soil applications-sustainability aspects, challenges and future prospects. Chemical Engineering Journal. 2022;**428**:131189

[66] Syuhada AB et al. Biochar as soil amendment: Impact on chemical properties and corn nutrient uptake in a Podzol. Canadian Journal of Soil Science. 2016;**96**(4):400-412

[67] Mukherjee A, Lal R. The biochar dilemma. Soil Research. 2014;**52**(3):217-230

[68] Diatta AA et al. Effects of biochar on soil fertility and crop productivity in arid regions: A review. Arabian Journal of Geosciences. 2020;**13**(14):1-17

[69] Igalavithana AD et al. The effects of biochar amendment on soil fertility. In: Agricultural and Environmental Applications of Biochar: Advances and Barriers. sssaspecpub63 ed. 2016. pp. 123-144

[70] He M et al. A critical review on performance indicators for evaluating

soil biota and soil health of biochar-amended soils. Journal of Hazardous Materials. 2021;**414**:125378

[71] Krishnakumar S et al. Impact of biochar on soil health. International Journal of Advanced Research. 2014;**2**(4):933-950

[72] Najar GR, Ganie MA, Tahir A. Biochar for sustainable soil health: A review of prospects and concerns. Pedosphere. 2015;**25**(5):639-653

[73] Das SK, Ghosh GK. Soil health management through low cost biochar technology. In: Biochar Applications in Agriculture and Environment Management. Springer; 2020. pp. 193-206

[74] Chan KY et al. Agronomic values of greenwaste biochar as a soil amendment. Soil Research. 2007;**45**(8):629-634

[75] Wu W et al. Chemical characterization of rice straw-derived biochar for soil amendment. Biomass and Bioenergy. 2012;**47**:268-276

[76] Chan K et al. Using poultry litter biochars as soil amendments. Soil Research. 2008;**46**(5):437-444

[77] Yadav A et al. Vacuum pyrolysed biochar for soil amendment. Resource-Efficient Technologies. 2016;**2**:S177-S185

[78] Nawkarkar P, Ganesan A, Kumar S. Carbon dioxide capture for biofuel production. In: Handbook of Biofuels. Elsevier; 2022. pp. 605-619

[79] Kral U. A new indicator for the assessment of anthropogenic substance flows to regional sinks. 2014

[80] Brückner R, Titgemeyer F. Carbon catabolite repression in bacteria: Choice of the carbon source and autoregulatory limitation of sugar utilization. FEMS Microbiology Letters. 2002;**209**(2):141-148

[81] Buss W et al. Mineral-enriched biochar delivers enhanced nutrient recovery and carbon dioxide removal. Communications Earth & Environment. 2022;**3**(1):1-11

[82] Das SK, Ghosh GK, Avasthe R. Biochar application for environmental management and toxic pollutant remediation. Biomass Conversion and Biorefinery. 2020:1-12

[83] Geden O, Schenuit F. Unconventional mitigation: Carbon dioxide removal as a new approach in EU climate policy. 2020

[84] Möllersten K, Naqvi R, Yan J. Qualitative Assessment of Classes of Negative Emission Technologies (NETs). Västerås: Mälardalen University; 2020

[85] Johnson K et al. Carbon dioxide removal options: A literature review identifying carbon removal potentials and costs. 2017

[86] Williamsona P et al. 10. Biologically-based negative emissions in the open ocean and coastal seas. 2021

[87] Yu O-Y, Raichle B, Sink S. Impact of biochar on the water holding capacity of loamy sand soil. International Journal of Energy and Environmental Engineering. 2013;**4**(1):1-9

[88] Ulyett J et al. Impact of biochar addition on water retention, nitrification and carbon dioxide evolution from two sandy loam soils. European Journal of Soil Science. 2014;**65**(1):96-104

[89] Abel S et al. Impact of biochar and hydrochar addition on water retention and water repellency of sandy soil. Geoderma. 2013;**202**:183-191

[90] Castellini M et al. Impact of biochar addition on the physical and hydraulic properties of a clay soil. Soil and Tillage Research. 2015;**154**:1-13

[91] Wang D et al. Impact of biochar on water retention of two agricultural soils—A multi-scale analysis. Geoderma. 2019;**340**:185-191

[92] Baronti S et al. Impact of biochar application on plant water relations in *Vitis vinifera* (L.). European Journal of Agronomy. 2014;**53**:38-44

[93] Novak JM et al. Impact of biochar amendment on fertility of a southeastern coastal plain soil. Soil Science. 2009;**174**(2):105-112

[94] Laird DA et al. Impact of biochar amendments on the quality of a typical Midwestern agricultural soil. Geoderma. 2010;**158**(3-4):443-449

[95] van de Ven DJ, Fouquet R. Historical energy price shocks and their changing effects on the economy. Energy Economics. 2017;**62**:204-216

[96] Xie T et al. Characteristics and applications of biochar for environmental remediation: A review. Critical Reviews in Environmental Science and Technology. 2015;**45**(9):939-969

[97] Mandal S et al. Progress and future prospects in biochar composites: Application and reflection in the soil environment. Critical Reviews in Environmental Science and Technology. 2021;**51**(3):219-271

[98] Shakoor MB et al. A review of biochar-based sorbents for separation of heavy metals from water. International Journal of Phytoremediation. 2020;**22**(2):111-126

[99] Shaheen SM et al. Wood-based biochar for the removal of potentially toxic elements in water and wastewater: A critical review. International Materials Reviews. 2019;**64**(4):216-247

[100] Inyang MI et al. A review of biochar as a low-cost adsorbent for aqueous heavy metal removal. Critical Reviews in Environmental Science and Technology. 2016;**46**(4):406-433

[101] Ltd ARP. Biochar Market Expected to Grow at 8.9% of High CAGR by Forecast 2028 With SWOT Analysis, Future Trends, Challenges and Growth Opportunities. USA: GlobeNewswire; 2022

[102] Bowling M et al. Machine learning and games. Machine Learning. 2006;**63**(3):211-215

[103] Samuel AL. Machine learning. The Technology Review. 1959;**62**(1):42-45

[104] El Naqa I, Murphy MJ. What is machine learning? In: Machine Learning in Radiation Oncology. Springer; 2015. pp. 3-11

[105] Gogas P, Papadimitriou T. Machine learning in economics and finance. Computational Economics. 2021;**57**(1):1-4

[106] Nilsson N. Learning Machines. Vol. 19652. New York: McGraw-Hill; 1965

[107] Clarke M. Pattern Classification and Scene Analysis. Wiley Online Library; 1974

[108] Abaimov S, Martellini M. Understanding machine learning. In: Machine Learning for Cyber Agents. Springer; 2022. pp. 15-89

[109] Bozinovski S. Reminder of the first paper on transfer learning in neural networks, 1976. Informatica. 2020;**44**(3)

[110] Mitchell TM. Artificial neural networks. Machine learning. 1997;**45**:81-127

[111] Mitchell TM. Machine Learning. New York: McGraw-Hill; 1997

[112] Harnad S. The annotation game: On Turing (1950) on computing, machinery, and intelligence. In: The Turing Test Sourcebook: Philosophical and Methodological Issues in the Quest for the Thinking Computer. Kluwer; 2006

[113] Patel J et al. Predicting stock and stock price index movement using trend deterministic data preparation and machine learning techniques. Expert Systems with Applications. 2015;**42**(1):259-268

[114] Soni S. Applications of ANNs in stock market prediction: A survey. International Journal of Computer Science & Engineering Technology. 2011;**2**(3):71-83

[115] Mohri M, Rostamizadeh A, Talwalkar A. Foundations of Machine Learning. MIT Press; 2018

[116] Ethem AM. Introduction to Machine Learning/Ethem Alpaydın. 2010. London: The MIT Press

[117] Angluin D. Computational learning theory: Survey and selected bibliography. In: Proceedings of the twenty-fourth annual ACM symposium on Theory of computing. 1992

[118] Rashidi HH et al. Artificial intelligence and machine learning in pathology: The present landscape of supervised methods. Academic Pathology. 2019;**6**:2374289519873088

[119] Moujahid A et al. Machine learning techniques in ADAS: A review. In: 2018 International Conference on Advances in Computing and Communication Engineering (ICACCE). IEEE; 2018

[120] Liu D et al. Optimizing wireless systems using unsupervised and reinforced-unsupervised deep learning. IEEE Network. 2020;**34**(4):270-277

[121] Jiang T, Gradus JL, Rosellini AJ. Supervised machine learning: A brief primer. Behavior Therapy. 2020;**51**(5):675-687

[122] Burkart N, Huber MF. A survey on the explainability of supervised machine learning. Journal of Artificial Intelligence Research. 2021;**70**:245-317

[123] Jaeger S, Fulle S, Turk S. Mol2vec: Unsupervised machine learning approach with chemical intuition. Journal of Chemical Information and Modeling. 2018;**58**(1):27-35

[124] Sutton RS. Introduction: The challenge of reinforcement learning. In: Reinforcement Learning. Springer; 1992. pp. 1-3

[125] Oyetunde T et al. Leveraging knowledge engineering and machine learning for microbial bio-manufacturing. Biotechnology Advances. 2018;**36**(4):1308-1315

[126] Lakshmi D et al. Artificial intelligence (AI) applications in adsorption of heavy metals using modified biochar. Science of the Total Environment. 2021;**801**:149623

[127] Zhou X et al. Hyperspectral technique combined with deep learning algorithm for detection of compound heavy metals in lettuce. Food Chemistry. 2020;**321**:126503

[128] Verma JK, Paul S, Johri P. Computational Intelligence and Its

Applications in Healthcare. Academic Press; 2020

[129] Meza JKS et al. Predictive analysis of urban waste generation for the city of Bogotá, Colombia, through the implementation of decision trees-based machine learning, support vector machines and artificial neural networks. Heliyon. 2019;**5**(11):e02810

[130] Golbayani P, Florescu I, Chatterjee R. A comparative study of forecasting corporate credit ratings using neural networks, support vector machines, and decision trees. The North American Journal of Economics and Finance. 2020;**54**:101251

[131] Ghavamzadeh M et al. Bayesian reinforcement learning: A survey. Foundations and Trends in Machine Learning. 2015;**8**(5-6):359-483

[132] Mitra T et al. Removal of Pb (II) ions from aqueous solution using water hyacinth root by fixed-bed column and ANN modeling. Journal of Hazardous Materials. 2014;**273**:94-103

[133] Zafar M et al. Ethanol mediated As (III) adsorption onto Zn-loaded pinecone biochar: Experimental investigation, modeling, and optimization using hybrid artificial neural network-genetic algorithm approach. Journal of Environmental Sciences. 2017;**54**:114-125

[134] Genuino DAD et al. Application of artificial neural network in the modeling and optimization of humic acid extraction from municipal solid waste biochar. Journal of Environmental Chemical Engineering. 2017;**5**(4):4101-4107

[135] Williams B et al. Data-driven model development for cardiomyocyte production experimental failure prediction. In: Computer Aided Chemical Engineering. Elsevier; 2020. pp. 1639-1644

[136] Misra S, Li H, He J. Noninvasive fracture characterization based on the classification of sonic wave travel times. In: Machine Learning for Subsurface Characterization. 2020. pp. 243-287

[137] Pradhan A. Support vector machine-a survey. International Journal of Emerging Technology and Advanced Engineering. 2012;**2**(8):82-85

[138] Yao Y et al. K-SVM: An effective SVM algorithm based on K-means clustering. Journal of Computers. 2013;**8**(10):2632-2639

[139] Tharwat A, Hassanien AE, Elnaghi BE. A BA-based algorithm for parameter optimization of support vector machine. Pattern Recognition Letters. 2017;**93**:13-22

[140] Natekin A, Knoll A. Gradient boosting machines, a tutorial. Frontiers in Neurorobotics. 2013;**7**:21

[141] Touzani S, Granderson J, Fernandes S. Gradient boosting machine for modeling the energy consumption of commercial buildings. Energy and Buildings. 2018;**158**:1533-1543

[142] Gong M et al. Gradient boosting machine for predicting return temperature of district heating system: A case study for residential buildings in Tianjin. Journal of Building Engineering. 2020;**27**:100950

[143] Fujiyoshi H, Hirakawa T, Yamashita T. Deep learning-based image recognition for autonomous driving. IATSS Research. 2019;**43**(4):244-252

[144] Deng L, Li X. Machine learning paradigms for speech recognition: An overview. IEEE Transactions on Audio,

Speech, and Language Processing. 2013;**21**(5):1060-1089

[145] Chen L et al. Machine learning-based product recommendation using apache spark. In: 2017 IEEE SmartWorld, Ubiquitous Intelligence & Computing, Advanced & Trusted Computed, Scalable Computing & Communications, Cloud & Big Data Computing, Internet of People and Smart City Innovation (SmartWorld/SCALCOM/UIC/ATC/CBDCom/IOP/SCI). IEEE; 2017

[146] Soni A et al. Design of a machine learning-based self-driving car. In: Machine Learning for Robotics Applications. Springer; 2021. pp. 139-151

[147] Kononenko I. Machine learning for medical diagnosis: History, state of the art and perspective. Artificial Intelligence in Medicine. 2001;**23**(1):89-109

[148] Richens JG, Lee CM, Johri S. Improving the accuracy of medical diagnosis with causal machine learning. Nature Communications. 2020;**11**(1):1-9

[149] Tenni J et al. Machine learning of language translation rules. In: IEEE SMC'99 Conference Proceedings. 1999 IEEE International Conference on Systems, Man, and Cybernetics (Cat. No. 99CH37028). IEEE; 1999

[150] De Prado ML. The 10 reasons most machine learning funds fail. The Journal of Portfolio Management. 2018;**44**(6):120-133

[151] Hong J-W, Wang Y, Lanz P. Why is artificial intelligence blamed more? Analysis of faulting artificial intelligence for self-driving car accidents in experimental settings. International Journal of Human–Computer Interaction. 2020;**36**(18):1768-1774

[152] Stilgoe J. Who killed Elaine Herzberg? In: Who's Driving Innovation? Springer; 2020. pp. 1-6

[153] Becker A. Artificial intelligence in medicine: What is it doing for us today? Health Policy and Technology. 2019;**8**(2):198-205

[154] Yampolskiy RV. Artificial intelligence safety engineering: Why machine ethics is a wrong approach. In: Philosophy and Theory of Artificial Intelligence. Springer; 2013. pp. 389-396

[155] Nath R, Sahu V. The problem of machine ethics in artificial intelligence. AI & Society. 2020;**35**(1):103-111

[156] Sandel MJ. The Case against Perfection: Ethics in the Age of Genetic Engineering. Harvard University Press; 2007

[157] Lucivero F. Ethical Assessments of Emerging Technologies. Cham: Springer; 2016

[158] Hinton G et al. Deep neural networks for acoustic modeling in speech recognition: The shared views of four research groups. IEEE Signal Processing Magazine. 2012;**29**(6):82-97

[159] Kobielus J. Gpus continue to dominate the ai accelerator market for now. 2019

[160] Ray T. AI Is Changing the Entire Nature of Compute. ZDNet; 2019. p. 30

[161] Chua L, Sirakoulis GC, Adamatzky A. Handbook of Memristor Networks. Springer Nature; 2019

[162] Zhu X, Wang Q, Lu WD. Memristor networks for real-time neural activity analysis. Nature Communications. 2020;**11**(1):1-9

[163] David R et al. TensorFlow lite micro: Embedded machine learning for tinyml systems. Proceedings of Machine Learning and Systems. 2021;**3**:800-811

[164] Gualtieri M. The Forrester Wave™: Predictive Analytics and Machine Learning Solutions, Q1. Forrester Research; 2017. p. 2017

[165] Dwivedi S, Kasliwal P, Soni S. Comprehensive study of data analytics tools (RapidMiner, Weka, R tool, Knime). In: 2016 Symposium on Colossal Data Analysis and Networking (CDAN). IEEE; 2016

[166] Grimes S. Text Analytics 2014: User Perspectives on. 2014

[167] Mazanetz MP et al. Drug discovery applications for KNIME: An open source data mining platform. Current Topics in Medicinal Chemistry. 2012;**12**(18):1965-1979

[168] Khan A et al. A review of machine learning algorithms for text-documents classification. Journal of Advances in Information Technology. 2010;**1**(1):4-20

[169] Mahesh B. Machine learning algorithms—A review. International Journal of Science and Research (IJSR). 2020;**9**:381-386

[170] Ibrahim I, Abdulazeez A. The role of machine learning algorithms for diagnosing diseases. Journal of Applied Science and Technology Trends. 2021;**2**(01):10-19

[171] Goeschel K. Reducing false positives in intrusion detection systems using data-mining techniques utilizing support vector machines, decision trees, and naive Bayes for off-line analysis. In: SoutheastCon 2016. IEEE; 2016

[172] Abdulrahman SA et al. Comparative study for 8 computational intelligence algorithms for human identification. Computer Science Review. 2020;**36**:100237

[173] Afridi R. List of Machine Learning Algorithms. 2020

[174] Hong S-H et al. Synthesis of Fe-impregnated biochar from food waste for selenium (VI) removal from aqueous solution through adsorption: Process optimization and assessment. Chemosphere. 2020;**252**:126475

[175] Hodgson E et al. Optimisation of slow-pyrolysis process conditions to maximise char yield and heavy metal adsorption of biochar produced from different feedstocks. Bioresource Technology. 2016;**214**:574-581

[176] Zhu X, Li Y, Wang X. Machine learning prediction of biochar yield and carbon contents in biochar based on biomass characteristics and pyrolysis conditions. Bioresource Technology. 2019;**288**:121527

[177] Sun Y et al. The application of machine learning methods for prediction of metal immobilization remediation by biochar amendment in soil. Science of the Total Environment. 2022:154668

[178] Cao H, Xin Y, Yuan Q. Prediction of biochar yield from cattle manure pyrolysis via least squares support vector machine intelligent approach. Bioresource Technology. 2016;**202**:158-164

[179] Li S et al. Predicting biochar properties and functions based on feedstock and pyrolysis temperature: A review and data syntheses. Journal of Cleaner Production. 2019;**215**:890-902

[180] Pyo J et al. Estimation of heavy metals using deep neural network with visible and infrared spectroscopy of

soil. Science of the Total Environment. 2020;**741**:140162

[181] Zhang H et al. Promotional effect of NH3 on mercury removal over biochar thorough chlorine functional group transformation. Journal of Cleaner Production. 2020;**257**:120598

[182] Hu B et al. Modelling bioaccumulation of heavy metals in soil-crop ecosystems and identifying its controlling factors using machine learning. Environmental Pollution. 2020;**262**:114308

[183] Razzaghi M et al. Phenol removal by HRP/GOx/ZSM-5 from aqueous solution: Artificial neural network simulation and genetic algorithms optimization. Journal of the Taiwan Institute of Chemical Engineers. 2018;**89**:1-14

[184] Wong YJ et al. Comparative study of artificial neural network (ANN), adaptive neuro-fuzzy inference system (ANFIS) and multiple linear regression (MLR) for modeling of Cu (II) adsorption from aqueous solution using biochar derived from rambutan (*Nephelium lappaceum*) peel. Environmental Monitoring and Assessment. 2020;**192**(7):1-20

[185] El Hanandeh A, Mahdi Z, Imtiaz M. Modelling of the adsorption of Pb, Cu and Ni ions from single and multi-component aqueous solutions by date seed derived biochar: Comparison of six machine learning approaches. Environmental Research. 2021;**192**:110338

[186] Talebkeikhah F et al. Investigation of effective processes parameters on lead (II) adsorption from wastewater by biochar in mild air oxidation pyrolysis process. International Journal of Environmental Analytical Chemistry. 2020:1-21

[187] Hafsa N et al. A generalized method for modeling the adsorption of heavy metals with machine learning algorithms. Water. 2020;**12**(12):3490

[188] Yaseen ZM. An insight into machine learning models era in simulating soil, water bodies and adsorption heavy metals: Review, challenges and solutions. Chemosphere. 2021;**277**:130126

[189] Wanyonyi FS et al. Role of pore chemistry and topology in the heavy metal sorption by zeolites: From molecular simulation to machine learning. Computational Materials Science. 2021;**195**:110519

[190] Li M et al. EDTA functionalized magnetic biochar for Pb (II) removal: Adsorption performance, mechanism and SVM model prediction. Separation and Purification Technology. 2019;**227**:115696

[191] Chakraborty V, Das P. Synthesis of nano-silica-coated biochar from thermal conversion of sawdust and its application for Cr removal: Kinetic modelling using linear and nonlinear method and modelling using artificial neural network analysis. Biomass Conversion and Biorefinery. 2020:1-11

[192] Zhao Y et al. Application of kernel extreme learning machine and kriging model in prediction of heavy metals removal by biochar. Bioresource Technology. 2021;**329**:124876

[193] Zhu X, Wang X, Ok YS. The application of machine learning methods for prediction of metal sorption onto biochars. Journal of Hazardous Materials. 2019;**378**:120727

Chapter 15

Applications and Data Analysis Using Bayesian and Conventional Statistics in Biochar Adsorption Studies for Environmental Protection

Obey Gotore, Tirivashe Phillip Masere, Osamu Nakagoe, Vadzanayi Mushayi, Ramaraj Rameshprabu, Yuwalee Unpaprom and Tomoaki Itayama

Abstract

The use of low-cost agricultural waste-derived biochar in solving water and environmental challenges induced by climate change was investigated and sound conclusions were presented. Water reuse strategies can diminish the impact of climate change in rural and remote areas of developing countries. The novel biochar materials from three agro-waste biomass (Matamba fruit shell, Mushuma, and Mupane tree barks) were investigated and characterized to attest to their capacity to remove iodine from the aqueous solution. Their surface morphologies were assessed using Field Emission Scanning Electron Microscopy with Energy Dispersive X-Ray Spectroscopy (FESEM-EDX) which exhibited their structural phenomena to purge environmental pollutants. The Fourier-transform infrared spectroscopy (FTIR) was conducted to show surface functional groups of the biochar materials and Matamba fruit shell exhibited hydroxyl (-OH), carbonyl groups (C=O), C=C stretches of aromatic rings, and the carboxylate (C–O–O–) groups on its surface with corresponding data from the Isotherm and Kinetic models, statistically analyzed by the conventional and Bayesian methods. These surface mechanisms are said to be induced by weak van der Waals forces and - and -stacking interaction on the biochar surface. These adsorbents promised to be potential materials for environmental-ecosystem-protection and water re-use approach.

Keywords: adsorption, Bayesian statistics, Matamba fruit shell, Mushuma bark biochar, Mupane bark biochar

1. Introduction

Global warming and climate change is triggering some drastic global environmental complications and developing countries are facing cumulative water insufficiency and such problems are subsequently increasing [1]. Some developing countries like Africa, South-East Asia, and South America are facing water deficits triggered by climate changes [2]. These countries are very vulnerable because of extremes of climatic change, which are increasing, their magnitude and frequency are making the availability of water a challenge to their societal livelihood's sustainability [3]. Portable safe water availability is becoming scarce due to high toxicity contaminants in water sources with a variety of constituents such as dyes [4] just to mention a few. Innovations, adaptations, and developments are being put in place to alleviate such burdens, and paramount measures are being employed to make sure that water is available and accessible to all in developing countries.

This study aims to remove micropollutants and recommend better wastewater reuse technology for unserved rural communities in an off-the-grid system to achieve socio-economic development using physiochemical properties of Mushuma, Mupani barks, and Matamba fruit shells, by analyzing their characteristics to evaluate the kinetic mechanism of adsorption from different models and statistical methods for the determination of equilibrium analysis. Thus, waste or residual biomass utilization such as biochar production has been given substantial attention because of its potential for carbon sequestration, waste management, and environmental remediation of pollutants [5]. A lot of technologies have been employed to mitigate challenges of water pollution globally, from a variety of sources and types of industries, for example, photocatalytic degradation [6], photooxidative degradation [7], Fenton reagent [8], and adsorption, wherein it is highly efficient in the removal of dyes and pigments from the liquid phase [9].

During the adsorption process, activated carbon is normally used due to its large specific surface area, well-developed pore structure, increased adsorption efficiency as well as good chemical stability [3]. [10] reiterated that adsorption takes place in the mesopores which act as conduits for adsorbate particles, and capillary condensation takes place to adsorb these macromolecules. The adsorption process has proved to be one of the best wastewater treatment technologies in the world and activated carbon acts as the universal adsorbent for the removal of different types of water pollutants. Most materials that come from carbon have great surface areas, which are stable with extensive functional groups, interconnected pore structures, and shapes [11]. To add to that, the Matamba fruit shell adsorption capacity, commonly found in Zimbabwe, can be a potential solution for water reuse techniques for the local people, and its kinetic adsorption was recently tested [11]. This can be attributed to adsorption as the most lucrative treatment technique [12], Langmuir and Freundlich's isotherms are common models which are extensively used since they are simple to use because of their empirical mathematical expressions [13].

The plant bark adsorbents for pollution reduction have widely been utilized in numerous studies which have been conducted recently [11]. Plant species analyzed included three eucalyptus, African border, flamboyant pods, and sycamore [14] among others. Biochar that is derived from waste biomass has also been considered one of the efficient adsorbents for wastewater pollutant removal because of its cost-effective merits, easy obtainability, and beneficial physicochemical properties [3]. The activated carbon materials, as adsorbents, has merits but are not limited to the adequate surface area, porosity phenomena, and thermal stability [15].

Different researchers are studying the application of biochar for wastewater treatment [5], and various materials are being pyrolyzed under different conditions and they can affect the physicochemical properties of the product [16]. The use of chemicals to modify the biochar by acids, bases, or polymers seems to give better adsorption effectiveness because of enlarged surface area, modified chemical functionality, and availability of high-affinity adsorption sites [16]. Matamba, Mushuma, and Mupane tree barks as a novel and recent research for the removal of micropollutants as a wastewater re-use strategy or technology for unserved rural communities in an off-the-grid system to achieve socio-economic development.

2. Production of biochar

2.1 Biomass materials and pyrolysis conditions

Biochar is a material rich in carbon, produced from cracking several biomasses, such as wood, sludge, plants, food waste, and animal waste after carbonization [17, 18]. Such waste biomass materials exhibited to be effective for the removal of harmful substances from the aqueous solutions considering their less production cost and economic benefits, wide availability of raw material, and conducive surface properties. Furthermore, as noted by [5, 11, 19, 20], the use of biomass or residual waste has now been prioritized due to several advantages including environmental remediation, waste management, carbon sequestration, and ameliorating the greenhouse gas effect.

Pyrolysis is one of the most used technology in the production of biochar. It involves the carbonization of organic materials in limited or no oxygen conditions [21]. It is a thermochemical decomposition process taking place at temperatures above 300°C. In addition, the pyrolysis process may also produce volatile liquids and gases (e.g., carbon monoxide, carbon dioxide, hydrogen, methane, and biogas [22]. Pyrolysis may be categorized into four groups based on temperature conditions, reaction time, and heating rate. These are: slow, fast, flash, and intermediate, and of these slow and fast are the most common types [22]. In fast pyrolysis, the temperature and heating rates are higher than in slow pyrolysis. As such the process can be done in seconds and the resulting product consists mainly of bio-oils [23].

Conversely, in slow pyrolysis, the process can go on for hours and the heating rate and temperature are lower; a temperature under 450°C is commonly used and the resulting product is mainly biochar [22]. Biochar may be chemically modified using acids, bases, or polymers to have better adsorption efficiency due to the increased surface area, modified chemical functionality, and the presence of high-affinity adsorption sites [16]. The adsorption mechanism of the biochar after pyrolysis is shown on **Figure 1** where both positive and negative charges do exist on the surface due to thermal decomposition. This property enhancement process makes biochar a cost-effective choice hazardous material removal from the environment.

2.2 Availability of Matamba fruit shell

Matamba (monkey orange - Strychnos spp.) are widely distributed in Southern Africa and particularly in Zimbabwe, where they are generally found throughout the country, but more so in the Midlands Province [11, 24]. These fruits proliferate in semi-arid areas of Zimbabwe, with limited rainfall water, and produce the fruit in abundance [25, 26]. Depending on the season, excess production of the fruit varies and sometimes leads to its underutilization, and this can be seen in the highveld around Zimbabwe where fruit remains and disturbs the environment [24]. The Strychnos spp. fruit is extensively found in Zimbabwe, it is underutilized, and little or no consideration has been raised for potential commercialization due to limited

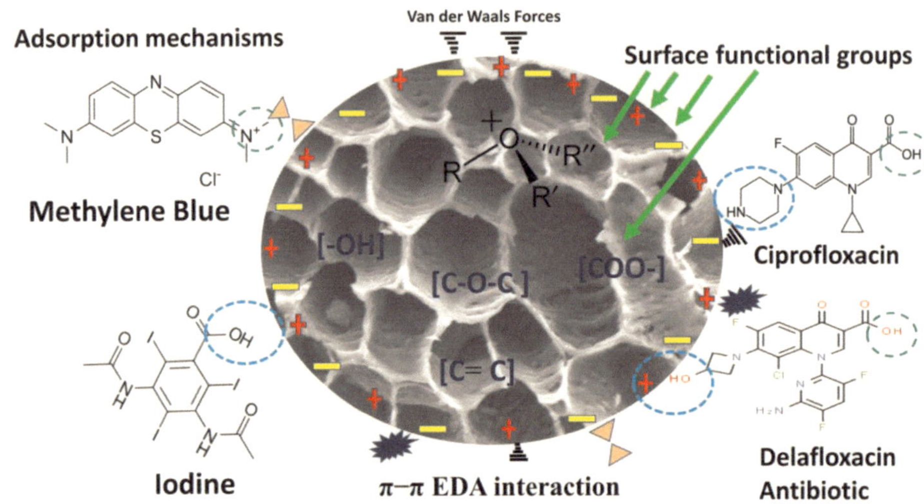

Figure 1.
The adsorption mechanism of adsorbates onto the biochar and surface characteristics after pyrolysis.

knowledge and dissemination of information about its propagation, agronomic practices, and product processing techniques for business [24].

The Matamba fruits begin to develop and grow during the autumn season and ripen in winter up to the spring season [11, 24]. The fruit is spherical with a hard thick shell, and the seeds are around 2–3 cm in diameter [11]. It is these seeds that are edible by humans and animals. To access these seeds the hard shells must be broken first, usually by hitting the shells on hard stone surfaces. After consumption of the seeds, the empty hard shells are often thrown away or littered around the veld or homesteads [24].

2.3 Agro-waste biomass

Agriculture and its related sectors like forestry generate massive volumes of biomass residues generated in the forestry in most developing countries. However, these residues should not be treated as waste given that a greater proportion of 'waste' is usable. As already discussed, pyrolysis is an important and more beneficial alternative to the usual farmer practice of burning, burying, or storing agricultural biomass residues [21, 27]. There is a large range of waste materials that could be suitable for pyrolysis and biochar production. However, for this study, agro-waste in the form of Mushuma, Mupane barks, and Matamba fruit shells were considered.

Mushuma tree, an African native species, is dominant in the Midlands province of Zimbabwe. The Shuma fruits (Jackle-berry, Diospyros mespiliformisare) [28] are syrup-like juice and smooth with a soft-transparent-jelly inside. The tree has a medium to huge tree stem with the outer bark peeling off naturally as the tree grows as well as the season changes. The Mupane tree (Colophospermum mopane) [28] is a legume family vegetation abundantly found in the Midlands of Zimbabwe in hot, dry low-lying areas with an altitude ranging from 200 to 1150 meters above sea level.

The Mupane tree is also prevalent in South Africa, particularly in the Northwestern part of that country [29]. Tree barks of Mushuma and Mupane are usually used to start fires because of their common availability in the province as well as their affinity to fires. It is very quick to start fires and the tree wood itself takes a long time without extinguishing. Generally, these trees' bark is either left in the forest after the tree ages or the outer barks peeled off or used as fuel by the local communities. However due to high rural to urban movements, the availability of tree barks is increasing, and the rest is getting decomposed in the bush with no value to the community.

3. Characterization of biomass materials

3.1 FESEM-EDX

In efforts to understand the thermal transformation and the structural setup of Mupane, Mushuma tree bark as well as the Matamba fruit shell biochar, it was necessary to characterize their surfaces with field emission scanning electron microscopy (FESEM) and energy-dispersive X-ray spectroscopy (EDX) after pyrolysis. These materials were characterized by using FESEM (JEOL JSM-7500FAM Tokyo, Japan) for surface morphology and image generation and EDX for element composition of biochar with a low vacuum. The pretreatment of the samples was conducted, where the biochar samples were dried at 105°C for 4 hours, stuck on the copper plate using a black double-sided tape, vacuumed for 12 hours, and analyzed for surface transformation.

The outcome of the EDX conducted revealed the purity of the elemental composition of Mushuma, Mupane barks, and the Matamba fruit shell biochar. Principally, the Matamba fruit shell biochar exhibited to be made up of 72.68 wt% C and some significant elements such as 10.35 wt%, 14.14 wt%, 0.97 wt%, 0.46 wt%, 0.37 wt%, and 0.31 wt% of O, N, K, Mg, Ca, P respectively with some trace compounds of Si and S as shown in **Table 1**.

It was revealed that adequate content of the C element remained after pyrolysis greatly influenced the adsorption capacity (44.071 mmol/g) of the biochar as ascribed by the Elovich kinetic model. Furthermore, the available O composition also offers enough polarization capability for high adsorption of the iodine used (43.65 mmol/g) as observed in the experimental data.

3.2 FTIR measurement

The Fourier transform infrared spectroscopy (FTIR) analysis was carried out using a wave number scanning range between 400 and 4000 cm^{-1}. Before that, the content of the moisture and ash that can be available in these materials was measured following the ASTM D1762–84 guide. The elemental compositional analysis of C, H, and N was executed accordingly. Acetanilide was used as a standard. Approximately 2 mg of biochar was used for each measurement, and each measurement was carried out in triplicate. The oxygen content (O) was then determined by the difference between the original dried sample and the sum of C, H, N, and ash content.

3.3 Surface area estimation using iodine solution

The results from the EDX and FESEM elemental presentation show a high content of Caborn with rigid skeleton structures of Matamba fruit shell, Mushuma, and Mupane bark which would be ascribed to the residual lignin from incomplete pyrolysis of the materials. Moreover, the weak van der Waals forces played a role in the removal of Iodine due to these high C, C/N, and O/C ratios which are inferred in **Table 2** and augmented biochar surface meso pore filling.

Material	Element	C*	N	O*	Na	Mg*	Al*	Si	P*	S*	Cl*	K*	Ca*	Mn*	Zn*
Mushuma biochar	ms%	74.76	nd*	19.6	nd*	0.45	0.03	nd*	0.11	0.03	0.06	0.23	4.51	0.07	0.15
	mol%	81.93	nd*	16.12	nd*	0.24	0.02	nd*	0.05	0.01	0.02	0.08	1.48	0.02	0.03
Mupane biochar	ms%	80.92	nd*	14.25	nd*	0.25	nd*	0.04	0.02	0.07	nd	0.07	4.18	0.04	0.16
	mol%	86.91	nd*	11.49	nd*	0.13	nd*	0.02	0.01	0.03	nd	0.02	1.34	0.01	0.03
Matamba biochar	ms%	72.68	14.14	10.35	nd*	0.46	0.04	0.02	0.31	0.22	0.02	0.97	0.37	nd*	nd*
	mol%	77.71	12.97	8.3	nd*	0.24	0.02	0.01	0.13	0.09	0.01	0.32	0.12	nd*	nd*

**nd not detected.*

Table 1.
The elemental composition of Mushuma, Matamba, and Mupane biochar.

DOI: http://dx.doi.org/10.5772/intechopen.105868

MODEL	Adsorbent			Bayesian	statistical	analysis					Conventional	analysis
		Model Parameter	Mean (±SD)	2.50%	25%	50%	75%	97.50%	Rhat	MAP	Mean	AICc
Langmuir	Matamba BC	KL (L /mol)	218.5 ± 81.8	110.2	171.5	206.03	248.42	400.1	1.001	195.6	206.4 ± 49.11	0.527
	Matamba BC	qmax (mmol/g)	2.12 ± 0.12	1.886	2.05	2.118	2.188	2.363	1.001	2.11	2.12 ± 0.089	
Freundlich	Matamba BC	KF (mmol L /mol)	3.061 ± 0.40	2.3	2.8	3.1	3.3	3.892	1.001	3.08	3.08 ± 0.295	5.377
	Matamba BC	mf [−]	0.194 ± 0.04	0.11	0.2	0.2	0.2	0.282	1.027	0.2	0.196 ± 0.032	
	Mushuma BC	k_2 (g mmol/min)	0.014 ± 0.002	0.011	0.013	0.014	0.016	0.019	1.002			
PSO	Mushuma BC	qt (mmol/g)	40.712 ± 0.986	38.776	40.106	40.703	41.308	42.715	1.001			
	Mupane BC	k_2 (g mmol/min)	0.014 ± 0.002	0.011	0.013	0.014	0.015	0.019	1.002			
	Mupane BC	qt (mmol/g)	41.64 ± 0.997	39.621	41.051	41.64	42.228	43.62	1.002			
	Mushuma BC	α (mmol/g/ min)	112.85 ± 40.2	55.621	86.492	105.82	131.39	212.627	1.002			
Elovich	Mushuma BC	β (g/ mmol)	0.16 ± 0.012	0.135	0.152	0.159	0.167	0.185	1.001			
	Mupane BC	α (mmol/g/ min)	120.2 ± 52.72	53.172	84.764	108.84	142.87	256.495	1.001			
	Mupane BC	β (g/ mmol)	0.156 ± 0.015	0.129	0.146	0.156	0.166	0.187	1.002			

Table 2.
Shows both the Bayesian and conventional statistical analysis results for the biochar investigated using isotherm and kinetic models.

The remaining alkaline elements such as Ca, Mg, and K with inorganic basic minerals present might be ascribed to the main component of ash established from the pyrolysis process of the biochar [29, 30]. In summary, the Iodine adsorption mechanisms onto the investigated biochar materials made it a probable choice for environmental contamination option, water reuse possibility, and global warming reduction due to high C, C/N, and enough polarization propensity.

Regarding the biochar produced from Mushuma and Mupane barks, it was from the FESEM images above that surface texture can be influenced by biomass type even under identical pyrolysis conditions. Biochar produced from Mushuma bark has large surface pores (10–15 m in diameter), uniformly distributed and separated by a thick carbon wall (2–3 m) than Mupane bark. Biochar from Mupane bark had smaller and heterogeneously distributed pores (3–5 m in diameter).

Similarly, the kinetic results from Iodine adsorption indicated that biochar from Mupane had higher qt values than biochar from Mushuma bark. As found by [29], larger pores tend to correlate to the limited surface area than small pores, thus there is greater adsorption on smaller pores than on larger pores. Further, the small pores are associated with high porosity and void volume.

4. Equilibrium mechanisms of adsorbents and data analysis

4.1 Adsorption kinetics of Matamba fruit shell and the tree bark adsorbents

In principle, [31] elucidated that adsorption is known as the mass transfer method that entails some time for the adsorbate to diffuse from the bulk solution of the aqueous phase, through the solid–liquid film into the material's pore spaces and onto the available active sites. Therefore, based on the results obtained from the experiments kinetic models like pseudo-first-order (PFO), pseudo-second-order (PSO), intraparticle diffusion (IPD), and Elovich models are shown in **Table 3** and plotted as shown in **Figure 2a**.

MPNBC advocated more adsorption for Iodine than MSHBC as exhibited in **Figure 3c** and **d** correspondingly, however, Matamba fruit shell outperformed both tree barks. Subsequently, the Iodine kinetic adsorption mechanism on these materials could be divided into three stages: rapid adsorption stage, slow adsorption stage, and adsorption equilibrium stage as elucidated by [32] as well.

The first 12 hours were observed to be a rapid Iodine adsorption stage on both biochars. The graph for MPNBC seems to be steeper than MSHBC. The adsorption capacity of the prepared Mupane and Mushuma barks were estimated to be 40.38 and 39.78 mmol g^{-1} respectively, from the experimental data. From conventional statistical analysis of the Pseudo-second order model, Mushuma and Mupani biochar exhibited adsorption capacity of 40.01 and 40.29 mmol g^{-1} which were slightly lower than the Bayesian outcome of 40.712 and 41.639 mmol g^{-1} as shown in **Table 2**.

This reveals the strength of Bayesian analysis against classical statistics as different quantile ranges revealed different estimations and the 50% (median) was so close to the actual mean for each parameter. The adsorption rate constants of the two biochar also exhibited the above phenomenon where the conventional method indicated a homogeneous reaction rate (0.014 min^{-1}), so as the Bayesian statistics as shown in **Table 2**. The figures also elucidated that the linear relationship is presented not as a continuous straight line but in two stages of least and enormous adsorption before and after 4 hr. of adsorption respectively.

Kinetic model	Equations	Model parameter	Matamba	Mushuma	Mupane	Reference
PFO	$q_t = q_e(1 - exp(-k_1 t))$ (1)	qt (mmol/g)	40.08	37.28	38.2	Lagergren, 1898 (Eq. (1))
		k_1 (1/min)	0.232	0.4	0.41	
		AIC	59.05	51.52	51.14	
PSO	$q_t = \frac{q_e^2 k_2 t}{1+q_e k_2 t}$ (2)	qt (mmol/g)	44.071	40.01	40.29	Ho and Makay, 1999 (Eq. (2))
		k_2 (g mmol/ min)	0.0079	0.01	0.01	
		AIC	51.95	37.76	38.03	
IPD	$q_t = k_p t^{\frac{1}{2}} + C$ (3)	α (mmol/g/ min)	45.41	110.7	117.88	Weber and Morris, 1963 (Eq. (3))
		β (g/ mmol)	0.127	0.16	0.16	
		AIC	38.26	41.52	44.92	
		k_p (mmol/ min)	7.72	3.64	3.66	
Elovich	$q_t = \left(\frac{1}{\beta}\right) ln(\alpha\beta) + \left(\frac{1}{\beta}\right) ln(t)$ (4)	C [−]	14.303	18.53	19.23	Bedin et al., 2016 (Eq. (4))
		AIC	46.56	54.29	55.94	

The AIC scores were used for non-linear model selection instead of the coefficient of determination (R^2).

Table 3.
The kinetic adsorption equations and estimated parameters from biochar materials after the experiment.

The higher adsorption could be attributed to smaller biochar particle sizes (0.25–1.00 mm) used in this experiment which started to diffuse into the pores later since IPD is a slow process. The Elovich model revealed that the initial adsorption was 110.701 and 117.88 mmol g^{-1} min^{-1}, 112.847 and 120.214 mmol g^{-1} min^{-1} for Mupane and Mushuma biochar from Conventional and Bayesian methods respectively.

Different adsorption mechanisms could have been encountered during the 48-hour contact time, but the adsorption rate gradually decreased until the adsorption reaches the equilibrium state as [28] elaborated. For Matamba biochar, Elovich (Eq. (3)) and IPD (Eq. (4)) better describe the kinetic adsorption of biochar through iodine adsorption than PFO (Eq. (1)) and PSO (Eq. (2)) models. Generally, the iodine adsorption rate decreases exponentially as the amount of iodine adsorbed increases on the heterogeneous surfaces of the Mushuma, Mushuma, and Matamba fruit shell biochar. Several adsorption experiments have been reported to follow the Elovich kinetic model [33, 34].

However, the adsorption kinetic results from conventional statistics on the tree bark revealed that the PSO kinetic model better described the adsorption behaviors of the biochar for Iodine adsorption [35, 36]. The model selection AICc scores for MPNBC and MSHBC were 38.03 and 37.76 respectively, away below other models used. AICc is a strong tool for model selection than using the correlation coefficient on non-linear model functions. This can be theoretically supported by the equilibrium adsorption capacity values from both statistical methods were also close to the experimental equilibrium adsorption capacity, signifying that the pseudo-second-order kinetic model could better describe the Iodine adsorption [36, 37]. From this point of

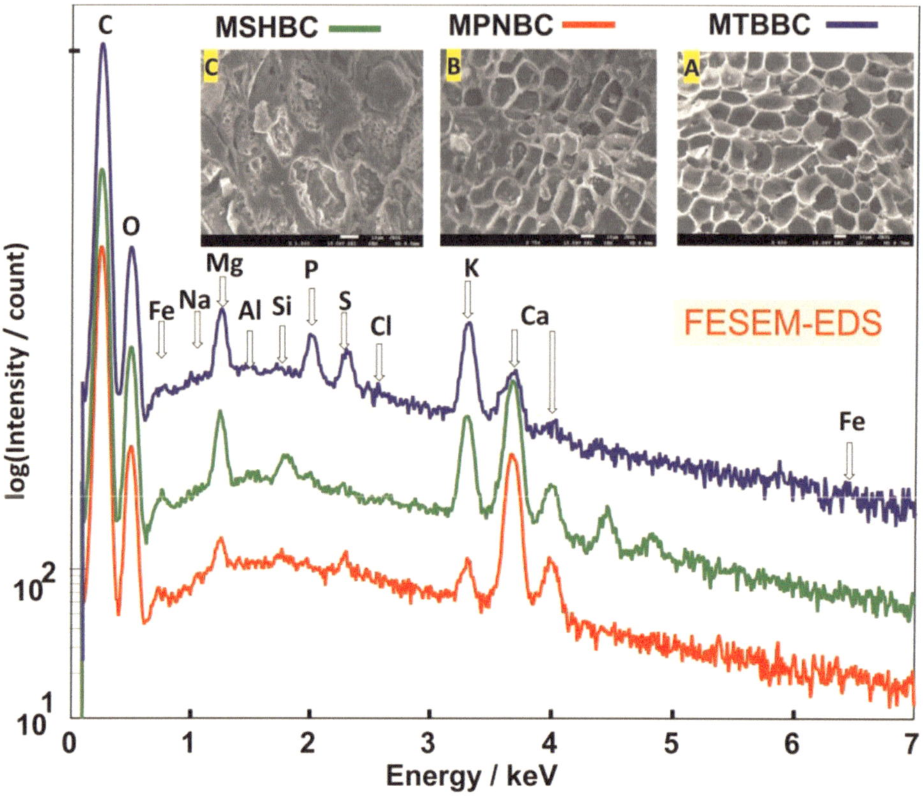

Figure 2.
Elemental composition and FESEM analysis of (a) Matamba biochar (MTBBC), (b) Mupane tree bark (MPNBC), and (c) Mushuma tree bark (MSHBC).

view, it can be inferred that both conventional and Bayesian approaches to estimations are well established and seem hard to justify if one of the two is preferred over the other [38, 39]. It is thought that the π-π electron donor-acceptor (EDA) interaction is the main player with a major role in the iodine - adsorbent interaction since the adsorption capacity after 2 days of investigation. The strong interaction of π-donor and π-acceptor compounds full fills the EDA theory taking into consideration the FTIR results. As given in **Figure 4e**, the biochar materials also show various surface functional groups. Regarding **Figure 4e**, the peaks at 3334 cm^{-1} and 1764–1710 cm^{-1}, can be ascribed to the hydroxyl groups (-OH) and the carbonyl groups (C=O) correspondingly. The shallow peak at 1385 cm^{-1} and deep and wide peak at 1568 cm^{-1} are due to C=C stretches of aromatic rings. Furthermore, the 1223 cm^{-1} peak can be ascribed to the C=O stretching in ethers, alcohols, and/or phenols. The FTIR outcomes clarify that the condition of pyrolysis has a great impact on the adsorption capacity of Iodine in terms of the hydrogen bond capacity created on the biochar materials.

Furthermore, the hydrophobic sites could be originated from the graphitic structure of biochar which is assumed to be interacting with hydrophobic molecules of the biochar. However, the adsorption isotherm results can corroborate this phenomenon. For the tree bark materials, the adsorption kinetic results are shown in **Figure 3c** and **d** and **Table 3** and **Table 3** revealed that the kinetic model fits follow the order PSO > Elovich > PFO > IPD yet for the Matamba fruit shell, Elovich model fitted the adsorption data better than other kinetic models.

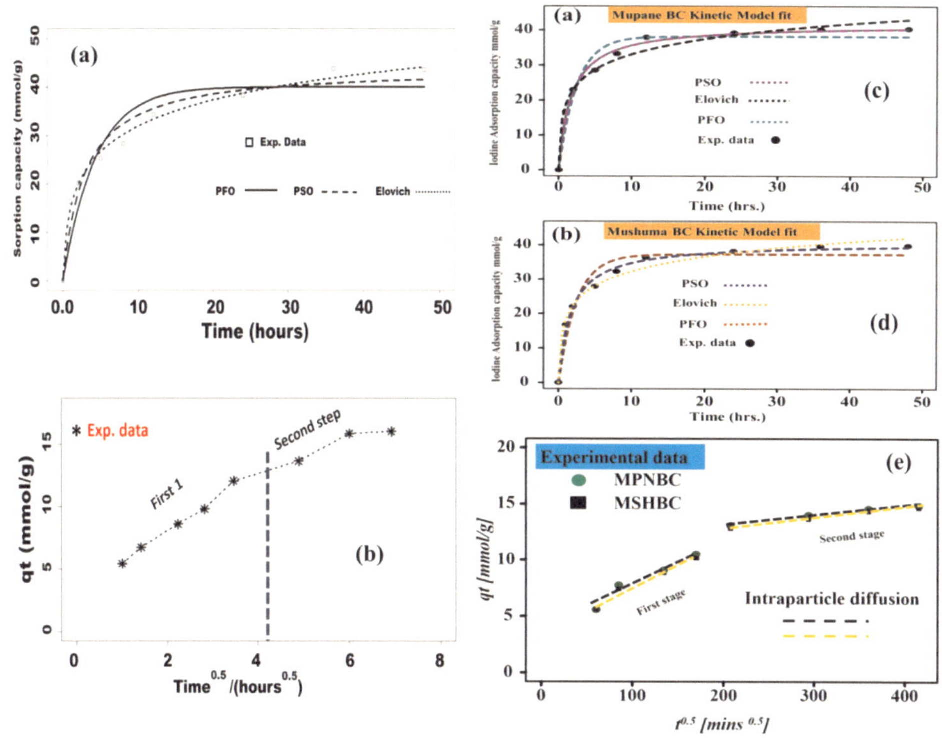

Figure 3.
The results from the kinetic adsorption experimental model analysis of biochar materials (a), (c), (d) are PFO, PSO, and Elovich models respectively, and (b), (e) are IPD models correspondingly.

4.2 Langmuir and Freundlich isotherms on Matamba fruit shell

Langmuir and Freundlich isotherm models (Eq. (5)) and (Eq. (6)) were used to examine and investigate the adsorption mechanisms of iodine onto the biochar surface. The Langmuir model described well the removal of iodine with the AICc of 0.527 (lower than 5.377 of the Freundlich model), which exhibited monolayer sorption on the Biochar surface with determinate indistinguishable adsorption sites.

Additionally, Bayesian statistics exhibited a clear difference between the two models from the ggplot2 since Freundlich (**Figure 4d**) shows a wider prediction band than Langmuir (**Figure 4c**). The maximum capacity of adsorption deliberated from the Langmuir model was so vital in biochar surface area estimation. The Matamba fruit shell biochar surface area was estimated to be 267.9 $m^2\,g^{-1}$ and 267.6 $m^2\,g^{-1}$ from NLS and Bayesian approaches respectively. The biochar surface area was estimated from Iodine adsorption using (Eq. (7)), whereas the Langmuir and Freundlich models reiterate that:

$$q_e = \frac{q_m\, k_L c_e}{1 + k_L c_e} \tag{1}$$

$$q_e = k_f c_e^{mf} \tag{2}$$

$$SAr = qt * 10 - 3 * NA * \omega I \tag{3}$$

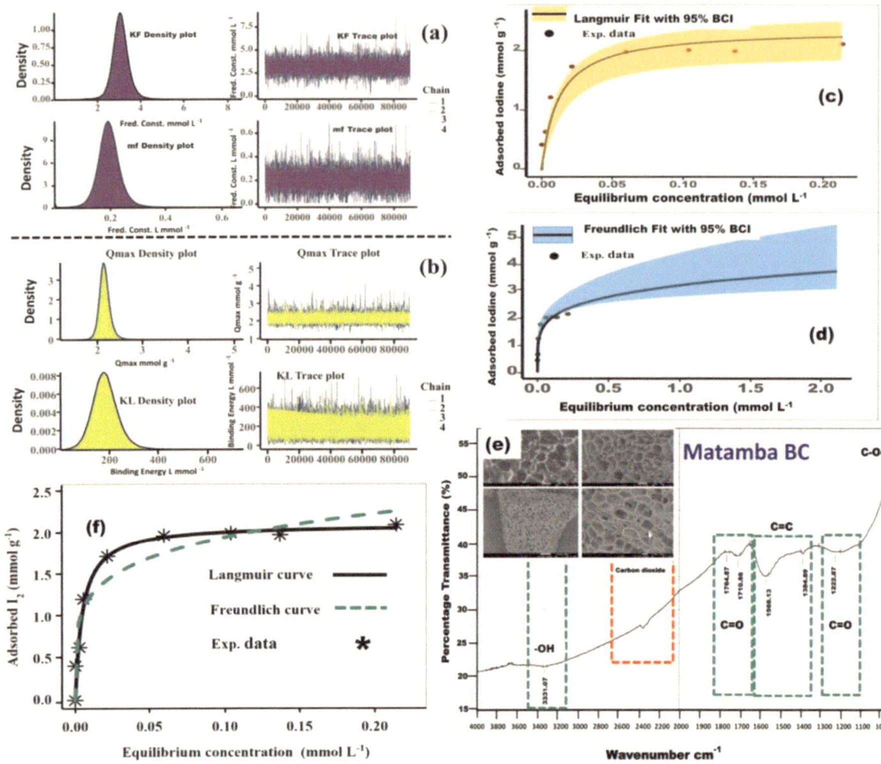

Figure 4.
(a) Freundlich and (b) Langmuir model density curves, (c) 95% Bayesian C.I. analysis of Langmuir and (d) Freundlich adsorption models, (e) FTIR and FESEM for Matamba biochar, (f) Langmuir and Freundlich adsorption models with conventional analysis method.

Where q*m* is the maximum adsorption capacity (mmol g^{-1}), k_L is Langmuir constant (L $mmol^{-1}$, Ce is the equilibrium concentration (mmol L^{-1}), k_f (mmol L $mmol^{-1}$), and m_f are Freundlich constants. From this qt is the maximum capacity of adsorption at equilibrium mmol/g, N_A is the Avogadro constant, $N_A = 6.023 \times 10^{23}$ mol^{-1}, and ω_I is the surface area occupied by one iodine molecule ($0.2096 * 10^{-18}$ m^2).

The surface area estimated from both Bayesian and Conventional statistics is insignificant since the *qmax* parameter (**Figure 5a–d**) only underscores the capacity of the biochar to adsorb the adsorbate yet has less substantial than the K_L as explained by [40]. The high value of the K_L parameter from the MCMC in **Figure 5** is directly proportional to the observed surface area because iodine molecules are small enough and strong to be attached to the biochar surface with minimum effects of desorption.

4.3 Statistical analysis using a Bayesian framework

The Bayesian statistics obtains *qmax* of 2.12 mmol g^{-1} and Conventional statistics resulted in the maximum adsorption capacity (*qmax*) of 2.122 mmol g^{-1}. Moreover, the median value was estimated to be 2.12 mmol g^{-1}, whereas the MAP value of 2.11 mmol g^{-1} was obtained and there were no significant differences with the *qmax*. The Rhat between 1.05 and 0.9 is acceptable and helps in the rejection of the Markov

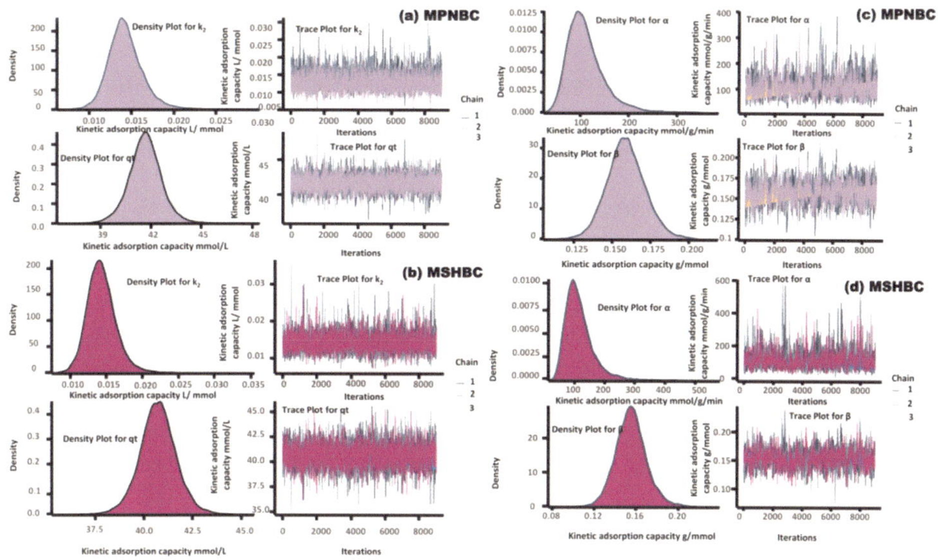

Figure 5.
Shows the MCMC density presentation of MPNBC and MSHBC from the Bayesian simulation of PSO and Elovich model's posterior probability distribution mean parameters.

chain Monte Carlo (MCMC) data simulation as it is far away from this range. The Bayesian statistics estimated that the energy binding strength (K_L) to be higher than the NLS, this is shown in **Table 2** where 218.5 L $mmol^{-1}$ and 206.43 L $mmol^{-1}$ were observed for the Bayesian and NLS observed respectively.

The K_L results exhibited a stronger evaluation as depicted by the Bayesian method than conventional statistics, so, Bayesian statistics seem to have a great capacity to estimate isotherm and kinetics parameters with consistency and supporting evidence than the former. The K_L is more significant as estimated by the Bayesian analysis and designated the degree of interface among iodine solution and the biochar surface property. Higher values of the K_L relatively resemble a strong interaction or sorption affinity of the adsorbate concentration onto the adsorbent as large values of K_L reflect the greater force of binding on the biochar material's surface [41, 42].

5. Conclusions

The pyrolysis condition at 600°C revealed the surface characteristics and adsorption mechanisms of the biochar materials to be sufficient in generating adequate biochar for the purpose. These agro-biomass materials used in this study were the first to be investigated for their potential application as low-cost adsorbents in rural areas of Zimbabwe for environmental protection. Easy access to these materials as well as lower production cost makes them fit to solve the water shortages and remove unwanted substances from the environment through adsorption. Elovich and PSO models fitted the data in this study, and this exhibits a heterogeneous surface characteristic of the biochar materials with significant chemisorption mechanisms developed during pyrolysis of the agro-biobased biochar. Bayesian statistical analysis has exhibited slightly higher qt estimations of 40.712 and 41.639 mmol/g when compared to the conventional statistics with 40.01 and 40.29 mmol/g for Mushuma and Mupane biochar. The Elovich model subsequently described the results very well, henceforth representing a heterogenous surface property with chemisorption phenomena. FESEM-EDX Spectroscopy also revealed that C (81.93 mol% and 86.91 mol %) and O (16.12 mol% and 11.49 mol%) for Mushuma and Mupane respectively. These percentages agreed with the FTIR results where the surface physical properties designated a rich surface with fundamental functional groups and, as a recommendation with the cost for future research, activating these materials could make them enduring adsorbents. The investigation outcomes unveiled the competence and potential of the locally obtainable and produced biochar in removing Iodine solution as affordable materials that can be established for other emerging contaminants and unwanted pollutants from the environment as water reuse and recycling strategy in developing countries and unserved communities and as a climate change mitigatory measure. Matamba, Museum, and Mupane biochar materials are locally available, no costs are required to obtain them, and the benefits of wastewater recycling strategy should be adopted with a proper design fit for rural communities as off-the-grid technology.

Acknowledgements

The authors would like to thank Nagasaki University, Graduate School of Engineering, for the provision of resources, chemicals, and analytical equipment that made this work a success. The authors appreciate the referees, editors, and reviewers for their effort in their correction and suggestions to improve the quality and content of this paper.

Conflict of interest

The authors declare that they have no known competing interests.

Notes/thanks/other declarations

Most acknowledgment is given to the JICA and HONJO Scholarship Foundation for the provision of these Scholarships to be in Japan for the study period.

Author details

Obey Gotore[1]*, Tirivashe Phillip Masere[2], Osamu Nakagoe[1], Vadzanayi Mushayi[3], Ramaraj Rameshprabu[4], Yuwalee Unpaprom[4] and Tomoaki Itayama[1]

1 Nagasaki University, Nagasaki, Japan

2 Midlands State University, Gweru, Zimbabwe

3 Harare Polytechnic College, Harare, Zimbabwe

4 Maejo University, Chiang Mai, Thailand

*Address all correspondence to: gotoreobey@gmail.com

References

[1] Abedin M, Collins AE, Habiba U, Shaw R. Climate change, water scarcity, and health adaptation in southwestern coastal Bangladesh. International Journal of Disaster Risk Science. 2019;**10**(1): 28-42. DOI: 10.1007/s13753-018-0211-8

[2] Hanna R, Oliva P. Implications of climate change for children in developing countries. The Future of Children. 2016;**26**:115-132. DOI: 10.1353/foc.2016.0006

[3] Liu X, Tian J, Li Y, Sun N, Mi S, Xie Y, et al. Enhanced dyes adsorption from wastewater via Fe3O4 nanoparticles functionalized activated carbon. Journal of Hazardous Materials. 2019;**373**: 397-407. DOI: 10.1016/j.jhazmat.2019.03.103

[4] Bekçi Z, Özveri C, Seki Y, Yurdakoç K. Sorption of malachite green on chitosan bead. Journal of Hazardous Materials. 2008;**154**(1–3):254-261. DOI: 10.1016/j.jhazmat.2007.10.021

[5] Zhou L, Liu Y, Liu S, Yin Y, Zeng G, Tan X, et al. Investigation of the adsorption-reduction mechanisms of hexavalent chromium by ramie biochars of different pyrolytic temperatures. Bioresource Technology. 2016;**218**: 351-359. DOI: 10.1016/j.biortech.2016.06.102

[6] Manea YK, Khan AM, Wani AA, Saleh MA, Qashqoosh MT, Shahadat M, et al. In-grown flower-like Al-Li/Th-LDH@ CNT nanocomposite for enhanced photocatalytic degradation of MG dye and selective adsorption of Cr (VI). Journal of Environmental Chemical Engineering. 2022;**10**(1):106848. DOI: 10.1016/j.jece.2021.106848

[7] Ghatge S, Yang Y, Ko Y, Yoon Y, Ahn JH, Kim JJ, et al. Degradation of sulfonated polyethylene by a bio-photo-Fenton approach using glucose oxidase immobilized on titanium dioxide. Journal of Hazardous Materials. 2022; **423**:127067. DOI: 10.1016/j.jhazmat.2021.127067

[8] Kodavatiganti S, Bhat AP, Gogate PR. Intensified degradation of acid violet 7 dye using ultrasound combined with hydrogen peroxide, Fenton, and persulfate. Separation and Purification Technology. 2021;**279**:119673. DOI: 10.1016/j.seppur.2021.119673

[9] Gotore O, Rameshprabu R, Itayama T. Adsorption performances of corn cob-derived biochar in saturated and semi-saturated vertical-flow constructed wetlands for nutrient removal under erratic oxygen supply. Environmental Chemistry and Ecotoxicology. 2022;**4**:155-163. DOI: 10.1016/j.enceco.2022.05.001

[10] Prajapati AK, Mondal MK. Comprehensive kinetic and mass transfer modeling for methylene blue dye adsorption onto CuO nanoparticles loaded on nanoporous activated carbon prepared from waste coconut shell. Journal of Molecular Liquids. 2020;**307**: 112949. DOI: 10.1016/j.molliq.2020.112949

[11] Obey G, Adelaide M, Ramaraj R. Biochar derived from non-customized Matamba fruit shell as an adsorbent for wastewater treatment. Journal of Bioresources and Bioproducts. 2022;**7**(2): 109-115. DOI: 10.1016/j.jobab.2021.12.001

[12] Castiglioni M, Rivoira L, Ingrando I, Meucci L, Binetti R, Fungi M, et al. Biochars intended for water filtration: A comparative study with activated

carbons of their physicochemical properties and removal efficiency towards neutral and anionic organic pollutants. Chemosphere. 2022;**288**: 132538. DOI: 10.3390/en14248472

[13] Lesaoana M, Mlaba RP, Mtunzi FM, Klink MJ, Ejidike P, Pakade VE. Influence of inorganic acid modification on Cr (VI) adsorption performance and the physicochemical properties of activated carbon. South African Journal of Chemical Engineering. 2019;**28**:8-18. DOI: 10.1016/j.sajce.2019.01.001

[14] Cong L, Feng L, Wei X, Jin J, Wu K. Study on the adsorption characteristics of Congo red by sycamore bark activated carbon. In: ENVIRONMENT. TECHNOLOGIES. RESOURCES. Proceedings of the International Scientific and Practical Conference. Republic of Latvia: The Scientific Journal of Rezekne Academy of Technologies; Vol. 1. 2017. pp. 64-69. DOI: 10.17770/etr2017vol1.2600

[15] Gautam UK, Panchakarla LS, Dierre B, Fang X, Bando Y, Sekiguchi T, et al. Solvothermal synthesis, cathodoluminescence, and field-emission properties of pure and N-doped ZnO nano bullets. Advanced Functional Materials. 2009;**19**(1):131-140. DOI: 10.1002/adfm.200801259

[16] Talha MA, Goswami M, Giri BS, Sharma A, Rai BN, Singh RS. Bioremediation of Congo red dye in im7obilized batch and continuous packed bed bioreactor by Brevibacillusparabrevis using coconut shell bio-char. Bioresource Technology. 2018;**252**:37-43. DOI: 10.1016/j.biortech.2017.12.081

[17] Fu MM, Mo CH, Li H, Zhang YN, Huang WX, Wong MH. Comparison of physicochemical properties of biochars and hydrochars produced from food wastes. Journal of Cleaner Production. 2019;**236**:117637. DOI: 10.1016/j.jclepro.2019.117637

[18] Sriburi T, Wijitkosum S. Biochar amendment experiments in Thailand: Practical examples. In: Bruckman VJ, Klinglmüller M, editors. Potentials to Mitigate Climate Change Using Biochar. Cambridge, United Kingdom: Cambridge University Press; 2016

[19] O'Connor D, Peng TY, Zhang JL, Tsang DCW, Alessi DS, Shen ZT, et al. Biochar application for the remediation of heavy metal polluted land: A review of *in situ* field trials. Science of the Total Environment. 2018;**619**(620):815-826. DOI: 10.1016/j.scitotenv.2017.11.132

[20] Chandra S, Bhattacharya J. Influence of temperature and duration of pyrolysis on the property heterogeneity of rice straw biochar and optimization of pyrolysis conditions for its application in soils. Journal of Cleaner Production. 2019;**215**:1123-1139. DOI: 10.1016/j.jclepro.2019.01.079

[21] Kambo HS, Dutta A. A comparative review of biochar and hydrochar in terms of production, physico-chemical properties and applications. Renewable and Sustainable Energy Reviews. 2015; **45**:359-378. DOI: 10.1016/j.rser.2015.01.050

[22] Gotore O, Osamu N, Rameshprabu R, Arthi M, Unpaprom Y, Itayama T. Iodine adsorption isotherms on Matamba fruit shell stemmed biochar for wastewater re-use strategy in rural areas owing to climate change. Chemosphere. 2022;**303**:135126. DOI: 10.1016/j.chemosphere.2022.135126

[23] Morgan TJ, Turn SQ, George A. Fast pyrolysis behavior of banagrass as a

function of temperature and volatiles residence time in a fluidized bed reactor. PLoS One. 2015;**10**:e0136511. DOI: 10.1371/journal.pone.0136511

[24] Ngadze RT, Verkerk R, Nyanga LK, Fogliano V, Linnemann AR. Improvement of traditional processing of local monkey orange (Strychnos spp.) fruits to enhance nutrition security in Zimbabwe. Food Security. 2017;**9**:1-13. DOI: 10.1007/s12571-017-0679-x

[25] Mwamba CK. Monkey Orange: Strychnos cocculoides Crops for the Future. Vol. 8. Southampton Centre for Underutilised Crops, UK; 2006

[26] National Research Council. Lost crops of Africa. In: Fruits. Washington DC: The National Academies Press; 2008

[27] Mwampamba TH, Owen M, Pigaht M. Opportunities, challenges and way forward for the charcoal briquette industry in sub-Saharan Africa. Energy for Sustainable Development. 2013;**17**:158-170. DOI: 10.1016/j.esd.2012.10.006

[28] Ajayi OC, Mafongoya PL. Indigenous knowledge systems and climate change management in Africa. Africa Report; Technical Center for Agricultural and Rural Cooperation; 2017. Available online: https://scholar.google.com/scholar_lookup?hl=en&publication_year=2017&author=P.L.+Mafongoya&author=O.C.+Ajayi&title=Indigenous+Knowledge+Systems+and+Climate+Change+Management+in+Africa

[29] Gotore O, Mushayi V, Rameshprabu R, Gochayi L, Itayama T. Adsorption studies of iodine removal by low-cost bioinspired Mushuma and Mupane bark derived adsorbents for urban and rural wastewater reuse. International Journal of Human Capital in Urban Management. 2022;**7**(3): 297-308. DOI: 10.22034/IJHCUM.2022.03.01

[30] Taha SM, Amer ME, Elmarsafy AE, Elkady MY. Adsorption of 15 different pesticides on untreated and phosphoric acid treated biochar and charcoal from water. Journal of Environmental Chemical Engineering. 2014;**2**:20-25. DOI: 10.1016/j.jece.2014.09.001

[31] Hevira L, Ighalo JO, Aziz H, Zein R. Terminalia catappa shell as low-cost biosorbent for the removal of methylene blue from aqueous solutions. Journal of Industrial and Engineering Chemistry. 2021;**97**:188-199. DOI: 10.1016/j.jiec.2021.01.028

[32] Zazycki MA, Godinho M, Perondi D, Foletto EL, Collazzo GC, Dotto GL. New biochar from pecan nutshells as an alternative adsorbent for removing reactive red 141 from aqueous solutions. Journal of Cleaner Production. 2018;**171**: 57-65. DOI: 10.1016/j.jclepro.2017.10.007

[33] Dotto GL, Pinto LA. Adsorption of food dyes onto chitosan: Optimization process and kinetic. Carbohydrate Polymers. 2011;**84**(1):231-238. DOI: 10.1016/j.carbpol.2010.11.028

[34] Cheung CW, Porter JF, McKay G. Elovich equation and modified second-order equation for sorption of cadmium ions onto bone char. Journal of Chemical Technology & Biotechnology. 2000; **75**(11):963-970. DOI: 10.1002/1097-4660(200011)75:11<963::AID-JCTB302>3.0.CO;2-Z

[35] Dang BT, Gotore O, Ramaraj R, Unpaprom Y, Whangchai N, Bui XT, et al. Sustainability and application of corncob-derived biochar for removal of fluoroquinolones. Biomass Conversion and Biorefinery. 2022;**12**(3):913-923. DOI: 10.1007/s13399-020-01222-x

[36] Doumer ME, Arízaga GG, da Silva DA, Yamamoto CI, Novotny EH, Santos JM, et al. Slow pyrolysis of different Brazilian waste biomasses as sources of soil conditioners and energy, and for environmental protection. Journal of Analytical and Applied Pyrolysis. 2015;**113**:434-443. DOI: 10.1016/j.jaap.2015.03.006

[37] Al-Wabel MI, Al-Omran A, El-Naggar AH, Nadeem M, Usman AR. Pyrolysis temperature induced changes in characteristics and chemical composition of biochar produced from conocarpus wastes. Bioresource Technology. 2013;**131**:374-379. DOI: 10.1016/j.biortech.2012.12.165

[38] Furia CA, Feldt R, Torkar R. Bayesian data analysis in empirical software engineering research. IEEE Transactions on Software Engineering. 2019;**47**(9):1786-1810. DOI: 10.1109/TSE.2019.2935974

[39] Gelman A, Carlin JB, Stern HS, Rubin DB. Bayesian Data Analysis Chapman & Hall. Boca Raton, FL, USA: CRC Texts in Statistical Science; 2004

[40] MacDermid-Watts K, Pradhan R, Dutta A. Catalytic hydrothermal carbonization treatment of biomass for enhanced activated carbon: A review. Waste and Biomass Valorization. 2021; **12**(5):2171-2186. DOI: 10.1007/s12649-020-01134-x

[41] Bandara T, Xu J, Potter ID, Franks A, Chathurika JB, Tang C. Mechanisms for the removal of Cd (II) and Cu (II) from aqueous solution and mine water by biochars derived from agricultural wastes. Chemosphere. 2020;**254**:126745. DOI: 10.1016/j.chemosphere.2020.126745

[42] Devi P, Saroha AK. Improvement in performance of sludge-based adsorbents by controlling key parameters by activation/modification: A critical review. Critical Reviews in Environmental Science and Technology. 2016;**46**(21–22):1704-1743. DOI: 10.1080/10643389.2016.1260902

Chapter 16

PAHs, PCBs and Environmental Contamination in Char Products

Karl Williams, Ala Khodier and Peter Bentley

Abstract

Biochar can have unique benefits to terrestrial and aquatic ecosystems. Investigations of biochar effectiveness within these environments often come from homogenous feedstocks, such as plant biomass, which have simple thermochemical processing methods and produce physically and chemically stable biochar. Current methods to increase biochar production include the addition of oil-derived products such as plastics, which produces a more heterogenous feedstock. This feedstock is similar to materials from waste recycling streams. The adoption of more heterogenous feedstocks produces additional challenges to biochar production and use. This can result in pollution contained within the feedstock being transferred to the biochar or the creation of pollutants during the processing. With the current climate emergency, it is essential to eliminate environmental contamination arising from biochar production. It is critical to understand the physiochemical composition of biochar, where detailed analysis of contaminants is often overlooked. Contamination is common from heterogenous feedstocks but on commercial scales, even homogeneous biochar will contain organic pollutants. This chapter investigates biochar produced from various waste feedstocks and the challenges faced in thermochemical processing. Using Automotive Shredder Residue (ASR) as an example of a heterogeneous feedstock, the levels of contamination are explored. Potential solutions are reviewed while assessing the environmental and economic benefits of using biochar from mixed sources.

Keywords: persistent organic pollutants, heterogeneous feedstock pyrolysis, biochar secondary processing, automotive shredder residue

1. Introduction

Biochar has been promoted as a solution to enhance soils as a conditioner and as an additive to enhance contaminated land remediation. For many of these proposed applications the positive properties of the biochar in the environment are championed, however, there is little investigation into their negative impacts on the environment. The main area of concern is the presence of persistent organic pollutants (POP), polyaromatic hydrocarbons (PAH) and polychlorinated biphenyl (PCB) within the biochar itself. Much of the research on the sources of material for biochar is carried out on small scale laboratory test rigs with carefully chosen homogenous feed sources. This does not represent the potential commercial application where a more heterogeneous feed would be present. There is also a drive to enhance and improve the production of biochar by the combination of organic and plastics. This again can give rise to contamination with undesirable by-products.

It is well known that soils already contain POPS however, there are concerns over these levels [1]. The addition of biochar containing POPs would increase the concentration. The main barrier to analysis of POP in soil is the variability of the soil and a methodology is complex and there is no specified guidelines and corresponding legislation [2]. Consequently, there is no incentive to analyze for organic contamination and the main analysis reports metal levels. For the threshold levels of POP in soil under UK legislation a risk assessment-based approach is required [3].

Many research projects investigating biochar from plant biomass assess the chemical status via the evaluation of organic elemental composition and the biochar porosity only [4, 5]. Although this is a useful method to understand how the char will develop in soil and its potential to absorb nutrients, further analysis of the inorganic metal concentration and the organic pollutants (such as PAHs) contained within the product may provide further information on the environmental contamination from the feedstock that is being added to the soil. Currently, there are limited regulations surrounding biochar reuse from organic products such as biomass, as it is assumed that plants are inert. However, bio-uptake from energy crops contaminated land sources, such as miscanthus, could be a result in a significant amount of pollution retained within biochar following thermal processing. Advanced chemical analysis of biochar is required to ensure that pollution from initial feedstock sources do not cause further pollution.

Biochar is the solid residue obtained during the thermochemical conversion of biomass in an oxygen limiting environment. Unlike combustible ash residues, biochar is a stable solid, rich in pyrogenic carbon. Biochar resides from biomass feedstock, of which there are 6 main sources: agricultural waste, forestry waste, animal waste, industrial residues, and municipal solid waste. Re-use of biochar from waste materials could have many positive environmental and economic effects for the waste recycling industry, including a reduction in waste to landfill and the provision of a circular economy from waste recycling. However, the chemical consistency of the feedstock can have significant implications on the quality of biochar and its potential re-use in certain applications. The use of non-homogeneous feed stocks such automotive shredder residue (ASR) is a good example of a mixture of organic material with oil derived plastic.

There are many applications of biochar, however it is commonly applied to agricultural systems as a soil improver. Addition of biochar to agroecosystems can have significant benefits to soil properties and plant health [6], where carbon sequestration, water retention, microbial activity and herbicide suppression is increased [7–10], whilst nutrient leaching is decreased [11]. It has been calculated that biochar addition can increase soil organic carbon (SOC) stocks by 29% (13 Mg ha^{-1}) [6]. Biochar can be added to soil via different methods, it can be mixed directly into soil, or used as an additive to other processes, such as compost, manures and fertilizers, where the biochar acts as a carrier for the nutrients. Through biochar applications increasing carbon storage within soil, the carbon footprint caused by thermal processing of biomass waste for energy is reduced [12, 13]. Little investigation has been carried out on the level of POP that come from the processing process and different feedstocks. This omission means that we do not have the full picture on what we are depositing onto the land.

Alongside the addition of biochar to agroecosystems to increase soil fertility and improve crop growth, the adsorbent properties of char make it a useful product for removal of contaminants in remediated soil sites and in aquatic environments. In soil, biochar can be used to immobilize contaminants such as lead, cadmium, arsenic and atrazine [14–16]. In water, biochar can be used to adsorb and remove metal ions such as cadmium, copper and zinc [17] and phenolic compounds [18]. Biochar can also be used to depollute wastewater, removing ammonia [19], dyes such as methylene blue [20] and toxic heavy metals [21, 22]. The chemical consistency and physical structure of biochar determines the pollutants that it can adsorb, where high aromaticity and porosity increase the sorption of organic contaminants and oxygen-containing functional groups increase the sorption of metals [23]. Feedstock type and pyrolysis conditions can alter the char chemical and physical consistency, which has an impact on its use for depollution. Outside of environmental applications, biochar is often used as an additive in the construction sector; where the porous structure acts as a micro-filler within concrete composites [24, 25]. Processing of heterogenous feedstocks to make char as a filler in concrete could increase carbon sequestration and reduce the carbon footprint of the concrete [26]. Biochar can also be added as an asphalt binder, increasing its high temperature performance and its resistance to aging [24]. Biochar from waste can also be added to epoxy resins, used in microelectronic, automotive and aircraft industries [27–31]. The adoption of biochar from more complex heterogenous sources such as municipal solid waste (MSW), contaminated wood and ASR could become a more viable option for the recycling industry. The caveat being that these types of products would retain any hazardous chemicals contained in the char and could pose problems at the end of life.

To use solid waste residues as a biochar for soil modification either depollution of feedstock may be required or that of the produced biochar. This will be required in some cases to meet the environmental requirements set by different governmental organizations [32, 33]. There are three main types of regulated contaminants that concern biochar these are: (i) PAHs, (ii) PCBs and (iii) heavy metals. PAHs is a term used for a large group of compounds which have multiple benzene rings in their chemical structure. PAHs are large compounds which are difficult to degrade in the environment. Many PAHs are non-toxic, yet some PAHs with specific chemical structures are carcinogenic and human exposure should be avoided [34, 35]. Current exposure limits to PAHs set by the UK government are $0.25 ng/m^3$ in air and < 0.2 ppb in water [36]. PCBs is a term used for a group of compounds which have two or more chlorine bonds within their hydrocarbon structure. PCBs are highly toxic and are banned in the UK and Europe [37]. Heavy metals that are regulated include lead, mercury and arsenic [37, 38]. There are different exposure limits set dependent on the location (inhalation, ingestion, skin contact). An example of the more common organic pollutants found within biochars is presented in **Figure 1**.

Char samples are typically analyzed for PAHs and PCB by chemical solvent extraction followed by GC–MS (Gas Chromatography—Mass Spectrometer) were extracted from cone and quartered samples of the ASR and pyrolysis solid residues. A common sample preparation method is ultrasonic-enhanced solvent extraction, based on the EPA 3550 method [39]. An example method for PAH analysis is shown in **Table 1**, where anhydrous sodium sulphate is added to a 5g biochar sample, which is extracted using ultrasonic extraction with a 50:50 mixture of hexane/acetone. In this example (**Table 1**), PAHs, PCBs, TPHs and BTXs were detected using Agilent 7890 and 6890 gas chromatographs, in various configurations.

Biochar produced from the thermal processing of organic solid residues is a growing technology which may be used to enhance processing a depollution of waste. However, waste streams are often a complex heterogenous mixture of material,

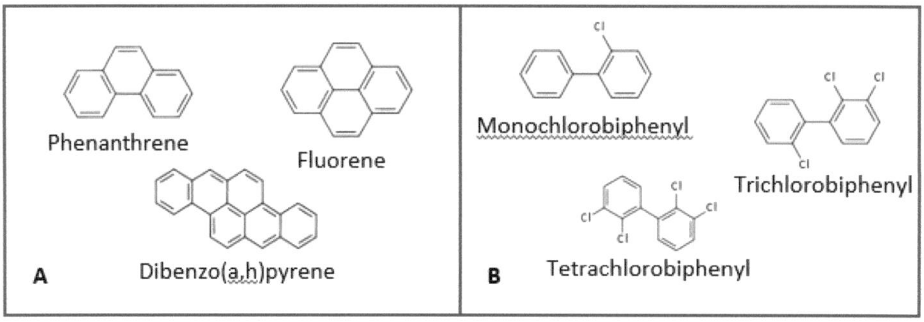

Figure 1.
The chemical structures of some common (A) PAHs, and (B) PCBs detected in solid residue products.

Pollutants	Agilent instrument	Injection volume µl	Detector	Column	Temperature program	Carrier gas
PAH	7890	2.0	GC/MS[a]	DB-5 ms	40°C for 1 min to 120°C at 25°C min^{-1}, then 160°C at 10°C min^{-1} and finally to 300°C at 5°C min^{-1}, final hold time of 15 min.	He
PCB	7890	2.0	GC/ECD[b]	HP-5 ms	75°C for 3 min, to 150°C at 15°C min^{-1}, then to 260°C at 6°C min^{-1}, finally to 300°C at 20°C min^{-1} rate held for 5 min	N_2
TPH	6890	1.0	GC/FID[c]	DB-5 ms	40°C for 1 min to 320°C at 10°C min^{-1}, final hold of 40 min.	He
BTEX	6890	1.0	GC/FID	DB-642	30°C for 1 min, to 100°C at 5 °C min^{-1} to 220°C at 8°C min^{-1}, final hold of 5 min.	He

[a]*GC/MS: gas chromatography equipped with high resolution mass spectrometry.*
[b]*GC/ECD: gas chromatography equipped with electron capture detector.*
[c]*GC/FID: gas chromatography equipped with flame ionization detector.*

Table 1.
Organic analysis operating conditions. Sourced from ref. [40].

making thermal processing methodology more complex. This chapter will define thermal processing methods and the effect on production of biochar in complex heterogenous waste streams.

2. Processes

Solid residues can be processed to produce biochar using two thermal processing methods: pyrolysis and gasification. Pyrolysis is the thermal processing of a material at an elevated temperature (400–1000°C) in the absence of oxygen. Pyrolysis produces three main products: syngas, oil, and biochar. Pyrolysis instruments vary in design with the main differences being in the type of kiln used to heat the feedstock and whether post-pyrolysis the gas is being distilled to remove any oil. Pyrolyser designs are often tailored by the feedstock, industry and components such as condensers and distillation systems can be added. Common pyrolyser designs are presented in **Figure 2**.

In addition to pyrolyser design, its operating parameters (temperature, residence time) can have a significant impact on the end products. The temperature and time that the waste is exposed to heat influences the breakdown of compounds and

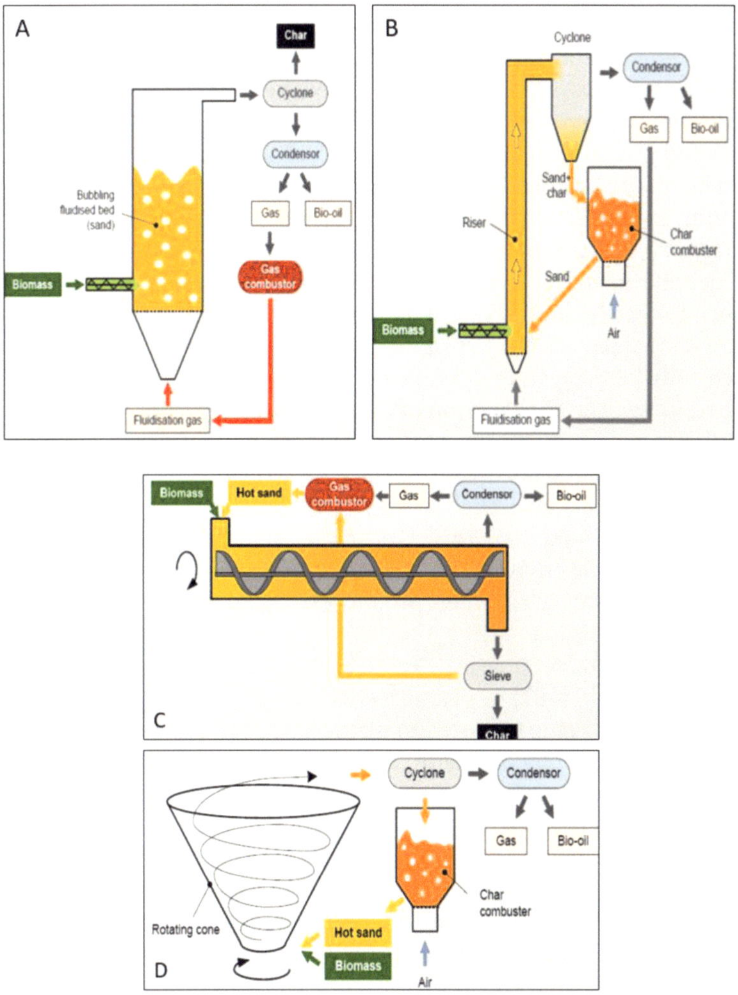

Figure 2.
Schematic diagrams of pyrolysis reactors used in waste processing. A = bubbling fluidised bed; B = circulating fluidised bed; C = screw reactor; and D = rotating cone reactor. Adapted from Khodier [41]; original source: [42].

the development and chemical consistency of the end-products. Often a higher temperature (800–1000°C) can increase char and syngas production, where lower temperatures (400–800°C) increase oil production [41]. Lower pyrolysis temperature does result in a reduction in contamination due to lower activation energy for larger compounds. Biochar is often produced as a byproduct, with the energy produced from pyrolysis of feedstock influencing the methodology. Often, this results in higher pyrolysis temperatures, causing pollutants to be contained within the biochar products, which requires clean-up.

Pyrolysis operating parameters can have significant impacts on the quality and yield of biochar. It is widely acknowledged that increased pyrolysis temperature and residence time can reduce the reactivity of the char produced [43, 44]. The effect of pyrolysis temperature (range 500–900°C) on char chemical structure was analyzed by Zhao et al. [45] where the pyrolysis temperature is greater than 700°C there was a significant reduction in the carbonyl groups within the aromatic structure. As there was a corresponding increase in oxygen within quinine compounds. Benzene ring condensation increased at 900°C with char having >6 benzene rings within the carbon structures. This was seen to be lower within the higher temperature chars. However, this reduced the char's chemical volatility and contributed to a larger pore size within the particles. Therefore, biochar produced under lower pyrolysis temperature had increased oxygen content and lower particle size, with the higher temperature biochar had increased chemical stability. This has an impact on what market the biochar can be utilized in and as we will see later the types of organic compounds present within the biochar structure.

Biochars that are lower in chemical reactivity may not be suitable for products within the depollution sector (chemical absorbent in water and air depollution) or as a feedstock for gasification. The chemical structure changes in char with pyrolysis temperature (explained above) has significant effects on the adsorption capability in water systems [46]. Pyrolysis temperatures above 500°C an increase the hydrophobicity of biochar, increasing the sorption of organic pollutants [47]. The reduction in biochar pore size and increase in oxygen content within hydrocarbon compounds in lower pyrolysis temperatures (<500°C) can encourage the sorption of inorganic pollutants from water systems, such as heavy metals [46]. It will also influence the retention of POP within the structure. Optimizing pyrolysis methodology to improve biochar utilization in the environment is crucial to meeting environmental targets. Types of biochar feedstocks and their products from different processing routes is presented in **Table 2**.

Investigations of biochar physical structure is focused on effects of homogenous biochar from biomass [84–87] or plastic feedstocks [28–30, 88]. The influence of co-feedstocks is currently being explored to increase syngas quality ($CO:H_2$ ratio), utilize waste, and improve byproducts, where research is still developing. Current findings suggest that co-feeding biomass with plastic feedstocks could have a synergistic effect on the quality of pyrolysis byproducts, where the lower oxygen concentration in plastic feedstocks lower oxygen concentrations and increase hydrogen and carbon concentration [58, 89–92]. However, most studies focus on the production of bio-oil and there is limited research on the impact on biochar structure and any corresponding update of POPS. Biochar from plastics requires pyrolysis at higher temperatures (900°C) [93] to fully decompose, therefore future research should investigate the effects of co-feeding at higher temperatures to determine the impact on solid residue products. This may reduce the potential for organic pollutants within the biochar structure.

Feedstock	Catalyst	Pyrolysis Temperature (°C)	Experimental Scale	Biochar type	Commercial Application
Lignin biomass	$FeSO_4$ [48] Char and metal oxide [49] Metal loaded Zeolite [50]	300–800	Laboratory	Biochar	Magnetic biochar [48] Biofuel [49, 50]
Cellulose Biomass	Synthetic Zeolites [51–53] Pyrolysis char [54]	600–800	Laboratory	Biochar	Hydrogen gas [51] Biofuel [52–54]
Wood Biomass	Biochar with metal catalysts [55, 56] Synthetic Zeolite [57]	550–600	Laboratory	Charcoal	Heating and Syngas [55, 57]; Activated carbon [56]
Pine Sawdust	Biochar with steam activation [58–60]; Treated steel slag [61]	850–1200	Laboratory	Charcoal	Heating [54, 112, 135] Fuel [61]
Straw Residues	CO_2 [62]	800	Laboratory	Biochar	Energy and Fuel [62]
Seaweed	CO [63]	800	Laboratory	Biochar	Energy and Fuel [63]
Papermill sludge	Fe_3O_4 [64]; $CaCO_3$ and Fe3O4 [65]	800	Laboratory	Magnetic biochar	Activated carbon [64, 65]
Municipal Sludge	Metal loaded zeolite with char [66]; Bentonite [67] HCl and Na_2CO_3 pretreatment [68]; biochar [69]	800–900	Laboratory	Pyrolysis char	Energy [66, 67] Activated carbon and hydrogen production [69, 68]
Cattle Manure	Biochar with metal oxides [54]	800	Laboratory	Charcoal	Energy production [54]
Food waste	Pyrolysis Char [70, 71]	700	Laboratory	Activated carbon	Air depollution [70, 71]
Waste Plastic	Synthetic Zeolite [72–74] with CaO [75] Dolomite [76] Modified pillared clays (MPCs) with Fe [77]	325–880	Laboratory	Pyrolysis char	Hydrogen production [72–77] Biofuel [73, 77]
Plastics & Biomass	HZSM-5 [146, 147]	500–600	Laboratory	Biochar	Biofuel production [78, 79]
Plastic from ELVs	Metal loaded Zeolites [80–82]	325–485	Laboratory	Pyrolysis char	Heating and Hydrogen production [80–82]
Biochar	Pyrolysis Char [83];	800	Laboratory	Pyrolysis char	Hydrogen production [83]

Table 2.
Types of feedstock that can be used in pyrolysis and the various products they can make under specific catalysts.

Biochar itself is currently being reused as a co-feed back into pyrolysis and gasification systems. Gasification is often conducted at a higher temperature than pyrolysis (800–1200°C) with controlled amounts of oxygen or steam to increase the rate of reaction [94]. Gasification does not produce bio-oil as one of the products. In some gasification systems, biochar is used as the feedstock, so the different techniques can complement one another [95]. Biochar produced at lower pyrolysis temperatures can be re-used back within the gasification system and reduce the activation energy required in syngas production. Biochar has been used as a co-feed for many pyrolysis and gasification feedstocks including biomass [49, 54, 57, 94, 95], sewage sludge and municipal waste [44, 66, 70, 71] and coal [61, 96–98]. Addition of biochar as a co-feed could also enhance the quality of the secondary biochar, causing a reduction in the inorganic components within the material. Recycling biochar back into the system will reduce pyrolysis impacts on waste to landfill. Addition of renewable biomass could improve secondary biochar quality and its effectiveness as a product. However, the impact towards the production of POPs is less understood due to limited research in this area.

A significant challenge for the waste industry is the complexities in processing material, which often results in heterogenous biomass feedstocks and the challenge to produce a usable biochar. This results in material often ending up in landfill. A heterogenous feedstock whose use as a biochar is being explored is ASR. The following section will discuss some of the hurdles of heterogeneous feedstocks and will use ASR as one of the worst-case materials.

3. Heterogeneous biochar feedstock: automotive shredder residue

To highlight the potential contamination within bio-chars a particular example has been chosen. This example will address some of the worst case for mono-source and mixed source feedstocks, ASR is a heterogenous organic waste produced at the end of the waste recycling process of ELVs (End-of-Life Vehicles). ASR makes up approximately 25% of the components of an ELV and is a mixture of organic biomass (textiles, wood) mixed with other waste (consisting of foams, plastics, fibers, glass and residual metals) [40, 93] (**Figure 3**). Recent ELV legislative targets in the UK and Europe require 95% of an ELV is required to be recycled or recovered by 2030 [99]. Currently ASR is sent to landfill; to meet legislative requirements further recovery or re-use is required therefore, the renewed interest in its conversion into a biochar as a potential processing route.

At present, there are only a small number of investigations into pyrolysis of ASR and its suitability for biochar production. As the trend increases to pyrolysis more heterogenous waste streams there will be an increase in the amounts of biochar which will require an end market. Studies indicated that carbon concentration within heterogeneous feedstocks such ASR char were not affected by temperature. This contrasts with crop-based feedstocks as mentioned in the earlier section [93, 100–103], however, the calorific value of the char did decrease with temperature [93, 103]. This could be caused by chemical structure changes within the char, previously seen in pyrolysis of other feedstocks. Further chemical analysis of carbon molecular structure of ASR pyrolyzed at different temperatures would be required to confirm this. The challenge being the heterogeneous nature of material and sampling errors. With governmental pressure to improve recycling activities and the environmental emergency requiring the elimination of fossil fuel energy production, research into this area is expected to expand over the next decade as more product types of biochar emerge.

Figure 3.
Image of ASR from waste recycling plant. Sourced from ref. [93].

Some of the key findings from research of ASR pyrolysis suggest that there is a significant effect of processing temperature on char particle size and chemical consistency. The biochar from ASR was produced in a 60 kg per hour pilot scale plant by Khodier and Williams [93]. The chars produced were subject to physio-chemically analysis under different temperature conditions (800–1000°C). Findings indicated that finer char was developed at higher pyrolysis temperature (1000°C), with a higher calorific value and lower oxygen content. The biochar produced at both 800 and 1000°C were separated into 'coarse' and 'fine' particle size fractions (coarse: > 0.1 mm diameter; fine = < 0.1 mm) see **Figure 4**. Therefore, allowing the biochar from different particle sizes having different applications depending on their characteristics. The larger particle sizes could be used in iron sintering [104] and to make H_2 through steam activation [69, 83, 91] Lower particle sized char, with its more irregular shape [100], which along with an increased microporosity has higher absorbent properties and would be more useful in environmental applications such as water storage in soils and water purification [105–107].

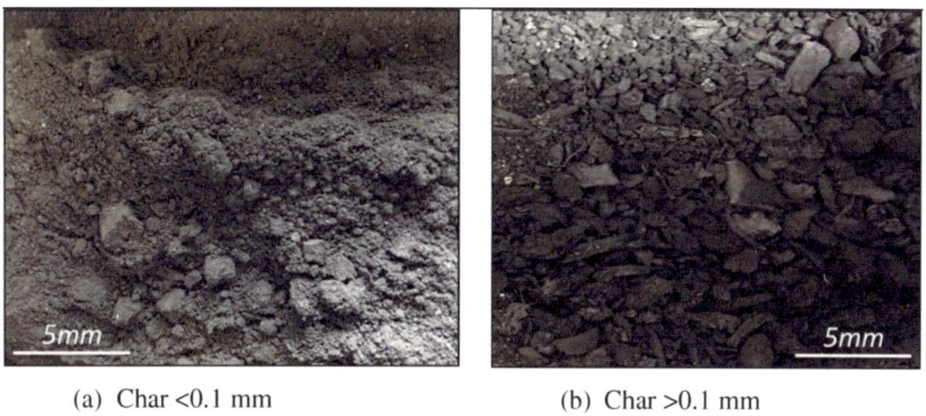

Figure 4.
Optical images of coarse (a) and fine (b) char. Source taken from ref. [92].

Biochar produced through pyrolysis of ASR and other heterogenous materials may have similar positive effects on soil properties and water purification as traditional homogenous feedstocks, however this still has to be proved as research into this area is limited. Recent laboratory studies indicated that coal residue biochar can increase SOC (Soil Organic Carbon) and TN (Total Nitrogen) concentration, when compared with maize biomass biochar, fresh residues and control soil [107], which could enhance crop growth. A supposition could be put forward that this would be true for heterogeneous feedstocks. However, it should be noted that with coal residue biochar no toxic contaminants within the chars (e.g., heavy metal concentration, PAH and PCB concentrations) were not studied or any impacts of leaching. As we will see later there are potential restrictions on the use of biochar produced from pyrolysis of heterogenous materials (such as ASR and waste sources such as contaminated wood etc.), due to the high concentrations of PAHs and dioxins in the char being over governmental limits for agricultural processes [36]. Increased chemical depollution of biochar from ASR and other heterogenous feedstocks will be required before use on land [108, 109].

It was found through more detailed analysis of the coarse and fine char fractions of ASR that there was a clear difference in organic pollutants. The fine particle sized fractions (<0.1 mm) had increased concentrations of PAHs and PCBs, which altered with temperature [40]. In contrast to the coarse char (>0.1 mm) which was determined to be inert with low contamination, (levels reported in **Tables 3** and **4**). There

Target Compounds	CAS	R.T. (min)	Char 800°C (mg kg^{-1})	Fit (%)	Char 1000°C (mg kg^{-1})	Fit (%)
Naphthalene	91–20-3	3.23	5010.00	99	46.60	99
Acenaphthylene	208–96-8	4.36	2040.00	99	91.00	99
Acenaphthene	83–32-9	4.48	56.80	73	<8.00	—
Fluorene	86–73-7	4.87	192.00	99	9.63	97
Phenanthrene	85–01–8	5.72	3980.00	99	429.00	99
Anthracene	120–12-7	5.77	724.00	97	101.00	98
Fluoranthene	206–44-0	7.07	2470.00	89	879.00	90
Pyrene	129–00-0	7.36	2870.00	87	1250.00	88
Benzo[a]anthracene	56–55-3	9.05	401.00	96	93.70	94
Chrysene	218–01–9	9.11	504.00	99	124.00	97
Benzo[b]fluoranthene	205–99-2	10.58	583.00	97	268.00	90
Benzo[k]fluoranthene	207–08-9	10.62	211.00	98	70.30	90
Benzo[a]pyrene	50–32-8	11.01	609.00	97	336.00	96
Indo[1,2,3-cd]pyrene	193–39-5	12.38	496.00	89	451.00	91
Dibenzo[a,h]anthracene	53–70-3	12.41	42.10	85	13.10	72
Benzo[g,h,i]perylene	191–24-2	12.68	524.00	93	627.00	95
Coronene	191–07-1	14.88	136.00	52	285.00	68
Total (USEPA16) PAHs			20712.90		<4797.33	

Table 3.
Concentrations of PAHs in fine char (at 800 and 1000 C) from ASR feedstock. Data sourced from ref. [40].

Compounds	Char 800 °C (mg kg^{-1})	Char 1000 °C (mg kg^{-1})
PCB28	<25.0	59.1
PCB52	<25.0	87.2
PCB101	<25.0	53.9
PCB118	<25.0	<25.0
PCB153	<25.0	<25.0
PCB138	<25.0	210.9
PCB180	<25.0	47.6
Benzene	13,100	420
Toluene	1220	<25
Ethylbenzene	167	<25
Xylenes	855	<75
m/*p*-xylenes	679	<50
o-xylene	176	<25
MTBE	<50	<50

Table 4.
Concentrations of PCBs (7 congeners) and BTEX in produced fine char fraction (at 800 and 1000°C). Data sourced from ref. [98].

were significant effects of pyrolysis temperature on the PAH and PCB levels within the fine char component, where PAHs decreased with higher pyrolysis temperature (**Table 3**) and concentrations of PCBs increased (**Table 4**). Further investigations on the effect of temperature on the formation and recreation of compounds from heterogenous mixtures is required to determine the best method to reduce environmental contaminants held within the feedstock through process control. It should be noted that the heterogeneous nature of the feedstock makes process control as the sole solution questionable. A more resilient solution would be secondary processing as an effective method to upgrade the biochar and reduce organic pollutants. The next section will define and evaluate current secondary processing of biochar from heterogenous waste sources.

4. Secondary processes to reduce organic pollutants in biochar from heterogeneous sources

Environmental contamination within biochar from heterogenous sources limits its use in other applications, therefore is often sent to landfill as hazardous waste [93]. Secondary processing of contaminated biochar could reduce the amount of waste to landfill and enable biochar from heterogenous sources to be used as de-pollutants in contaminated land and water systems. There are many methods to reduce pollutants held within biochar. Based on the example in Section 3 [40], a simple reduction in contamination would be size segregation by sieving. If the biochar was sieved to <0.1 mm particle size, the contaminated fine char could be segregated, and the coarse fraction could be re-used. Size segregation of chars would not fully eliminate waste to landfill, so further secondary processing to clean up finer fractions of biochar would be required.

A common secondary process of biochar is carbonization and activation [110]. Carbonization is where volatile and inorganic components of feedstock are removed through thermal treatment, such as a secondary pyrolysis or calcination. The carbon contained in biochar from the pyrolysis process has a disorganized physical structure. Activation is the upgrading of the carbon porosity to regulate the structure. This is conducted though steam or CO_2 activation and the addition or impregnation of a catalyst (such as $ZnCl_2$, H_3PO_4 or KOH) [68, 111]. Activated carbon is chemically stable, with good conductivity due to its' high surface area and can be used to generate EDLCs (Electrical Double Layer Capacitators) and used in depollution of water due to its high adsorption capacity [112]. However, organic waste containing heterogenous components, such as those from ASR, still produce substantial amounts of pollutants following activation [113, 114], so further post treatment is required. Nitric acid addition can be used to remove inorganic metals, followed by a base to neutralize. Studies suggest that this can significantly improve the conductivity of the EDLC without altering the porosity and char texture [115]. However, acid treatment results in excessive amounts of waste which then requires depolluting [116, 117], so may not be a cost-effective solution. Current research has explored molten salt post treatment as an alternative to acid treatment which removes the metal impurities [118]. Cleaner alternatives to activated carbon production for heterogenous feedstocks is required if this is to be economically viable.

In addition to activation and carbonization, another secondary processing method applied to biochar is magnetic synthesis, which can enhance its use as a water decontamination agent, due to the easy removal from the system post-adsorption [119, 120]. The use of Fe_3O_4 as a catalyst under CO_2 can encourage the formation of magnetic biochar (magnetite Fe_3O_4; saturation magnetization 28.4 emu g^{-1}), which has a high heavy metal adsorption [64]. Magnetization could increase removal of heavy metals from aquatic environments and improve water quality in polluted areas [120]. Impregnation of metal composites such as $FeSO_4$ into heterogeneous feedstock pre-pyrolysis can produce magnetic biochar. It has been found that an iron loading of 8% in the feedstock also enhanced biochar production [48]. Impregnation of iron composites within pyrolysis systems with heterogenous feedstocks, such as ASR, could enhance the utilization of the biochar as a magnetic activated carbon product.

Within heterogenous waste streams, further sorting of material pre-treatment could have significant effects on the contamination found within the biochar product. Using ASR as an example, the elimination of PVC from plastics within the material could significantly reduce the number of PCBs in the final product [93]. In addition to this, improved sorting could reduce the number of contaminants within the biochar, making secondary treatment more effective. Further sorting of the biomass (wood) and polymer (plastics, foams) materials of ASR may improve secondary depollution of biochar [108] and improve production of activated carbon [121]. Certain types of plastic removal from ASR would increasing the homogeneity of the feedstock to be pyrolyzed [122]. Development of feedstock sorting practices is possible; however, this would require significant changes to waste management practices which may not be practical.

If feedstock sorting is not a viable homogenization option, pre-treatment of feedstock by calcination could increase homogenization of the feedstock by reducing the particle size without causing depolymerization of hydrocarbons and devolatilization of plastic components [123]. It should be noted that typical feedstocks have not been tested at larger pilot scales, so it is difficult to evaluate the impact of scaling on the outputs. Torrefaction may be another suitable method of homogenizing feedstocks without fine metal sorting [124]. The process of calcination is a thermal pre-treatment conducted under limited oxygen, whereas torrefaction is conducted in the absence of oxygen. Further tests are required to determine the differences between torrefaction and calcination on the chemical consistency of the improved heterogenous feedstock to provide information on the optimum conditions. The economic impacts of an extra thermal pre-treatment step on the overall pyrolysis process requires careful evaluation to determine if increase product quality and yield are enough to promote investment.

Another method of reducing POP contamination of biochar is the reprocessing of it back into a thermal process. Biochar can be utilized back within the pyrolysis system to upgrade and clean the syngas, where the absorbent properties of char can increase H_2S removal [70] and can improve production of other byproducts such as ethylbenzene [71], where its catalytic properties can crack hydrocarbon chains. Alongside directly altering syngas properties, addition of char as a co-catalyst can increase regeneration of catalysts, improving production costs [57]. Utilization of char in other pyrolysis systems where the feedstock is more oxygenated, such as plant biomass, can have a deoxygenation effect, improving the quality of bio-oil products and increasing the syngas value [125]. Added to syngas systems, biochar can be used to clean up combustion systems by adsorbing CO2 emissions, reducing the negative industrial impact on global warming. Upgrading biochar through addition of metal composites such as Fe_2O_3 and Al_2O_3 can increase the adsorption through increasing char surface area and sorption capacity [57]. This is a low-cost CO_2 adsorption method, where catalyst desorption and regeneration temperature occurs at 120°C. Reprocessing biochar developed from heterogeneous feedstocks could be a viable option of creating more homogenous products which can be more effectively utilized. If using this approach, it is essential that contaminants within biochar are monitored to ensure that the addition of a catalyst in the gasification process can increase H_2 production [126] and limit PAH formation within the biochar [127]. Using Ca/Na compounds as a catalyst reduces the production of aromatic structures and increase the formation of more active intermediates beneficial to gasification [127]. However, research into the effects of biochar as a feedstock for gasification and H_2 production is focused on char derived from homogenous feedstocks [59, 96, 97, 126–129]. ASR derived biochar has a more volatile carbonaceous structure than homogenous chars [130], meaning it could be more effective in gasification processes under steam activation. However, it could also be more difficult to select the correct catalyst with a wide range of pollutants present in the char (**Tables 3** and **4**). Further information on the physiochemical structure of ASR-derived biochars and effects of catalyst addition under steam activation is required ensure that depollution of char in this process is effective.

Alternative methods to processing contaminated biochar from heterogenous waste sources is to look at the sequestration of this product in composite materials. This will be in areas where the physical structure of the biochar improves the physical composition of the material and the pollutants are contained; reducing their effects. Containing polluted char within concrete may be a sustainable method for their use whilst at the same time reducing natural resource depletion from production concrete materials. Although, there is significant potential to utilize char in concrete materials, where an increased cement hydration and the immobilization of contaminants has been determined, there is still significant research required before commercial products can be manufactured and sold. The effect of char particle size, feedstock type and dosing amount can influence the tensile strength of the concrete, where if not correct, micro cracking can be caused [131]. Caeteno et al., [132] highlighted how finer fractions of heterogeneous bio char (ASR) when added to concrete had a beneficial effect. From the earlier sections highlighting that POPs were associated with certain size fractions for biochar from ASR sieved finer char could be used as a concrete agent and the inert coarser biochar for other applications. However, size separation of bulk material could be a time-consuming and expensive process.

A significant limiting factor of research into biochar production and its' secondary processing is the lack of pilot scale projects. Many initial pyrolysis trials of biochar production and applications were conducted at a laboratory-scale (**Table 2**). Upscaling of laboratory scale to pilot scale systems is required to increase accuracy in the effectiveness of a catalyst on product yield and quality. To gain accurate results from a laboratory scale experiment a large amount of replication is required due to the small sample size (often 1-10 g feedstock), where upscaling to a pilot reactor can process 1000× more material, providing more realistic results. This would benefit heterogenous feedstocks by reducing error in sampling due to missing of potential contamination. This is also true of homogeneous feedstocks with added plastic material to improve yield. The next step in the development of products from heterogeneous feedstocks such as ASR will be to test effective catalysts on a pilot scale reactor. This would provide more accurate information on the effects of catalysis on a commercial scale, improving depollution of complex feedstocks. Due to the heterogeneity of the material, more than one catalyst may be required to target specific components of the feedstock. Two-stage catalysis of plastics has been investigated [133, 134], which might be a viable option for heterogenous biochar production.

5. Economic and environmental impacts driving biochar depollution

Producing biochar from heterogenous feedstocks and the potential contamination from POPs will be decided by two conflicting economical drivers: (i) whether biochar is being developed to stop feedstock going to landfill, or (ii) whether biochar is being produced for a specific application (such as activated carbon). If reducing material to landfill was the business focus, then biochar production from heterogenous sources (such as ASR) will be the driver for secondary processing development. Although secondary processing of biochar will reduce environmental contamination, the energy and resources required to implement these changes may outweigh other costs, such as landfill. If it is to produce biochar for specific applications then pre-processing technology would be the focus. In future, plastic components within waste streams including ASR will be classed as hazardous [40]. Many plastics already contain significant amounts of POPs, increasing the cost of landfill tax and the expense of disposal [3, 135], (EU Regulation on persistent organic pollutants (2019/1021) was adopted on 20 June 2019). This may lead to unforeseen consequences where businesses are making biochars that cannot be used because of elevated levels of contamination. The economics of processes will lead to a trade-off between reduction in waste to landfill and the creation of contaminated char with or without secondary processing. An analogous situation also arises with the use of crops being used to depollute contaminated soil systems and then used for energy. The biochar produced will contain pollutants which is then spreading contaminated biochar as a conditioner. This section will assess the economic and environmental costs to a waste recycling business when introducing thermal processing systems and the challenges and opportunities faced during commercialization.

Waste biochar produced from heterogenous sources such as ASR and MSW can produce a circular economy from waste streams [136–138]. In addition to the environmental incentive of reducing waste to landfill, a reduction in landfill tax is a significant economic opportunity, where current UK rates are £98.60 per tonne [139]. However, the chemical contamination within biochar (**Tables 3** and **4**) means that biochar from certain waste streams (such as ASR) could be classified as hazardous [40] which would increase landfilling costs. This could deter waste recycling industries from investing in biochar production, where a large financial investment is already required upon the purchase of a pyrolysis plant (**Table 5**) Upgraded biochar from secondary processing methods could produce a viable product that would promote a circular economy. However, the addition in business costs from development and maintenance of a secondary processing system might outweigh the costs of landfilling contaminated biochars. Long-term lifecycle assessment studies are required to investigate the payback and carbon/energy balances of these systems, which will determine whether secondary processing is an appropriate method in the future. There is no simple solution, and we are potentially creating legacy problems for the future.

Plant size ($t\,d^{-1}$)	Feedstock	Capital investment (M $)	Annual operating costs (M $)	Feed costs ($/t)	Production costs ($/gal)
2000	Forest residues	427	154	69	6.25
2205	Woody biomass	546	25.41	80	3.46
2205	Woody biomass	700	37.66	80	3.39
2000	Corn stover	200	12.3	83	0.26
1650	Wood pellet	180	12	—	0.24
1000	Dry wood	68	10.6	44	0.41
1000	Wet wood	72	11.3	30	0.60
1000	Peat	76	10.2	20	0.61
1000	Straw	82	10.2	42.5	0.64
900	Wet wood	46	9.9	34	0.50
550	Dry wood	48.2	9.6	45	0.71
400	Wet wood	14.3	8.8	36	1.02
250	Dry wood	14	8.92	44	0.55
200	Wet wood	8.8	4.84	36	1.11
100	Wet wood	6.6	2.84	36	1.48
24	Rice husk	3.89	0.170	22	0.82
2.4	Rice husk	0.97	0.34	22	1.73

Table 5.
Summary of reported pyrolysis plant cost. Sourced from ref. [39].

6. Conclusion

Production of biochar from heterogenous materials is likely to increase over the next decade as governments attempt to reach environmental targets for 2030 following COP26. The use of waste biomass for energy sources will be a driver in future energy production as the world resorts towards cleaner energy and away from fossil fuels. As highlighted in this chapter, utilization of biochar produced from thermal recycling of heterogenous waste feedstocks pose many challenges due to prominent levels of POPs and heavy metals within the feedstock. The amounts and types of persistent organic present is discussed. Secondary processing is a potential solution to remove contamination from biochars but the economics and readiness for the market are currently the limiting factors. Future opportunities to upgrade biochar through secondary processing are being adopted within the sector but are yet to be commercially available.

Acknowledgements

The authors gratefully acknowledge the following financial support: Higher Education Innovation Fund UKRI, Innovate UK's support through the Knowledge Transfer Partnership (KTP). As well as both the University of Central Lancashire and Recycling Lives Limited, Preston, UK for their financial support and access to their facilities.

Conflict of interest

The authors declare no conflict of interest.

Author details

Karl Williams*, Ala Khodier and Peter Bentley
University of Central Lancashire, Preston, UK

*Address all correspondence to: kswilliams@uclan.ac.uk

References

[1] Nam JJ, Thomas GO, Jaward FM, Steinnes E, Gustafsson O, Jones KC. PAHs in background soils from Western Europe: Influence of atmospheric deposition and soil organic matter. Chemosphere (Oxford). 2008;**70**(9):1596-1602. DOI: 10.1016/j.chemosphere.2007.08.010

[2] Kalbe U, Lehnik-Habrink P, Bandow N, Sauer A. Validation of European horizontal methods for the analysis of PAH, PCB and dioxins in sludge, treated biowaste and soil. Environmental Sciences Europe. 2019;**31**:1. DOI: 10.1186/s12302-019-0211-3

[3] UK Government. 2017. Landfill Tax rates for 2022 to 2023 . Landfill Tax rates for 2022 to 2023-GOV.UK. Available from: www.gov.uk

[4] Mimmo T, Panzacchi P, Baratieri M, Davies C, Tonon G. Effect of pyrolysis temperature on miscanthus (Miscanthus × giganteus) biochar physical, chemical and functional properties. Biomass & Bioenergy. 2014;**62**:149-157. DOI: 10.1016/j.biombioe.2014.01.004

[5] Singh A, Nanda S, Guayaquil-Sosa JF, Berruti F. Pyrolysis of Miscanthus and characterization of value-added bio-oil and biochar products. Canadian Journal of Chemical Engineering. 2021;**99**(1):S55-S68. DOI: 10.1002/cjce.23978

[6] Schulz H, Dunst G, Glaser B. Positive effects of composted biochar on plant growth and soil fertility. Agronomy for Sustainable Development. 2013;**33**(4):817-827

[7] Jones DL, Edwards-Jones G, Murphy DV. Biochar mediated alterations in herbicide breakdown and leaching in soil. Soil Biology and Biochemistry. 2011;**43**:804-813

[8] Quilliam RS, Marsden KA, Gertler CH, Rousk J, De Luca TH, Jones DL. Nutrient dynamics, microbial growth and weed emergence in biochar amended soil are influenced by time since application and reapplication rate. Agriculture, Ecosystems and Environment. 2012;**158**:192-199

[9] Han F, Ren L, Zhang X-C. Effect of biochar on the soil nutrients about different grasslands in the loess plateau. Catena. 2016;**137**:554-562

[10] Raclavská H, Růžičková J, Škrobánková H, Koval S, Kucbel M, Racklavský K, et al. Possibilities of the utilisation of char from the pyrolysis of tetrapak. Journal of Environmental Management. 2018;**219**:231-238

[11] Gross A, Bromm T, Glaser B. Soil organic carbon sequestration after biochar application: A global meta-analysis. Agronomy. 2021;**11**:2474

[12] Smith P. Soil carbon sequestration and biochar as negative emission technologies. Global Change Biology. 2016;**22**:1315-1324

[13] Fuss S, Lamb WF, Callaghan MW, Hilaire J, Creutzig F, Amann T, et al. Negative emissions, part 2: Costs, potentials and side effects. Environmental Research Letters. 2018;**13**:063002

[14] Cao XD, Ma LN, Gao B, Harris W. Dairy manure derived biochar effectively sorbs lead and atrazine. Environmental Science and Technology. 2009;**43**:3285-3291

[15] Cao XD, Ma LN, Liang Y, Gao B, Harris W. Simultaneous immobilization of lead and atrazine in contaminated soils using dairy manure biochar. Environmental Science and Technology. 2011;**45**:4884-4889

[16] Liu L, Xiu L, Wang D, Lin H, Huang L. Removal and reduction of Cr (VI) in simulated waste water using magnetic biochar prepared by co-pyrolysis of nano-zero-valent iron and sewage sludge. Journal of Cleaner Production. 2020;**257**:120562

[17] Bogusz A, Oleszczuk P, Dobrowlowski R. Application of laboratory prepared and commercially available biochars to adsorption of cadmium, copper and zinc ions from water. Bioresource Technology. 2015;**196**:540-549

[18] Kuśmierek K, Świątowski A, Kotkowski T, Cherbański R, Molga E. Adsorption properties of activated tire pyrolysis chars for phenol and chlorophenols. Chemical Engineering and Technology. 2020;**43**(4):770-780

[19] Tang Y, Alam MS, Konhauser KO, Alessi DS, Xu S, Tian W, et al. Influence of pyrolysis temperature on production of digested sludge biochar and its application for ammonium removal from municipal wastewater. Journal of Cleaner Production. 2019;**209**:927-936

[20] Hu M, Deng W, Hu M, Chen G, Zhou P, Zhou Y, et al. Preparation of binder-less activated char briquettes from pyrolysis of sewage sludge for liquid-phase adsorption of methylene blue. Journal of Environmental Management. 2021;**299**:113601

[21] Calugaru LL, Neculita CM, Genty T, Zagury GJ. Removal efficiency of As(V) and Sb(III) in contaminated neutral drainage by Fe-loaded biochar. Environmental Science and Pollution Research. 2019;**26**:9322-9332

[22] Liu Y, Huang J, Xu H, Zhang Y, Hu T, Chen W, et al. A magnetic macro-porous biochar sphere as vehicle for activation and removal of heavy metals from contaminated agricultural soil. Chemical Engineering Journal. 2020;**390**:124638

[23] Braghiroli FL, Bouafif H, Neculita CM, Koubaa A. Activated biochar as an effective sorbent for organic and inorganic contaminants in water. Water, Air and Soil Pollution. 2018;**229**(7):1-22

[24] Gupta S, Kua HW, Low CY. Use of biochar as a carbon sequestering additive in cement mortar. Cement and Concrete Composites. 2018;**87**:110-129

[25] Cuthbertson D, Berardi M, Briens C, Berruti F. Biochar from residual biomass as a concrete filler for improved thermal and acoustic properties. Biomass and Bioenergy. 2019;**120**:77-83

[26] Wijesekara DA, Sargent P, Ennis CJ, Hughes D. Prospects of using chars derived from mixed post waste plastic pyrolysis in civil engineering applications. Journal of Cleaner Production. 2021;**317**:128212

[27] Kumar A, Choudhary R, Kumar A. Aging characteristics of asphalt binders modified with waste tire and plastic pyrolytic chars. PLoS One. 2021;**16**(8):e0256360

[28] Ahmetli G, Kocaman S, Ozaytekin I, Bozkurt P. Epoxy composites based on inexpensive char filler obtained from plastic waste and natural resources. Polymer Composites. 2013;**34**:500-509

[29] Sogancioglu M, Yel E, Ahmetli G. Pyrolysis of waste high density polyethylene (HDPE) and low density

polyethylene (LDPE) plastics and production of epoxy composites with their pyrolysis chars. Journal of Cleaner Production. 2017;**165**:369-381

[30] Sogancioglu M, Yel E, Ahmetli G. Behaviour of waste polypropylene pyrolysis char-based epoxy composite materials. Environmental Science and Pollution Research. 2020;**27**:3871-3884

[31] Kahrizsangi AG, Shariatpanahi H, Neshati J, Akbarinezhad E. Corrosion behaviour of modified nano carbon black/epoxy coating in accelerated conditions. Applied Surface Science. 2015;**331**:115-126

[32] European Commission. The Environmental Liability Directive, Directive 2004/35/EC. 2004. Liability - Legislation - Environment - European Commission. Available from: europa.eu

[33] Environmental Protection Agency. Environmental Justice Strategy. 1995. Available from: ej_strategy_1995.pdf (epa.gov)

[34] Culp SJ, Beland FA. Comparison of DNA (Deoxyribonucleic Acid) Adduct Formation in Mice Fed Coal Tar or Benzo[a]Pyrene. Carcinogenesis (New York). OXFORD: Oxford University Press. 1994;**15**(2):247-252. DOI: 10.1093/carcin/15.2.247

[35] Culp SJ, Gaylor DW, Sheldon WG, Goldstein LS, Beland FA. A comparison of the tumours induced by coal tar and benzo[a]pyrene in a two-year assay. Carcinogenesis. 1998;**19**(1):117-124. DOI: 10.1093/carcin/19.1.117

[36] European Commission. 1996. Council Directive 96/59/EC of 16 September 1996 on the disposal of polychlorinated biphenyls and polychlorinated terphenyls (PCB/PCT), Official Journal L 243, 24 September 1996. pp. 31-35. Available from: eur39237.doc(live.com)

[37] European Commission. 2008. Directive 2008/50/EC of the European Parliament and of the Council of 21 May 2008 on ambient air quality and cleaner air for Europe. Available from: EUR-Lex-32008L0050-EN-EUR-Lex(europa.eu)

[38] European Commission. 2004. Directive 2004/107/EC of the European Parliament and of the council of 15 December 2004 relating to arsenic, cadmium, mercury, nickel and polycyclic aromatic hydrocarbons in ambient air

[39] US EPA. 2007. EPA 3550C, Ultrasonic Extraction, Office of research and development. Available from: www.epa.gov/sites/production/files/2015-12/documents/3550c.pdf

[40] Williams KS, Khodier A. Meeting EU ELV targets: Pilot-scale pyrolysis automotive shredder residue investigation of PAHs PCBs and environmental contaminants in the solid residue products. Waste Management. 2020;**105**:233-239

[41] Khodier A. Automotive shredder residue (ASR) for clean energy systems (pyrolysis and gasification) to produce sustainable green energy. Masters Thesis, University of Central Lancashire. 2019. Available from: clok.uclan.ac.uk

[42] Ronsse F, Dickinson D, Nachenius R, Prins W. Biomass pyrolysis and biochar characterization. In: Presented at the 1st FOREBIOM workshop. Vienna: Austrian Academy of Sciences; 2013

[43] Tushar MSHK, Mahinpey N, Khan A, Ibrahim H, Kumar P, Idem R. Production, characterization and reactivity of chars produced by the isothermal pyrolysis of flax straw. Biomass and Bioenergy. 2012;**37**:97-105

[44] Wang X, Zhai M, Guo H, Panahi A, Dong P, Levendis YA. High temperature pyrolysis of biomass pellets: The effect of ash melting on char structure. Fuel. 2021;**285**:119084

[45] Zhao Y, Feng D, Zhang Y, Huang Y, Sun S. Effect of pyrolysis temperature on char structure and chemical speciation of alkali and alkaline earth metallic species in biochar. Fuel Processing Technology. 2016;**141**(1):54-60

[46] Enaime G, Baçaoui A, Yaacoubi A, Lübken M. Biochar for wastewater treatment: Conversion technologies and applications. Applied Sciences. 2020;**10**(10):34912

[47] Keiluweit MM, Nico PS, Johnson MG, Kleber M. Dynamic molecular structure of plant biomass derived black carbon (biochar). Environmental Science and Technology. 2010;**44**:1247-1253

[48] Han T, Yang W, Jönsson PG. Pyrolysis and subsequent steam gasification of metal dry impregnated lignin of the production of H2-rich syngas and magnetic activated carbon. Chemical Engineering Journal. 2020;**394**:124902

[49] Waheed QMK, Wu C, Williams PT. Hydrogen production from high temperature steam catalytic gasification of bio-char. Journal of the Energy Institute. 2016;**89**(2):222-230

[50] Ellison CR, Boldor D. Mild upgrading of biomass pyrolysis vapors via ex-situ catalytic pyrolysis over an iron montmorillonite catalyst. Fuel. 2021;**291**:120226

[51] Widyaningrum RN, Church TL, Zhao M, Harris AT. Mesocellular-foam-silica-supported Ni catalyst: Effect of pore size on H_2 production from cellular pyrolysis. International Journal of Hydrogen Energy. 2012;**37**(12):9590-9601

[52] Rezaei PS, Shafaghat H, Daud WMAW. Suppression of coke formation and enhancement of aromatic hydrocarbon production in catalytic fast pyrolysis of cellulose over different zeolites: Effects of pore structure and acidity. Royal Society of Chemistry Advances. 2015;**5**:65408-65414

[53] Huynh VN, Dang NT, Truong TT, Van T. Catalytic upgrading and enhancing the combustion characteristic of pyrolysis oil. International Journal of Green Energy. 2021;**18**(12):1277-1288

[54] Zhou Y, Chen Z, Gong H, Wang X, Yu H. A strategy of using recycled char as a co-catalyst in cyclic in-situ catalytic cattle manure pyrolysis for increasing gas production. Waste Management. 2020;**107**:74-81

[55] Singh S, Kumar Bhaumik S, Dong L, Li C-Z, Vuthaluru H. An integrated two-step process of reforming and adsorption using biochar for enhanced tar removal in syngas cleaning. Fuel. 2022;**307**:121935

[56] Williams PT, Horne PA. The influence of catalyst type on the composition of upgraded biomass pyrolysis oils. Journal of Analytical and Applied Pyrolysis. 1995;**31**:39-61

[57] Creamer AE, Gao B, Wang S. Carbon dioxide capture using various metal oxyhydroxide-biochar composites. Chemical Engineering Journal. 2016;**283**:826-832

[58] Tong W, Liu Q, Yang C, Cai Z, Wu H, Ren S. Effect of pore structure on CO_2 gasification reactivity of biomass chars under high temperature pyrolysis. Journal of Energy Institute. 2020;**93**(3):962-976

[59] Yan F, Luo S-Y, Hu Z-Q, Cheng G. Hydrogen-rich gas production by steam

gasification of char from biomass fast pyrolysis in a fixed bed reactor: Influence of temperature and steam on hydrogen yield and syngas composition. Bioresource Technology. 2010;**101**(14):5633-5637

[60] Yang S, Zhang X, Chen L, Sun L, Xie X, Zhao B. Production of syngas from pyrolysis of biomass using Fe/CaO catalysts: Effect of operating conditions on the process. Journal of Analytical and Applied Pyrolysis. 2017;**125**:1-8

[61] Wu Y, Yu H, Chao H, Chen D. A novel nickel catalyst supported on activated steel slags for syngas production and tar removal from biomass pyrolysis. International Journal of Hydrogen Energy. 2021;**46**(75):37268-37280

[62] Kwon EE, Jeon EC, Castaldi MJ, Jeon YJ. Effect of carbon dioxide on the thermal degradation of lignocellulosic biomass. Environmental Science and Technology. 2013;**47**:10541-10547

[63] Kwon EE, Cho SH, Kim S. Synergetic sustainability enhancement via utilisation of carbon dioxide as carbon neutral chemical feedstock in the thermo-chemical processing of biomass. Environmental Science and Technology. 2015;**49**:5028-5034

[64] Cho D-W, Kwon G, Yoon K, Tsang YF, Ok YS, Kwon EE, et al. Simultaneous production of syngas and magnetic biochar via pyrolysis of paper mill sludge using CO_2 as a reaction medium. Energy Conversion and Management. 2017;**145**:1-9

[65] Rocha LS, Sousa ÉML, Gil MV, Oliveira JABP, Otero M, Esteves VI, et al. Producing magnetic nanocomposites from paper sludge for the adsorptive removal of pharmaceuticals from water—A fractional factorial design. Nanomaterials (Basel, Switzerland). 2021;**11**(2):1-20. DOI: 10.3390/nano11020287

[66] Wang S, Shan R, Gu J, Zhang J, Yuan H, Chen Y. Pyrolysis of municipal sludge char supported Fe/Ni catalysts for catalytic reforming of tar model compound. Fuel. 2020;**279**:118494

[67] Qin J, Jiao Y, Li X, Liu Y, Lei Y, Gao J. Sludge char-to-fuel approaches based on the catalytic pyrolysis II: Heat release. Environmental Science and Pollution Research International. 2018;**25**(36):36581-36588. DOI: 10.1007/s11356-018-3596-4

[68] Alvarez J, Lopez G, Amutio M, Bilbao J, Olazar M. Preparation of absorbents from sewage sludge pyrolytic char by carbon dioxide activation. Process Safety and Environmental Protection. 2016;**103**:76-86

[69] Chun YN, Lim MS, Yoshikawa K. Characteristics of the products from steam activation of sewage sludge. Journal of Industrial and Engineering Chemistry. 2012;**18**(2):839-847

[70] Hervy M, Pham M, Gérente C, Weiss-Hortala E, Nzihou A, Villot A, et al. H_2S removal from syngas using waste pyrolysis chars. Journal of Chemical Engineering. 2018;**334**:2179-2189

[71] Hervy M, Villot A, Gérrente C, Pham M, Weiss-Hortala E, Nzihou A, et al. Catalytic cracking from ethylbenzene as tar surrogate using pyrolysis chars from wastes. Biomass and Bioenergy. 2018;**117**:86-95

[72] Yao D, Yang H, Chen H, Williams PT. Investigation of nickel-impregnated zeolite catalysts for hydrogen/syngas production from the catalytic reforming of waste polyethylene. Applied Catalysis B: Environmental. 2018;**227**:477-487

[73] Oh D, Lee HW, Kim Y-M, Park Y-K. Catalytic pyrolysis of polystyrene and polyethylene phthalate over Al-MSU-F. Energy Procedia. 2018;**144**:111-117

[74] Lee HW, Park Y-K. Catalytic pyrolysis of polyethylene and polypropylene over desilicated beta and Al-MSU-F. Catalysts. 2018;**8**(501):1-15

[75] Dou B, Wang K, Jiang B, Song Y, Zhang C, Chen H, et al. Fluidized bed gasification combined continuous sorption-enhanced steam reforming system to continuous hydrogen production from waste plastic. International Journal of Hydrogen Energy. 2016;**41**:3803-3810

[76] Sonaware YB, Shindikar MR, Khaladkar MY. High calorific value fuel from household plastic waste by catalytic pyrolysis. Nature, Environment and Pollution Technology. 2017;**16**(3):879-882

[77] Li K, Lei J, Yuan G, Weerachanchai P, Wang J-Y, Zhao J, et al. Fe-, Ti-, Zr-, and Al-pillared clays for efficient catalytic pyrolysis of mixed plastics. Chemical Engineering Journal. 2017;**317**:800-809

[78] Muneer B, Zeeshan M, Qaisar S, Razzaq M, Iftikhar H. Influence of in-situ and ex-situ HZSM-5 catalyst on co-pyrolysis of corn stalk and polystyrene with a focus on liquid yield and quality. Journal of Cleaner Production. 2019;**237**:117762

[79] Xuanjun J, Lee JH, Choi JW. Catalytic co-pyrolysis of woody biomass with waste plastics: Effects of HZSM-5 and pyrolysis temperature on producing high value pyrolytic products and reducing wax formation. Energy. 2022;**235:A**:121739

[80] Jung S, Lee T, Lee J, Lin K-YA, Park Y-K, Kwon EE. Catalytic pyrolysis of plastics derived from end-of-life vehicles (ELVs) under the CO_2 environment. International Journal of Energy Research. 2021;**45**(11):16781-16793

[81] Miskolczi N, Juzsakova T, Sója J. Preparation and application of metal loaded ZSM-5 and y-zeolite catalysts for thermo-catalytic pyrolysis of real end of life vehicles plastic waste. Journal of the Energy Institute. 2019;**92**:118-127

[82] Miskolczi N, Sója J, Tulok E. Thermo-catalytic two-step pyrolysis of real waste plastics from end-of-life vehicle. Journal of Analytical and Applied Pyrolysis. 2017;**128**:1-12

[83] Chaudhari ST, Dalai AK, Bakhshi NN. Production of hydrogen and/or syngas (H_2 + CO) via steam gasification of biomass-derived chars. Energy & Fuels. 2003;**17**(4):1062-1067

[84] Xiao R, Yang W. Influence of temperature on organic structure of biomass pyrolysis products. Renewable Energy. 2013;**50**:136-141

[85] Guizani C, Jeguirim M, Valin S, Limousy Y, Salvadour S. Biomass chars: The effects of pyrolysis conditions on their morphology, structure, chemical properties and reactivity. Energies. 2017;**10**(6):1-18

[86] Yu J, Sun L, Berrueco C, Fidalgo B, Paterson N, Millian M. Influence of temperature and particle size on structural characteristics of chars from beechwood pyrolysis. Journal of Analytical and Applied Pyrolysis. 2018;**130**:127-134

[87] Huang Y, Liu S, Aktar MA, Li B, Zhou J, Zhang S, et al. Volatile char-interactions using biomass pyrolysis: Understanding the potential origin of char activity. Bioresource Technology. 2020;**316**:123938

[88] Mastral F, Esperanza E, Garcia P, Juste M. Pyrolysis of high-density polyethylene in a fluidised bed reactor: Influence of temperature and residence time. Journal of Analytical and Applied Pyrolysis. 2002;**63**:1-15

[89] Al-Rumaihi A, Shahbaz M, McKay G, Mackey H, Al-Ansari T. A review of pyrolysis technologies and feedstock: A blending approach for plastic and biomass towards optimum char yield. Renewable and Sustainable Energy Reviews. 2022;**167**:112715

[90] Ke L, Wu K, Zhou N, Xiong J, Yang Q, Zhang L, et al. Lignocellulosic biomass pyrolysis for aromatic hydrocarbons production: Pre and in-process enhancement methods. Renewable and Sustainable Energy Reviews. 2022;**165**:112607

[91] Vo TA, Tran QK, Ly HV, Kwon B, Hwang HT, Kim J, et al. Co-pyrolysis of lignocellulosic biomass and plastics: A comprehensive study on pyrolysis kinetics and characteristics. Journal of Analytical and Applied Pyrolysis. 2022;**163**:105464. DOI: 10.1016/j.jaap.2022.105464

[92] Likun Z, H., & Fan, Y. Improving hydrocarbons production via catalytic co-pyrolysis of torrefied-biomass with plastics and dual catalytic pyrolysis. Chinese Journal of Chemical Engineering. 2022;**42**:196-209. DOI: 10.1016/j.cjche.2020.09.074

[93] Khodier A, Williams KS, Dallison N. Pilot-scale thermal treatment of automotive shredder residue: Pyrolysis char is a resource or waste. WIT Transactions on Ecology and the Environment. 2017;**224**(1):439-450

[94] Mahinpey N, Gomez A. Review of gasification fundamentals and new findings: Reactors, feedstock, and kinetic studies. Chemical Engineering Science. 2016;**148**:14-31

[95] Esmaeili V, Ajalli J, Faramarzi A, Abdi M, Gholizadeh M. Gasification of wastes: The impact of the feedstock type and co-gasification on the formation of volatiles and char. International Journal of Energy Research. 2020;**44**(5):3587-3606. DOI: 10.1002/er.5121

[96] Liu H, Zhu H, Yan L, Huang Y, Kato S, Kojima T. Gasification rate of char with CO2 at elevated temperatures: The effect of heating rate during pyrolysis. Asia-Pacific Journal of Chemical Engineering. 2011;**6**(6):905-911

[97] Wang Q, Zhang R, Luo Z, Fang M, Cen K. Effects of pyrolysis atmosphere and temperature on coal char characteristics and gasification reactivity. Energy Technology. 2016;**4**(4):543-550

[98] Li R, Zhang J, Wang G, Ning X, Wang H, Wang P. Study on CO_2 gasification reactivity of biomass char derived from high-temperature rapid pyrolysis. Applied Thermal Engineering. 2017;**121**:1022-1031

[99] Cossu R, Lai T. Automotive shredder residue (ASR) management: An overview. Waste Management. 2015;**45**:143-151

[100] Galvagno S, Fortuna F, Cornacchia G, Casu S, Coppola T, Sharma VK. Pyrolysis process for treatment of automotive shredder residue: Preliminary experimental results. Energy Conversion and Management. 2004;**42**:573-586

[101] Haydary J, Susa D, Gelinger V, Cacho F. Pyrolysis of automobile shredder residue in a laboratory scale screw type reactor. Journal of Environmental Chemical Engineering. 2016;**4**:995-972

[102] Nortanicola M, Cornacchia G, De Gisi S, Di Canio F, Freda C, Garzone P, et al. Pyrolysis of automotive shredder residue in a bench scale rotary kiln. Waste Management. 2017;**65**:92-103

[103] Zolezzi M, Nicolella C, Ferrara S, Iacobucci C, Rovatti M. Conventional and fast pyrolysis of automobile shredder residue (ASR). Waste Management. 2004;**24**:691-699

[104] Chong Z, Yuan S, Ruimeng S. Particle size-dependant properties of a char produced using a moving-bed pyrolyzer for fuelling pulverized coal injection and sintering operations. Fuel Processing Technology. 2019;**190**:1-12

[105] Liu Z, Dugan B, Masiello CA, Gonnermann HM. Biochar particle size, shape, and porosity act together to influence soil water properties. PLoS One. 2017;**12**(6):e0179079

[106] Hameed R, Lei C, Lin D. Adsorption of organic contaminants on biochar colloids: Effects of pyrolysis temperature and particle size. Environmental Science and Pollution Research. 2020;**27**(15):18412-18422

[107] Shar AG, Peng JY, Tian X, Siyal TA, Shar AH, Yuhan J, et al. Contrasting effects of maize residue, coal gas residue and their biochars on nutrient mineralization, enzyme activities and CO_2 emissions in sandy loess soil. Saudi Journal of Biological Sciences. 2021;**28**(8):4155-4163

[108] Cunliffe AM, Williams PT. Properties of chars and activated carbons from the pyrolysis of used tyres. Environmental Technology. 1998;**19**(12):1177-1190

[109] Williams PT. Pyrolysis of waste tyres: A review. Waste Management. 2003;**33**(8):1714-1728

[110] Antoniou N, Stavropoulos G, Zabaniotou A. Activation of end of life tyres pyrolytic char for enhancing viability of pyrolysis—Critical review, analysis and recommendations for a hybrid dual system. Renewable & Sustainable Energy Reviews. 2014;**39**:1053-1073. DOI: 10.1016/j.rser.2014.07.143

[111] Makigrianni V, Giannakas A, Bairamis F, Papadaki M, Konstaninou I. Adsorption of Cr(VI) from aqueous solutions by HNO3-purified and chemically activated pyrolytic tire char. Journal of Dispersion Science and Technology. 2017;**38**(7):992-1002

[112] Qu DY, Shi H. Studies of activated carbons used in double-layer capacitator. Journal of Power Sources. 1998;**74**(1):99-107

[113] Joung H-T, Seo Y-C, Kim K-H, Seo Y-C. Effects of oxygen, catalyst and PVC on the formation of PCDDs, PCDFs and dioxin-like PCBs in pyrolysis products of automotive shredder residues. Chemosphere. 2006;**65**(9):1481-1489

[114] Joung H-T, Seo Y-C, Kim K-H. Distribution of dioxins, furans and dioxin-like PCBs in solid products generated by pyrolysis and melting of automotive shredder residue. Chemosphere. 2007;**68**(9):1636-1641

[115] Han Y, Zhao P-P, Dong X-T, Zhang C, Liu S-X. Improvement in electrochemical capacitance of activated carbon from scrap tires by nitric acid treatment. Frontiers in Material Science. 2014;**8**(4):391-398

[116] Shilpa K, Rudra SA. Morphologically tailored activated carbon derived from waste tires as high performance anode for Li-ion battery. Journal of Applied Electrochemistry. 2017;**48**(1):1-13

[117] Iraola-Arregui I, Van Der Gryp P, Görgens JF. A review on the demineralisation of pre- and post-pyrolysis biomass and Tyre wastes. Waste Management. 2018;**79**:667-668

[118] Tang H, Hu H, Li A, Yi B, Li X, Yao D, et al. Removal of impurities of waste pyrolysis char using the molten salt thermal treatment. Fuel. 2021;**301**:121019

[119] Yan LL, Kong L, Qu Z, Lo L, Shen GQ. Magnetic biochar decorated with ZnS nanocrystals for Pb (II) removal. ACS Sustainable Chemistry and Engineering. 2015;**3**:125-132

[120] Yang JP, Zhao YC, Ma SM, Zhu BB, Zhang JY, Zheng CG. Mercury removal by magnetic biochar derived from simultaneous activation and magnetization of sawdust. Environmental Science and Technology. 2016;**50**:12040-12047

[121] Hossain R, Al Mahmood A, Sahajwala V. Recovering renewable carbon materials from automotive shredder residue (ASR) for micro-supercapacitor electrodes. Journal of Cleaner Production. 2021;**304**:e127131

[122] Ruffino B, Panepinto D, Zanetti M. A circular approach for the recovery and recycling of automotive shredder residues (ASRs): Material and thermal valorization. Waste and Biomass Valorization. 2021;**12**:3109-3123

[123] Vijayan SK, Kibria MA, Uddin MH, Bhattacharya S. Pretreatment of automotive shredder residues, their chemical characterization and pyrolysis kinetics. Sustainability. 2021;**13**:1-19

[124] Jagodzińska K, Yang W, Jönsson PG, Forsgren C. Can torrefaction be a suitable method of enhancing shredder fines recycling? Waste Management. 2021;**128**:211-220

[125] Zhou Q, Zarei A, De Girolamo A, Yan Y, Zhang L. Catalytic performance of scrap Tyre char for the upgrading of eucalyptus pyrolysis derived bio-oil via cracking and deoxygenation. Journal of Analytical and Applied Pyrolysis. 2019;**139**:167-176

[126] Li N, Li Y, Ban Y, Song Y, Zhi K, Teng Y, et al. Direct production of high hydrogen syngas by steam gasification of Shengli lignite/chars: Remarkable promotion effect of inherent minerals and pyrolysis temperature. International Journal of Hydrogen Energy. 2017;**42**(9):5865-5872

[127] Bai Y, Lv P, Li F, Song X, Su W, Yu G. Investigation into Ca/Na compounds catalysed coal pyrolysis and char gasification with steam. Energy Conversion and Management. 2019;**184**:172-179

[128] Luo S, Xiao B, Hu Z, Liu S, Guo X, He M. Hydrogen-rich gas from catalytic steam gasification of biomass in a fixed bed reactor: Influence of temperature and steam on gasification performance. International Journal of Hydrogen Energy. 2009;**34**(5):2191-2194

[129] Xu MX, Wu YC, Nan DH, Lu Q, Yang YP. Effects of gaseous agents on the evolution of char physical and chemical structures during biomass gasification. Bioresource Technology. 2019;**292**:121994-121994

[130] Loftfian S, Ahmed H, El-Geassy A-HA, Samuelsson C. Alternative reducing agents in metallurgical processes: Gasification of shredder residue material. Journal of Sustainable Metallurgy. 2016;**3**(2):336-349

[131] Wang L, Chen L, Tsang DCW, Kua HW, Yang J, Ok YS, et al. The roles of biochar as a green admixture for sediment-based construction products.

Cement and Concrete Composites. 2019;**104**:103348

[132] Caetano JA, Schalch V, Pablos JM. Characterization and recycling of the fine fraction of automotive shredder residue (ASR) for concrete paving blocks production. Clean Technologies and Environmental Policy. 2020;**22**:835-847

[133] Azara A, Belbessai S, Abatzoglou N. A review of filamentous carbon nanomaterial synthesis via catalytic conversion of waste plastic pyrolysis products. Journal of Environmental Chemical Engineering. 2022;**10**(1):107049. DOI: 10.1016/j.jece.2021.107049

[134] Williams PT. Hydrogen and carbon nanotubes from pyrolysis-catalysis of waste plastics: A review. Waste and Biomass Valorization. 2020;**12**(1):1-28. DOI: 10.1007/s12649-020-01054-w

[135] European Commission. 2019. Regulation (EU) 2019/1021 of the European Parliament and of the Council of 20 June 2019 on persistent organic pollutants. Available from: EUR-Lex-32019R1021-EN-EUR-Lex(europa.eu)

[136] Elkhalifa S, Al-Ansari T, Mackey HR, McKay G. Food waste to biochars through pyrolysis. Resources, Conservation and Recycling. 2019;**144**:310-320. DOI: 10.1016/j.resconrec.2019.01.024

[137] Bolognesi S, Bernardi G, Callegari A, Dondi D, Capodaglio AG. Biochar production from sewage sludge and microalgae mixtures: Properties, sustainability and possible role in circular economy. Biomass Conversion and Biorefinery. 2019;**11**(2):289-299. DOI: 10.1007/s13399-019-00572-5

[138] Porshnov D. Evolution of pyrolysis and gasification as waste to energy tools for low carbon economy. Wiley Interdisciplinary Reviews. Energy and Environment. 2022;**11**(1):e421–n/a. DOI: 10.1002/wene.421

[139] UK Government. 2022. Landfill Tax rates 2022/23. Landfill Tax rates for 2022 to 2023- GOV.UK. Available from: www.gov.uk